PROTOCHLOROPHYLLIDE REDUCTION AND GREENING

ADVANCES IN AGRICULTURAL BIOTECHNOLOGY

Already published in this series

Akazawa T. et al., eds: The New Frontiers in Plant Biochemistry. 1983.
ISBN 90-247-2829-0

Gottschalk W. and Müller H.P., eds: Seed Proteins: Biochemistry, Genetics, Nutritive Value. 1983. ISBN 90-247-2789-8

Marcelle R., Clijsters H. and Van Poucke M., eds: Effects of Stress on Photosynthesis. 1983. ISBN 90-247-2799-5

Veeger C. and Newton W.E., eds: Advances in Nitrogen Fixation Research 1984.
ISBN 90-247-2906-8

Chinoy N.J., ed: The Role of Ascorbic Acid in Growth, Differentiation and Metabolism of Plants. 1984. ISBN 90-247-2908-4

Witcombe J.R. and Erskine W., eds: Genetic Resources and Their Exploitation
– Chickpeas, Faba beans and Lentils. 1984. ISBN 90-247-2939-4

Sybesma C., ed: Advances in Photosynthesis Research. 1984.
ISBN 90-247-2946-7

Protochlorophyllide Reduction and Greening

edited by

C. SIRONVAL and
M. BROUERS

Department of Botany, Laboratory of Photobiology
University of Liège
Sart Tilman
Liège
Belgium

1984 **MARTINUS NIJHOFF/DR W. JUNK PUBLISHERS**
a member of the KLUWER ACADEMIC PUBLISHERS GROUP
THE HAGUE / BOSTON / LANCASTER

Distributors

for the United States and Canada: Kluwer Boston, Inc., 190 Old Derby Street, Hingham, MA 02043, USA
for all other countries: Kluwer Academic Publishers Group, Distribution Center, P.O.Box 322, 3300 AH Dordrecht, The Netherlands

Library of Congress Cataloging in Publication Data

Main entry under title:

Protochlorophyllide reduction and greening.

   (Advances in agricultural biotechnology)
   Includes indexes.
   1. Protochlorophyllide--Congresses. 2. Chlorophyll
--Synthesis--Congresses. 3. Photosynthesis--Congresses.
I. Sironval, C. II. Brouers, M. III Series.
QK898.P84P76 1984   581.19'218   84-3983

ISBN-13: 978-94-009-6145-6   e-ISBN-13: 978-94-009-6143-2
DOI: 10.1007/978-94-009-6143-2   .

Katzuo SHIBATA
(November 15, 1918 - July 27, 1983)

The discoveries of Katzuo Shibata have had a great influence on most of the contributors to this book.

Using the opal glass technique, which he introduced to science, Shibata made a decisive step forward in recording for the first time, the exact absorption of a leaf *in vivo*. This technique enabled him to demonstrate that the green pigments were changing colour *in vivo* following defined rules.

Shibata understood that these changes played an active role in the development of the plastids from the etiolated to the green state. He was one of the first men, alongside C.S. French, J.H.C. Smith and J.B. Thomas who distinguished differing pigment forms by using specific characteristics of their *in vivo* spectra. Shibata was able to prove that these pigment forms, not only of protochlorophyll(ide) but also of chlorophyll(ide), were engaged in well defined reactions despite of the fact that at that time no clear physico-chemical interpretations could be made.

In this respect, the following quotation which conclude Shibata's remarkable paper entitled "Spectroscopic studies on chlorophyll formation in intact leaves", is particularly illustrative :

"It was found in this study that the transformation of protochlorophyll a into chlorophyll a is not a simple process. The process can be summarized in the following way :

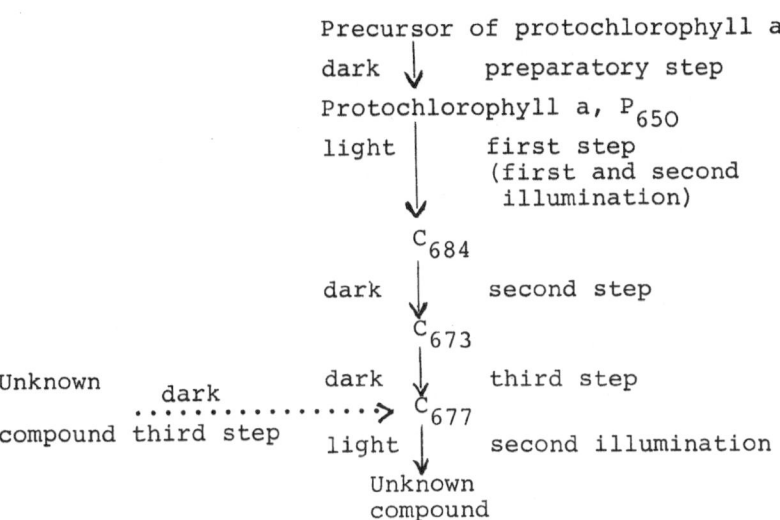

Precursor of protochlorophyll a

dark — preparatory step

Protochlorophyll a, P_{650}

light — first step
(first and second
illumination)

C_{684}

dark — second step

C_{673}

Unknown   dark   dark — third step
compound third step .....> C_{677}

light — second illumination

Unknown
compound

Once etiolated leaves are illuminated, many processes are initiated. The process proceeds in a series of reactions...".

This quotation does not only mention the "Shibata shift", it also remarks other shifts, some of which are discussed in this book. This paper provided the first clear data and a methodological basis for the opening of a new chapter of science.

Shibata never ceased in his research to open new doors to new horizons. His work on the shrinkage of chloroplasts, thermoluminescence, the effects of successive light flashes and his promotion of methods for separation of pigment-protein complexes when such techniques were in their infancy, demonstrate an uninterrupted series of significant discoveries to science.

It is to this pioneer that this book is dedicated.

CONTENTS

LIST OF PARTICIPANTS

to the meeting on PChlide reduction and greening held in Liège August 8-9, 1983.

ADAMSON H.
  Macquarie University - North Ryde - New South Wales- 2113
  AUSTRALIA.

AFFOLTER D.
  King's College London - 68 Half Moon Lane, London SE24 9JF
  ENGLAND.

AKOYUNOGLOU G.
  Biology Department - Nuclear Research Center "Demokritos"
  Aghia Paraskevi - Attikis - Athens - GREECE.

ARGYROUDI-AKOYUNOGLOU J.H.
  Biology Department - Nuclear Research Center "Demokritos"
  Aghia Paraskevi - Attikis - Athens - GREECE.

BEREZA B.
  Institute of Biochemistry - University of Wroclaw - POLAND

BOMBART P.
  Université de Liège - Département de Botanique, Sart Tilman,
  B4000 - Liège, BELGIQUE.

BROUERS M.
  Université de Liège - Département de Botanique, Sart Tilman,
  B4000 Liège, BELGIQUE.

BUSCHMANN C.
  Botanisches Institut Pflanzenphysiologie - Universität Karlsruhe - Kaiserstrasse 12, D7500 Karlsruhe- GERMANY (B.R.D).

CAHAY C.
  Université de Liège - Département de Botanique, Sart Tilman,
  B4000 - Liège, BELGIQUE.

DILOVA S.
  Popov Institute of Plant Physiology - Bulgarian Academy of
  Sciences. ML. Acad. G. Bontchev, BL. VI Sofia 1113 -
  BULGARIA.

DUJARDIN E.
  Université de Liège - Département de Botanique, Sart Tilman,
  B4000 - Liège, BELGIQUE.

EL HAMOURI B.
  Department of Biochemistry and Plant Physiology - Agronomic
  Institute Hassan II- BP 704 - Rabat - MAROC

FRANCK F.
  Université de Liège - Département de Botanique, Sart Tilman,
  B4000 - Liège, BELGIQUE.

FREIBERG A.
  Institut of Physics - 142 Rija Street - 202400 Tartu - USSR.

FUAD N.
  Department of Environmental Biology, Research School of Bio-

logical Sciences. The Australian National University. Post
Office Box 4, Canberra A.C.T. 2600, AUSTRALIA.

GERDAY C.
Université de Liège - Département de Botanique, Sart Tilman,
B4000 - Liège, BELGIQUE.

GRIFFITHS W.T.
Department of Biochemistry - University of Bristol, Medical
School, University Walk - Bristol BS8 1TD - ENGLAND.

HENDRICH W.
Institute of Biochemistry - University of Wroclaw - POLAND.

HILLER R.
Macquarie University - North Ryde, N.S.W. 2113 AUSTRALIA.

INOUE Y.
The Institute of Physical and Chemical Research - Wako -
Saitama 351, JAPAN.

KASEMIR H.
Biological Institute II, University of Freiburg - Schanzler-
strasse 1 - D7800 Freiburg, GERMANY (B.R.D.).

KIEKECZAVA J.
Institute of Biochemistry - University of Wroclaw - POLAND.

KLIMOV V.
Institute of Photosynthesis - USSR Academy of Sciences-
Pushcino - 142292 - U.S.S.R.

KLOCKARE B.
University of Göteborg - Botanical Institute - Department
of Plant Physiology - Carl Skottsbergs gata 22 - S-413 19
Göteborg, SWEDEN.

LÜTZ C.
Department of Botany II - University of Kölnn Gyrhofstrasse
15 - 5 Köln 41 - GERMANY.

MICHEL-WOLWERTZ M-R.
Université de Liège - Département de Botanique, Sart Tilman,
B4000 , Liège , BELGIQUE.

MURAKAMI S.
Department of Biology - University of Tokyo - Komaba-machi
3-8-1 Meguro-ku, Tokyo, JAPAN 153.

PACKER N.
Macquarie University - North Ryde - New South Wales - 2113
AUSTRALIA.

PAQUES M.
Université de Liège - Département de Botanique, Sart Tilman,
B4000 - Liège , BELGIQUE.

RÖPER U.
Department of Botany II - University of Köln, Gyrhofstrasse
15, 5 Köln 41 - GERMANY.

RUBIN A.
Dept. of Biophysics Fac. of Biology - Moscow State Univer-

sity - Moscow 117234, U.S.S.R.

RYBERG M.
  University of Göteborg - Botanical Institute , Dept. of plant
  Physiol.- Carl Skottsbergs gata 22 - S-413 19 Göteborg -
  SWEDEN.
ŠEŠTÁK Z.
  Institute of Experimental Botany - Czechoslovak Academy of
  Sciences, Flemingovo n.2 - CS-16000 Praha 6 -CZECHOSLOVAKIA.

SHUVALOV V.
  Institute of Photosynthesis USSR Academy of Sciences, Pushcino, 142292, U.S.S.R.

SIRONVAL C.
  Université de Liège - Département de Botanique - Sart Tilman,
  B4000 - Liège, BELGIQUE.

STRASSER R.J.
  Institute of Biology, Department of Bioenergetics - Universi-
  ty of Stuttgart, Ulmerstrasse 227, 7000 Stuttgart 60,
  GERMANY.

SURIN C.
  Université de Liège - Département de Botanique - Sart Tilman,
  B4000 - Liège, BELGIQUE.

VALCKE R.
  Limburgs Universitair Centrum - Department S.B.M. Universi-
  taire Campus, 3610 Diepenbeek, BELGIUM.

VAN BOCHOVE A.C.
  Department of Biophysics, State University Leiden - P.O.Box
  9504 NL 2300 RA Leiden, NETHERLANDS.

VAN DER CAMMEN J.
  Phys. Lab. Princetonplein 5, 3508 TA, Utrecht , NETHERLANDS.

ACKNOWLEDGMENTS

The organization of the meeting on PChlide reduction and greening, Liège 1983 and the publication of this book have been supported by:
- The "Fonds National de la Recherche Scientifique (F.N.R.S.)"
- The University of Liège, BELGIUM.

PREFACE

        This book gathers contributions presented during an In-
ternational meeting organized by the Laboratory of Photobiology of the University of Liège, Belgium, on 8 and 9 August
1983.

        The general topic of the discussions was protochlorophyl-
lide reduction and greening. Among the reasons for choosing
this topic were the recent advances in the field. These advances deal with:

        (1) The characterization of the basic constituents of
the photoenzymatic complex responsible for protochlorophyllide reduction. This complex is known to be ternary, comprising
the pigment: protochlorophyllide, NADPH and the enzyme protochlorophyllide oxidoreductase.

        (2) The discovery of short-lived intermediates in the
photoreduction process, and in particular, the recent findings
resulting from the progresses of the picosecond and nanosecond
spectrometry.

        (3) The obtention of new data on the components of the
plastids, on the changes they undergo during the first steps
of greening, and on the distribution of the pigment-protein
complexes between the various substructures of the etioplast.

        (4) The detection of early photoactivities apart from
protochlorophyllide reduction.

        These subjects have been extensively discussed during
the meeting and several sections of this book are devoted to
the presentation of the new data.

        The book brings together informations about the events
which lead to the formation of the various chlorophyll forms

(antenna and active centres) and to the onset of early photoactivities in the greening plastids. We feel that it is worth making these informations available for searchers in photosynthesis, photobiology, biophysics, plant physiology and biochemistry and we hope that this goal has been fulfilled.

A great liberty has been left to the authors with regard to the form of their contributions and to the expression of their opinions and interpretations. This decision causes some lack of uniformity in the presentation (in particular of the reference quotation lists).

One problem was the variety of denominations for the same protochlorophyllide protein complexes. This is due mainly to some variability of the position of the absorption and fluorescence maxima depending on leaf age, plant species, methods of growing or precision of the spectrophotometers used.
The denominations comprise (1) an abbreviation or a letter P for protochlorophyll(ide) or pigment - C for chlorophyll(ide), and (2) subscripts giving the wavelength of maximum absorption or maximum fluorescence in the red region of the spectrum.
The following tables may help to clarify the equivalences. It must be emphasized, however, that all denominations are empirical and that some of the complexes named in the tables might correspond to mixtures of complexes.

Most contributions to this book contain original data. Contradictory opinions may be found almost side by side when comparing the papers one with other. We consider these controversies as sources for future research.

We espress our deep gratitude to C.S. French who accepted to write the Introduction to this book.
We also thank the chairmen : G. Akoyunoglou, W.T. Griffiths, Y. Inoue and A. Rubin for their excellent leadership during the sessions.
Mrs F. Hayet-Brouers did an excellent job as secretary before and during the meeting.

C. Sironval, M. Brouers.

SOME ABBREVIATIONS USED

FOR DESIGNATING THE MAIN NATURAL PIGMENT FORMS

Table 1. General abbreviations

Protochlorophyllide :	PChlide, PChld
Protochlorophyll :	PChl
The mixture (protochlorophyllide + protochlorophyll) :	PChl(ide), Pchl(d)
Chlorophyllide :	Chlide, Chld
Chlorophyll :	Chl
The mixture (chlorophyllide + chlorophyll) :	Chl(ide), Chl(d)

Table 2. Protochlorophyllide forms.

Red absorp. max. (nm)	X_A	X_{A-F}	$X_{F,A}$	X_A^F	line n^r
628	Pchlide 628 P_{630}	Pchlide$_{628-632}$	$P_{630-628}$, $P_{633-628}$	P_{632} P_{628}	1
642	Pchlide 642 P_{642}				2
635-640	Pchlide 637 P_{637}, P_{635} $P_{637,650}^{*}$	Pchlide$_{636-657}$	$P_{645-637}$, $P_{645-640}$ $P_{657-650,638}^{*}$	$P(645)$ 640	3
650	Pchlide 650 P_{650}	Pchlide$_{650-657}$	$P_{657,650}$	P_{657} 650	4

Table 3. Chlorophyll(ide) forms.

668-670			$P_{675-670}$, $P_{676-670}$, $P_{675-668}$		5
672-673	Chlide 672 P_{673}	$C_{685-672}$	$P_{685-672}$, $P_{685-673}$, $P_{684-672}$		6

676-678	C_{678} $\quad$ P_{678}	$C_{688-677}, C_{688-678}$ $P_{688-676}, P_{688-678}$		7
682-684	Chlide 684 $\quad$ P_{684}	$C_{696-684}$ $P_{696-684}, P_{695-683}, P_{695-682}'$, $P_{694-682}$	$p696$ P_{682}	8

Table 4. Intermediate forms.

$\geqslant 690$	X_{690}	$P_{-}, 680 - 750$		9

In the X_A column, the pigments are designated by the location of the red absorption maximum. In the other columns, they are designated by the location of the red absorption maximum (A) and by the location of the main fluorescence band (F).

The letter P stays for "protochlorophyllide " (or "protochlorophyll") in table 2, and for "pigment" in table 3 and 4. The letter C stays for "chlorophyllide" (or "chlorophyll").

X in table 4 designates an unknown pigment.

Equivalent abbreviations for one pigment form are on a same line.

* These abbreviations designate a complex comprising pigments of both lines 3 and 4.

INTRODUCTION

C.S. FRENCH
Carnegie Institution of Washington. Department of Plant Biology. Stanford, California 94305. USA.

Chlorophyll is the most obvious colored substance on the earth. It is therefore not suprising that so much work has been done trying to find out how it is made. The editors have kindly asked me to prepare brief comments on the early studies of protochlorophyll and its conversion to chlorophyll emphasizing the times before most of the present day investigators were active. Certainly other authors in this volume will give adequate references to works of the more recent past. This discussion will be limited mainly to the "pre-historic" era when I had the pleasure of knowing many of the active individuals and is not in any sense a review of current events in chlorophyll formation. Therefore the significant contributions of many distinguished colleagues, who will I presume, write other chapters have been omitted here. They had best speak for themselves. To adequately review the contributions of just one of the many important investigators of those early times would more than fill the space allotted for this introduction, hence the citations given are primarily to the excellent review articles. Many significant discoveries are here omitted. I will however, discuss a few of the ideas then popular so we can trace back some present beliefs to the time when they were beginning to appear dimly out of the fog.

During the time covered by these comments the subject of chlorophyll formation was generally included in meetings, reviews, and books devoted

C. Sironval and M. Brouers (eds.); Protochlorophyllide Reduction and Greening. ISBN 90 247 2954 8
© 1984, Martinus Nijhoff/Dr W. Junk Publishers, The Hague/Boston/Lancaster.

to various other aspects of photosynthesis of which chlorophyll formation was considered a part.

The "Old Testament", H.A. Spoehr, 1926 (1) mentions neither protochlorophyll nor chlorophyll formation even though much work had been done by that time. The name "protochlorophyll" was given by Monteverde in 1893 according to the "New Testament", Rabinowitch, 1945 (2). That volume allows one page out of six hundred to our present subject. However, because of the complete contemporary coverage of chlorophyll formation along with all other aspects of photosynthesis, the second volume, part 2, Rabinowitch 1956 (3) is an excellent survey up to that time. A review on protochlorophyll was given by Paul Rothemund, a student of Hans Fischer, at the "Cold Spring Harbor Symposium on Quantitative Biology #3" in 1935 (4). This talk brought protochlorophyll and its transformation to the attention of U.S. investigators of photosynthesis, of photochemistry, and of related biological processes.

About that time the concept was still viable that photosynthesis might take place by water splitting with the liberation of O_2 while hydrogen went to protochlorophyll thus transforming it to chlorophyll. Just how that process could recycle was not brought out. Later Smith (5) found that the protochlorophyll photoconversion in a stream of purified hydrogen produced less than 2.5% of the amount of O_2 that would come from water splitting were it used as a source of H_2 for the transformation. This experiment disposed of one of the many early simple thoughts on the mechanism of photosynthesis. Particularly relevant to the present topic are the thorough reviews on chlorophyll formation up to 1951 by Smith and Young (6) and Smith, 1960 (7). They point out that the structural formula for protochlorophyll that was substantiated by Fischer and collaborators was first proposed by Stoll and Weidermann in 1938. Those and other

(7,36,37,44) reviews cover the main subject matter of chlorophyll formation
in what we may call the pre-Shibata era. That is before direct
spectrophotometric measurements of protochlorophyll and of recently formed
chlorophyll had been made in intact live leaves. One remarkable effect
discovered by Smith (8) still deserves further work. That is the rapid
accumulation of ether-soluble magnesium and phosphate compounds, other
than chlorophyll, early in the illumination of etiolated leaves. Many
times more organic magnesium was found shortly after illumination than
could be accounted for by the chlorophyll content. However, later, over
90% of the organic magnesium is in chlorophyll.

Extraction of leaf samples with water-miscible organic solvents
followed by transfer to ether for spectrophotometry was generally used in
studies of the transformation. However, Smith (9) and also Krasnovsky
and Kosobutskaya (10) were able to measure the transformation
spectrophotometrically in combined extracts of several leaves with
glycerol-water mixtures or alkaline buffers. This extraction of the active
holochrome was a great advance. The phototransformable aqueous extracts
contained protein as well as protochlorophyll. The protein went along
with protochlorophyll in ammonium sulfate fractionations behaving like
hemoglobin and the chlorophyll protein complexes of Stoll and Wiedermann
(36) and others. Smith proposed the name protochlorophyll-holochrome to
refer to the whole complex and isolated a moderately pure holochrome by
standard protein isolation methods. Much recent work on this substance
will no doubt be covered in this book. Boardman (18) reviews his own as
well as Smiths (7) isolation procedures and the older approximate molecular
weight determinations giving about 1-2 protochlorophyll molecules per
particle of about 10^6 daltons.

Because of the small amount of protochlorophyll in leaves it was believed impossible to measure the transformation by direct spectrophotometry in leaves. Shibata, 1958 (11), however found that with opal glass to roughly equalize the light scattering of the reference and measuring beams it was easy to see protochlorophyll in live leaves and thus to follow chlorophyll formation in real time. The first formed chlorophyll observed at 672 nm soon changes to 684 nm then back to 678 nm. This "Shibata shift" shows the details of the transformation from the immediately formed "C 670" to "C 678" which accumulates as was seen by Krasnovsky and Kosobutskaya 1953 (12). The quantum yield for the protochlorophyllide to chlorophyllide transformation in purified glycerol-water extracts of protochlorophyll holochrome was found to be 0.6, Smith, 1958 (13). Probably Professor Sironval and Dr. Brouers will bring us up to date on this subject.

An action spectrum for protochlorophyll transformation had been measured by Frank 1946 (14) in Selig Hecht's laboratory with color filters. At that time there were no interference filters and any suitable monochromator had to be home made. A precise action spectrum for chlorophyll formation in normal and in albino corn showed that protochlorophyll-holochrome has its _in vivo_ absorption at 650 nm and that carotenoids do not participate in active light absorption, but act as internal filters, Koski, French and Smith 1951 (15). That action spectrum for the albino gave the absorption spectrum of protochlorophyll _in vivo_ before it had been directly measured spectrophotometrically. Furthermore, those experiments showed that the remaining protochlorophyll after partial transformation, was not changed by light absorbed by the chlorophyll that had already been formed.

The central role of protochlorophyll was shown also by the findings of Ogawa, _et al._ (16) that the action spectra for formation of chlorophyll _a_ and _b_ and for β-carotene were identical with that for the transformation of protochlorophyll. Large amounts of protochlorophyll are found in pumpkin and squash seeds but that material does not transform to chlorophyll in light. Two variants of seed protochlorophyll lead to the idea that chlorophyll a and chlorophyll b might come from different protochlorophylls. Squash and related seeds produced large quantities of protochlorophyll for studies of its structural chemistry, but not for the investigation of chlorophyll formation. Leaf protochlorophyll after extraction by organic solvents cannot be transformed by light or other means (3,4,17,18). The non-transformability of squash seed protochlorophyll as well as the dark transformation in pine seeds without light pointed early on to the need for other substances than protochlorophyll in chlorophyll transformation in leaves. Godnev and Terent'eva 1953 (19) found that an extract of spruce seedlings caused the transformation of protochlorophyll to chlorophyll in the dark when infiltrated into etiolated corn seedlings. Eventually the discovery of NADPH-enzyme-protochlorophyll complex clarified the situation but that story belongs to recent, rather than to ancient history.

The use of radioactive tracers in etiolated plants, largely by Shlyk (20) and his collaborators, eventually proved that chlorophyll b came from freshly formed chlorophyll a and not from a previously postulated protochlorophyll b as had been suggested by Seybold (as discussed in ref. 18). While chlorophyll formation by light from protochlorphyll in etiolated plants was very evident there was some doubt that all chlorophyll in mature leaves was made by that process. Smith 1960 (7) discusses the meager evidence for the conclusion that this is indeed the case. However Virgin 1955 (21) found the initial rate of protochlorophyll formation in the dark at different temperatures to parallel the further accumulation of chlorophyll in the light. Furthermore protochlorophyll has been found in mature barley leaves harvested in the field. Another proof came from the experiments of Litvin, Krasnovsky and Rikhireva 1959 (22) who put fully green leaves in the dark then exposed some of them to light at -150 degrees C where protochlorophyll does not transform. These leaves, unlike those exposed to light at room temperature, showed protochlorophyll in the fluorescence spectrum. Smith believed that chlorophyll b could not come from chlorophyll a during greening because the a/b ratio did not change much. However Shlyks experiments with ^{14}C showed that only the freshly formed chlorophyll a makes chlorophyll b thus removing Smiths objection to that idea.

During much of the period we are discussing little distinction was made between protochlorophyll esterified with phytol, and protochlorophyllide without phytol. The difference in acid extractability of freshly formed chlorophyll and that accumulated later showed the freshly formed chlorophyll to lack phytol, Wolff and Price (23). Therefore, "protochlorophyll" was recognized to be actually protochlorophyllide in etiolated leaves. The literature of the early times usually speaks of protochlorophyll and in this discussion the term is used loosely to cover protochlorophyllide, or its esters with vinyl, phytol or geraniol. The rapidly formed C 683 absorption is due to non-phytylated chlorophyll while the C 672 formed from it does have phytol. The P 650 transformable protochlorophyll is not phylytated unlike the phylytated P 635 which does not transform in light, Sironval et al.,1965 (24).

Two lines of investigation on the various forms of chlorophyll in plants are now converging. One of these - the heterogenity of chlorophyll a complexes in green leaves and algae started with Lubimenko's studies 1927, (25) of leaves and their water macerates. Using only a microspectroscope and sharp eyes he found irregularities in the red absorption bands of in vivo chlorophyll. These he attributed to the presence of different forms of chlorophyll a. Krasnovsky and Kosobutskaya (26) clarified the existance of "C670" and "C680" as separate forms of chlorophyll a following up the ideas of Lubimenko. Since then a large amount of work has been done in many laboratories on the different forms of chlorophyll a occuring together in mature plants, French (27), Brown (28). Butler 1966 (29) wrote an early review of the discovery of chlorophyll a forms and of chlorophyll transformation during development. The other line of investigation that can well be tied in closely with the knowledge of various chlorophyll forms in mature leaves is the study of relations between esterification and in vivo spectroscopy of newly formed chlorophyll.

Thus the studies of protochlorophyll transformation may merge even more closely with such investigations of the various forms of chlorophyll a and of the Shibata shift as in the work of Virgin (30). Possibly more extensive correlations of spectral deconvolution of fractions of chloroplast material with studies of the chemistry of the chlorophyll extracted from the fractions may be profitable. It would be good to see the details of the biosynthetic pathways for the formation of individual chlorophyll a forms from etiolated to mature plans by further expanding the work of Wolff and Price (23), of Virgin 1969 (30,32), of Goedheer (31), and of Sironval et al.,1965 (24) and correlating such experiments with curve analyses of the low temperature absorption spectra.

The scope of contemporary studies on chlorophyll formation is evident from the 210 recent references in this departments computerized file on on photosynthesis (Dr. J. S. Brown) and by the two reviews that appeared while this introduction was in preparation: Kasemir (33) and Castelfranco and Beale (34).

The following reviews although not specifically referenced here were also used extensively in the preparation of this introduction, (35-44), where the many contributions of Seybold, Granick, Godnev, Virgin, Aronoff, and others may be found.

There are remarkable analogies between the protochlorophyll- holochrome and another water-soluble phototransformable chlorophyll-protein also found in small amounts in many plants. That is the CP668 of Yakushiji et al. (46) which transforms in light to a form with a 740 nm peak but whose function if any is not yet known. Each of these substances has its own independent groups of investigators. Strangely enough both protochlorophyll and the phototransformable water soluble chlorophyll protein have both been isolated from plants grown in the field and brought into a dark laboratory for extraction.

The literature is complicated by marital name changes, thus Koski became Young and Kosobutskaya became Vorobeva. This year we should pay special honor to Hemming Virgin who retires this fall and to the memory of Kazuo Shibata who died July 27, 1983.

REFERENCES
1. Spoehr, H. A. 1926. Photosynthesis. Am. Chem. Soc. Monograph #29, The Chemical Catalog Co. Inc., NY.
2. Rabinowitch, E. 1945. Photosynthesis and related processes. Vol. I. Interscience, NY.
3. Rabinowitch, E. 1951. Vol. II, part 1; Vol. II, part 2, 1956. Interscience, NY.
4. Rothemund, P. 1935. Protochlorophyll. Cold Spring Harbor Symposia on Quantitative Biology III, 71-79.
5. Smith, J. H. C. 1951. The formation of chlorophyll and the beginning of photosynthesis. Carnegie Inst. Yearbook, 50, 123-124.
6. Smith, J. H. C. and Young, V. M. K. 1951. Chlorophyll formation and accumulation in plants. Radiation Biology, A. Hollander ed., III, 393-442. McGraw Hill, NY.
7. Smith, J.H.C. 1960. Protochlorophyll transformation. In: Comparative biochemistry of photoreactive systems. M. B. Allen, ed., 259-277. Academic Press, NY.
8. Smith, J.H.C. 1949. Processes accompanying chlorophyll formation. In: photosynthesis in plants. J. Franck, W.E. Loomis, eds., pp. 209-217. Iowa State College Press.
9. Smith, J.H.C. 1952. Factors affecting the transformation of photochlorophyll to chlorophyll. Carnegie Inst. Yearbook, 52, 151-153.
10. Krasnovsky, A.A., Kosobutskaya, L.M. 1952. Spectral study of the state of chlorophyll during formation in plants and in colloyidal solutions of etiolated leaf substances. Dokl. Akad. Nauk. SSSR. 85,177-180.
11. Shibata, K. 1958. Spectrophotometry of intact biological materials.

J. Biochem. (Tokyo) 45, 599-623.

12. Krasnovsky, A.A., Kosobutskaya, L.M. 1953. Different conditions of chlorophyll in plant leaves. Dokl. Akad. Nauk. SSSR. 91, 343-346.

13. Smith, J.H.C. 1958. Quantum yield of the protochlorophyll-chlorophyll transformation. Brookhaven Symposia in Biology, 11, 296-302.

14. Frank, S. R., 1946. The effectiveness of the spectrum in chlorophyll formation. J. Gen. Physiol. 29, 157-179.

15. Koski, M., French, C.S., Smith, H. C. 1951. The action spectrum for the transformation of protochlorophyll to chlorophyll in normal and albino corn seedlings. Arch. Biochem. 31, 1-17.

16. Ogawa, T., Inoue, Y., Kitajima, M., Shitaba, K. 1973. Action spectra for biosynthesis of chlorophyll $\underline{a}$ and $\underline{b}$ andβ-carotene. Photochem. Photobiol. 18, 229-235.

17. French, C.S. 1960. The chlorophylls $\underline{in\ vivo}$ and $\underline{in\ vitro}$. In: Encyclopedia of Plant Physiology, W. Ruhland, ed., V, 252-297, Springer, Berlin.

18. Boardman, N.K. 1966. Protochlorophyll. In: The Chlorophylls. L. P. Vernon and G.R. Seely, eds., pp. 437-479. Academic Press, NY.

19. Godnev, T.N., Terent,'eva, M.V. 1953. Transformation of protochlorophyll in etiolated corn leaves by infiltration with an extract of spruce seedlings. Dokl. Akad. Nauk. SSSR. 88, 725-727.

20. Shlyk, A,A., et al. 1969. Relationship between chlorophyll metabolism and heterogeneity of pigment apparatus. in: Progress in Photosynthesis Research, H. Metzner, ed., II, 572-591, Tübingen.

21. Virgin, H. 1955. Pigment transformation in leaves of wheat after irradiation. Physiol. Plant. 13, 155-164.

22. Litvin, F.F., Krasnovsky, A., Rikhireva, G.T. 1959. Formation and transformation of protochlorophyll in green leaves. Dokl. Akad. Nauk. SSSR. 127, 699-701.

23. Wolff, J.B., Price, L. 1959. Terminal steps in chlorophyll $\underline{a}$ biosynthesis in higher plants. Arch. Biochem. Biophys., 72, 293-301.

24. Sironval, C., Michel-Wolwertz, M.R., Madsen, A. 1965. On the nature and possible function of the 673 and 684 nm forms $\underline{in\ vivo}$ of chlorophyll. Biochim. Biophys. Acta. 94, 344-354.

25. Lubimenko, V., 1927. Recherches sur les pigments des plastes et sur la photosynthesis. I. Les pigments des plastes et leur transformation dans les tissues vivants de la plante. 1. Les pigments des chloroplasts. Rev. gén. Bot. 39, 547-559.

26. Krasnovsky, A.A., Kosobutskaya, L.M. 1953. Different conditions of chlorophyll in plant leaves. Dokl. Akad. Nauk. SSSR. 91, 343-346.

27. French, C.S. 1971. The distribution and action in photosynthesis of several forms of chlorophyll. Proc. Nat. Acad. Sci. USA, 68, 2893-2897.

28. Brown, J.S. 1972. Forms of chlorophyll $\underline{in\ vivo}$. Ann. Rev. Plant Physiol. 23, 73-86.

29. Butler, W.L. 1966. Spectral characteristics of chlorophyll in green plants. In: The chlorophylls. L.P. Vernon and G.R. Seeley, eds., pp. 343-379. Academic Press, NY.

30. Virgin, H.I. 1960. Pigment transformations in leaves of wheat after irradiation. Physiol. Plant. 13, 155-164.

31. Goedheer, J.C. 1961. Effect of changes in chlorophyll concentration on photosynthetic properties. 1. Fluorescence and absorption of greening bean leaves. Biochim. Biophys. Acta. 51, 494-504.

32. Virgin, H. I., French, C. S. 1973. The light induced protochlorophyll-chlorophyll $\underline{a}$ transformation and the succeeding

interconversions of the different forms of chlorophyll. Physiol. Plant. 28, 350-357.

33. Kasemir,H. 1983 . (Yearly Review). Action of light on chlorophyll (ide) appearance. Photochem. Photobiol. 37, 701-708.

34. Castelfranco, P.A. and Beale, S. 1983. Chlorophyll biosynthesis: Recent advances and areas of current interest. Ann. Rev. Plant. Physiol. 34, 241-278.

35. Milner, H.W. 1949-1965. Russian Translations, Seven volumes, 2,000 pages. In: Library of Department of Plant Biology, Carnegie Institution of Washington, Stanford, CA 94305 and the John Crerar Library, 35 W 33rd St., Chicago, IL, 60616.

36. Stoll. A., Wiedermann, E. 1952. Chlorophyll. Fortschr. Chem. Forschung 2, 538-608.

37. Smith, J.H.C., Benitez, A. 1955. Chlorophylls: Analysis in plant materials. In: Modern Methods of Plant Analysis, K. Paech and M.V. Tracey, eds., 4, 142-196, Springer, Berlin.

38. Shibata, K. 1959. Spectrophotometry of translucent biological materials - opal glass transmission method. In: Methods of Biochemical Analysis, D. Glick, ed., 7, 77-109, Interscience, NY.

39. French, C.S. 1960. The chlorophylls in vitro. In: Encyclopedia of Plant Physiol, W. Ruhland, ed., V, 252-297, Springer, Berlin.

40. Bogorad, L. 1960. The biosynthesis of protochlorophyll. In: Comparative biochemistry of photoreactive systems, M.B. Allen, ed., pp. 227-256, Academic Press, NY.

41. Smith, J.H.C., French, C.S. 1963. The major and accessory pigments in photosynthesis. Ann. Rev. Plant Physiol. 14, 181-224.

42. Bogorad, L. 1966. The biosynthesis of chlorophylls. In: The Chlorophylls. L.P. Vernon and G.R. Seeley, eds., pp. 481-510, Academic Press, NY.

43. Bogorad, L. 1976. Chlorophyll biosynthesis. In: Chemistry and Biochemistry of Plant Pigments, T.W. Goodwin, ed., 1, 64-148, Academic Press, London.

44. Kupke, D.W., French. C.S. 1960. Relationship of chlorophyll to proteins and lipids; molecular and colloidal solutions. Chlorophyll units. In: Encyclopedia of Plant Physiology, W. Ruhland, ed., 5, 298-322, Springer, Berlin.

45. Gurinovich, G.P., Sevchenko, A.N., Solov'er, K.N. 1971. Spectroscopy of chlorophyll and related compounds. English translation by U.S. Joint Publication Research Service, Washington, D.C., U.S. Atomic Energy Commission, Division of Technical Information, S.E.C. tr- 7199 Chemistry, (TID-4500).

46. Yakushiji, C., Uchino, K., Segimura, Y., Takamiya, F. 1963. Isolation of water-soluble chlorophyll protein from the leaves of Chenopodium album. Biochim. Biophys. Acta 75, 193-298.

Carnegie Institution of Washington
Department of Plant Biology
Stanford CA 94305
Publication #835

PROTOCHLOROPHYLLIDE
PHOTO-OXIDOREDUCTASE;
PROPERTIES AND LOCALISATION.

A CRITICAL APPRAISAL OF THE ROLE AND REGULATION OF NADPH-
PROTOCHLOROPHYLLIDE OXIDOREDUCTASE IN GREENING PLANTS.

W.T. GRIFFITHS, R.P. OLIVER and S.A. KAY

Department of Biochemistry, University of Bristol, The Medical
School, Bristol, BS8 ITD U.K.

It is well known that angiosperms when germinated in
darkness produce etiolated chlorophyll-free plants in marked
contrast to the green tissues enriched with chlorophyll produced in the light. The reason for this difference is that
the biosynthesis of chlorophyll by angiosperms involves a
reaction that is obligatory light dependent i.e. the reduction of protochlorophyllide (pchlide) to chlorophyllide (chlide)
(see Figure 1). Gymnosperms and certain algae however can

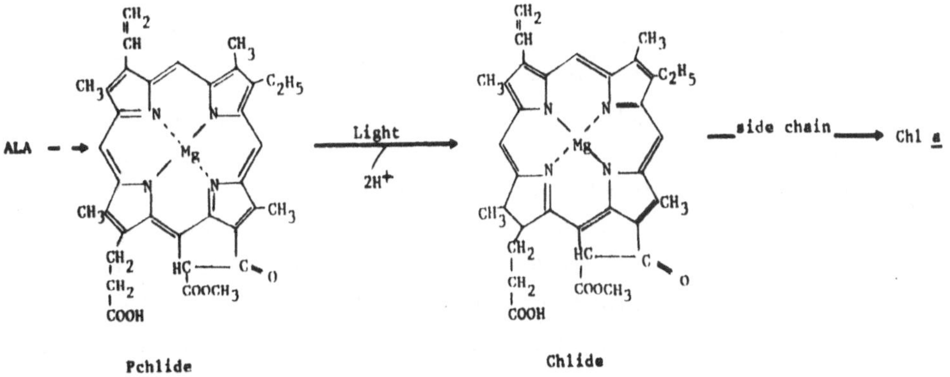

FIGURE 1. Chlorophyll formation in angiosperms.

presumably achieve this reduction without light since they
can synthesise chlorophyll and become green in complete
darkness.

The enzyme which catalyses the light dependent reduction
of pchlide, designated pchlide reductase, has been identified,
isolated and partially characterised (1,2,3). In the presence
of the substrates NADPH and pchlide, the enzyme forms a dark
stable ternary complex which on illumination produces

C. Sironval and M. Brouers (eds.); Protochlorophyllide Reduction and Greening. ISBN 90 247 2954 8
©1984, Martinus Nijhoff/Dr W. Junk Publishers, The Hague/Boston/Lancaster.

the products $NADP^+$ and chlide. Protochlorophyllide is very
abundant and active in membranes prepared from isolated etioplasts. Rather surprisingly however it is difficult to assay
or detect in extracts prepared from illuminated tissues (4,5).
In fact, the suggestion has even been made that the reductase
may not be involved with chlorophyll formation in illuminated
plants, reduction of pchlide here being assumed to be catalysed
by some as yet unidentified enzyme (6). In the present article
we wish to survey progress in our understanding of the physiological role of the enzyme together with the mechanism of its
regulation by light. In particular an attempt will be made to
try and establish the status of the enzyme in green plants
grown under normal illumination conditions.

The presence of pchlide reductase is unchallenged in
etiolated tissues. It occurs in dark grown leaves loosely
bound to the internal membrane system of the etioplasts and
can be solubilised by mild extraction with non-ionic detergents such as Triton X-100 (7,3). Various estimates for the
molecular size of the enzyme from different tissues have been
proposed. For example the enzyme has been described as a single
peptide of mol. wt. 36 kD from barley (3,8) whereas two closely
related peptides of mol. wt. 35 and 37 kD have been described
as the enzyme in oats (8). Whether these represent genuine
specie differences or reflect varying degrees of proteolysis
occuring during isolation remains to be unequivocally established. Further, the macromolecular organisation of the enzyme in the plastid membrane has also to be established. By
way of explanation it is perhaps worth mentioning that the
widely studied pchlide holochrome of earlier literature (9)
represents, according to current ideas, various mixed aggregates of detergent solubilised pchlide reductase.

The role of the enzyme in pchlide phototransformation is
depicted in Fig. 2 which is based on data obtained from spectroscopic studies carried out *in vitro* on isolated etioplast
membranes. The scheme involves several complexes of the reductase with its substrates and products, each characterised by

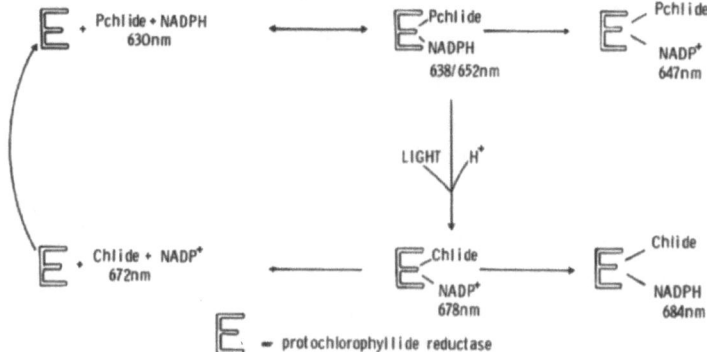

FIGURE 2. Scheme for Pchlide phototransformation.

its own distinct spectral properties (10). The assignment of
a spectral form to a particular complex is based upon reconstitution studies whereby specific absorbing froms can be generated *in vitro* under appropriate conditions. Thus "free" pchlide
or pigment that is non-specifically adsorbed shows maximum
light absorption at approximately 628nm. In the presence of
NADPH and active reductase free pchlide combines with the enzyme to form a complex absorbing at 638 and 652nm (Fig. 2).
Illumination of this complex results in hydrogen transfer
from the NADPH to the pchlide producing eventually a complex
of the reductase with the products, $NADP^+$ and chlide, such a
complex absorbing maximally at 678nm. Two alternative fates
are open to this form depending on conditions (Fig. 2). In
the presence of excess NADPH the nucleotide on the enzyme
can be reduced forming a "mixed" complex of the reductase with
NADPH and chlide -this absorbing maximally at approximately
684nm. In the situation with excess pchlide present however,
this serves to displace chlide from the enzyme, irrespective
of whether the coenzyme bound is oxidised or reduced, and
producing free or non-specifically bound chlide absorbing
maximally at 672nm. This step, designated the Shibata shift
(11) after its discovery by Professor Katsuo Shibata, regene-
rates the reductase for participation again in the cyclic production of chlide.

The progress of the Shibata shift can readily be studied
both *in vivo* (12) and *in vitro* in NADPH supplemented membranes

(10) by continuous monitoring of absorbance at 690nm relative
to 710nm. At 690nm a large difference in absorbance exists
between the chlide 684 and 672 forms whereas at 710nm there
is no specific pigment absorption. The role of pchlide in
inducing the Shibata shift, as demonstrated by this technique,
in membranes which have been photoconverted in the presence
of excess NADPH is shown in Fig. 3. Photoconversion of these

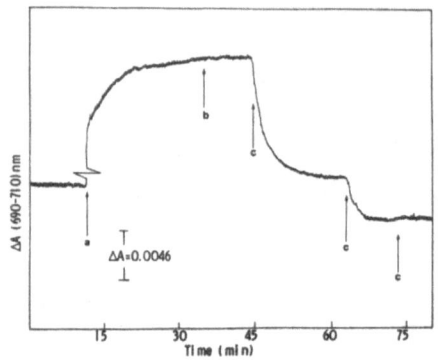

FIGURE 3. Effect of
pchlide on the Shibata
shift *in vitro*.

membranes produces an absorbance increase at 690-710nm due to
formation of chlide 684 (see Figs. 2 and 3). On addition of a
methanolic solution of pchlide (point C Fig. 3), the sample
undergoes a rapid decrease in absorbance at 690nm as a result
of the rapid Shibata shift giving chlide 672. The spectral
forms generated *in vitro* appear qualitatively similar to the
forms observed during photoconversion of pchlide in whole
leaves and on this basis it is assumed that the scheme shown
in Fig. 2 is also a true reflection of the *in vivo* photoconversion process.

That our interpretation of the Shibata shift holds good
for the *in vivo* situation can be seen from the whole leaf
measurements shown in Fig. 4. In this experiment, etiolated
bean leaves were allowed to accumulate different amounts of
pchlide by pre-feeding for varying lengths of time (0, 6 and
20 hours) with the porphyrin precursor δ-aminolevulinic acid
(ALA). Each sample was then photoconverted whilst monitoring
absorbance changes at 690-710nm as before. In the control
leaves the t 1/2 of the Shibata shift - measured as the time

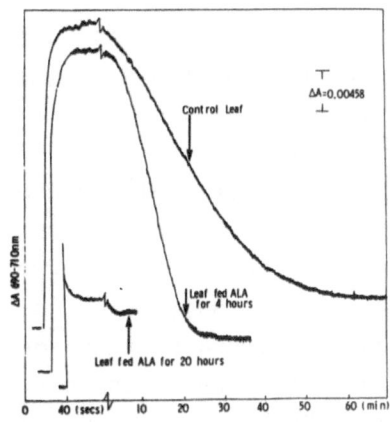

FIGURE 4. Effect of ALA feeding of bean leaves on the
Shibata shift *in vivo*.

taken for the absorbance at 690 to decrease to half the maximum extent after the initial photoconversion - was 18 minutes.
For the 6 hour ALA fed leaves, containing approximately twice
as much pchlide as the control leaves, the Shibata shift was
much more rapid, being half complete in 10 min. In the 20 hour
fed leaves, containing a large excess of pchlide, the Shibata
shift was too fast to measure and no red shifted chlide absorbing at 684nm accumulated in this sample (Fig. 4). This de-
monstrates therefore that whole leaves respond to excess
pchlide in causing stimulation of the Shibata shift in much
the same way as isolated membranes.

There is in fact already ample evidence in the literature
to support this suggestion. Many workers have already shown
that ALA feeding of etiolated leaves results in marked stimulation of the Shibata shift (12, 13, 14). A barley mutant
(alb f[17]) which contains no non-photoactive pchlide (15) does
not show the Shibata shift. Interestingly however, such a shift
can be induced in plastid membranes prepared from the mutant
by pchlide addition (Oliver and Griffiths, unpublished work)
further supporting our suggestion. It is also apparent from
the model put forward in Fig. 2 that the turnover rate of the
pchlide photoconversion cycle is governed by the availability
of pchlide, which of course, under normal conditions reflects
the flux through the porphyrin pathway providing this pigment.

One amazing property of pchlide reductase in etiolated

plants is the way it responds to illumination of the plants.
As already mentioned, such a treatment results in a rapid
decrease in activity of the enzyme due to a decrease in the
amount of enzyme protein (4,2). In this early work however it
was noted that sufficient of the enzyme was retained following
illumination to accommodate chlorophyll biosynthesis by the
greening plant (1). In the meantime however, despite this
report, claims of the complete absence of the reductase from
normally grown plants have appeared with the suggestion that
some other mechanism of pchlide reduction operates in green
plants (6). This has led to our reinvestigation of the status
of pchlide reductase in green tissues. In this study, young
oat and barley seedlings were grown under normal greenhouse
conditions except that immediately prior to harvesting the
plants were given either an extended (16 hours) dark or (30
hours) light periods respectively and designated dark and
light plants. Analysis of pchlide reductase activity in the
various samples (Table 1) indicates that the enzyme could be
detected in the darkened oat and barley plants whereas in the
light plants a considerably reduced level of activity could
be detected in the oat extracts only.

Table 1. Protochlorophyllide reductase activity of plastids
from dark and light oat and barley seedlings.

Seeds were germinated in a greenhouse under an 18 hours light
and 6 hours dark regime for 6 days (barley) and 11 days (oats).
Before harvesting the seeds were exposed to continuous light
for 30 hours (light) or darkness for 16 hours (dark) and plastids isolated and assayed for the reductase.

Sample	Protochlorophyllide reductase activity units min^{-1} mg protein^{-1}
'dark' oats	0.17
'light' oats	0.05
'dark' barley	0.15
'light' barley	0.01

On comparing the peptide profile of the plastids from the dark
and light barley (Fig. 5) the only difference found is in a
peptide with approx. mol. wt. 36 kD (arrowed in Fig. 5), pre-

sent in the dark sample but hardly visible amongst the peptides of the light sample. The identity of this band as pchlide
reductase has been ascertained using the ^{3}H-N-phenylmaleimide
labelling technique (2) thereby providing confirmation of the
data on the activity of the reductase in the dark and light
barley already presented (Table 1).

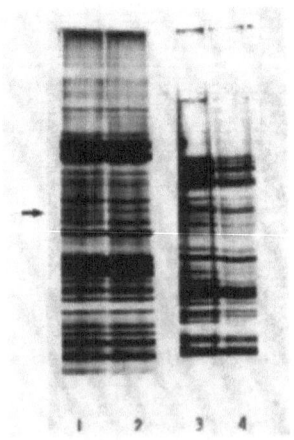

FIGURE 5. Peptides from chloroplasts
of : dark barley (1), light barley
(2), dark oats (3), light oats (4).

On application of the infinitely more sensitive technique
of immune blotting to protein extracts taken from the plant
samples (Fig. 6) it is apparent that traces of the reductase
rendered visible by this technique are present in all the
samples. Thus, even the extracts from the light barley plants,
plastids from which had no detectable reductase activity
(Table 1), gave a faint positive reaction for the reductase
by the immune blotting technique. On a quantitative basis it
appears from this data (Table 1 and Fig. 6) that illumination/
darkness affects the level of the reductase in green tissue,

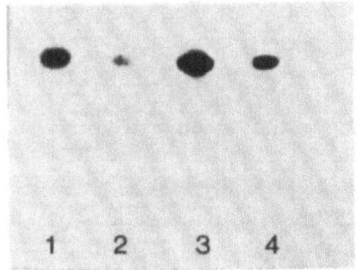

FIGURE 6. Immune blotting of
pchlide reductase from : (1) dark
barley, (2) light barley and (3)
dark oats, (4) light oats.

the amount of enzyme being low after a period of illumination
and increased after a period of darkness. This illumination
induced diurnal variation of the level of the reductase has
some important consequences in discussions on mechanisms of
regulation of the reductase by light (see below). The problem
remains as to the identity of the process(es) involved in the
light regulation of the level of pchlide reductase in normal
plants. This question must obviously consider the effect of
light on both breakdown and synthesis of the enzyme.

The effects of illumination on the breakdown of the
reductase comes dramatically into play during illumination of
etiolated plants as already mentioned. We have recently been
successful in reproducing this light induced breakdown in
isolated etioplast membranes and have furthermore demonstrated
that the process can be inhibited either by the presence of
added substrates (pchlide and NADPH) or by certain protease
inhibitors such as antipain (16). This data we have interpreted as implying the existence, on etioplast membranes, of a
protease capable of breaking down the reductase whenever the
latter exists free of its substrates e.g. after illumination.
The presence of the substrate presumably keeps the reductase
in some conformation which is resistant to proteolytic attack.

It is also possible to demonstrate substrate protection
of pchlide reductase against light induced breakdown in whole
leaves. The endogenous pchlide level of etiolated leaves can, as
already mentioned, be considerably elevated by prefeeding
with ALA. Illumination of such plants (with red light to avoid
photodestruction) results in reduced loss of the reductase,
activity and protein, compared with control plants (Fig. 7).
While we know of no way of artificially elevating the NADPH
levels in leaves it is interesting that ALA feeding appears
to increase the state of reduction of NADPH in the etioplasts
compared with the control plastids from unfed plants (Fig. 7).
This factor must also contribute to the substrate protection
of the reductase observed in these experiments.

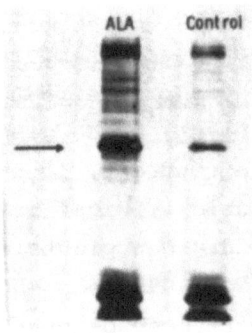

FIGURE 7. Effect of
ALA feeding of leaves
on illumination induced
loss of pchlide reduc-
tase from membranes
(arrow indicates posi-
tion of reductase).

	ALA	Control
Reductase activity (units mg protein^{-1})	2.01	0.82
Plastid NADPH/NADP^{+}	1.5	1.1

Establishing the existence and properties of the protease
has meant that a satisfactory explanation can now be made
for the loss of reductase which accompanies illumination of
etiolated plants. Illumination produces the reductase-products
complex. On release of the products free reductase is generated and this acts as a substrate for the protease. A certain
amount of enzyme survives breakdown however, even in the light,
this reflecting the steady state level of pchlide present in
the light. The existence of such a pool of pchlide in illuminated plants is already well established (17,18). In summary
therefore, light through the activity of the protease, controls the degradation of the reductase ensuring a steady state
level of the enzyme compatible with the prevailing level of
pchlide.

Another possible way light may regulate the level of
the reductase is obviously *via* its effect on the synthesis of
the enzyme. In this respect in etiolated barley plants, it
has been shown that illumination, acting through the phytochrome system, results in a dramatic decline in the level of
reductase mRNA, this disappearing completely after 24 hours
of illumination (6,19). This result implies that such plants

lose the ability to synthesise the reductase and, since light simultaneously also causes breakdown of the reductase it has prompted the suggestion that pchlide reduction during chlorophyll synthesis by green plants is carried out by some other mechanism (6). However, in contrast, it has been demonstrated in the current article (Table 1, Figs. 5 and 6) that active pchlide reductase exists in green plants of both oat and barley. Furthermore, this amount has been shown to increase after a dark period implying that these plants, at least when returned to darkness, have retained (or recovered) the ability to synthesise and accumulate the enzyme. Accumulation of the enzyme in darkness in these plants is undoubtedly due to stabilisation of the newly synthesised molecule by the simultaneously synthesised pchlide and endogenous NADPH. We do not know however if the rate of synthesis of the reductase in the light is similar to that occuring in darkness. If it is, then the lower level of enzyme accumulated in the light (Table 1, Figs 5 and 6) in green plants can be accounted for on the basis of enhanced proteolysis as a result of liberation by light of free protease susceptible enzyme.

This discussion on the light regulated diurnal variation in the reductase has ignored any contribution to the process from changes in transcription of the reductase mRNA. It may well be that the true situation in the living plant is far more complex than the simple picture presented here and may well involve in addition important contributions from a light regulated mRNA pool.

ACKNOWLEDGEMENT

The work described in the article was carried out with generous financial assistance from the Science and Engineering Research Council (UK) (Grant n°. GR/B/71992).

REFERENCES

1. Griffiths WT. 1978. Biochem. J. 174 : 681-692.
2. Oliver RP, Griffiths WT. 1980. Biochem. J. 191 : 277-280.
3. Apel K, Santel HK, Redlinger TE, Falk H. 1980. Eur. J. Biochem. 111 : 251-258.
4. Mapleston RE, Griffiths WT. 1977. Biochem. Soc. Trans. 5 : 319-321.
5. Mapleston RE, Griffiths WT. 1980. Biochem. J. 189 : 125-133.
6. Apel K. 1981. Eur. J. Biochem. 120 : 89-93.
7. Beer NS, Griffiths WT. 1981. Biochem. J. 195 : 83-92.
8. Oliver RP, Griffiths WT. 1981. Biochem. J. 195 : 93-101.
9. Boardman NK. 1962. Biochim. Biophys. Acta 62 : 63-79.
10. Oliver RP, Griffiths WT. 1982. Plant Physiol. 70 : 1019-1025.
11. Shibata K. 1957. J. Biochem. (Tokyo) 44 : 147-173.
12. Gassman ML. 1973. Plant Physiol. 52 : 295-300.
13. Klockare B, Sundqvist C. 1977. Photosynthetica 11 : 189-199.
14. Walter G. 1974. Photosynthetica 8 : 40-46.
15. Henningsen KW, Thorne SW. 1974. Physiol. Plant. 30 : 82-89.
16. Kay SA, Griffiths WT. 1983. Plant Physiol. 72 : 229-237.
17. Shlyk AA, Savchenko GY, Averina NG. 1969. Biofisika (English translation) 14 : 119.
18. Dujardin E, Correia M, Sironval C. 1981. In : Photosynthesis and Chloroplast development. Philadelphia Pa. Balaban International Science Services. Ed. G. Akoyunoglou.
19. Santel HJ, Apel K. 1981. Eur. J. Biochem. 120 : 95-103.

CHANGES IN THE PROPERTIES OF PROTOCHLOROPHYLLIDE REDUCTASE DURING EARLY GREENING OF ETIOLATED SQUASH COTYLEDONS

M. IKEUCHI* and S. MURAKAMI

Department of Biology, University of Tokyo, Komaba, Meguro-ku, Tokyo 153, JAPAN

INTRODUCTION

A photoconvertible complex of protochlorophyllide(Pchld) and protein, called Pchld holochrome has been isolated from various etiolated seedlings with(1-6) or without(7-9) the aid of detergents. The molecular size of the Pchld holochrome has been extensively studied, but is still controversial(10) ranging widely from 600,000 to 63,000. Changes in size, and spectroscopic and physicochemical properties have also been reported. Illumination of the isolated Pchld holochrome induces a decrease of its apparent molecular size, probably indicating partial dissociation of the components of the Pchld holochrome(11-13).

It is now generally accepted that the Pchld holochrome is an aggregate of the photoactive Pchld-Pchld reductase complex. Pchld reductase(PR) which forms a ternary complex with Pchld and NADPH has a single polypeptide with a molecular weight of 36,000 (5, 6) and is located mostly in the prolamellar body of etioplasts (14). In our previous work(6) the Pchld-PR complex was solubilized with Triton X-100 and fractionated by glycerol density gradient centrifugation originally used by Apel et al.(5). The Pchld-PR complex thus obtained was an oligomeric form of the complex. Therefore, Triton X-100-solubilized preparation can be used as a material to study light-induced changes in physicochemical and biochemical properties of the Pchld-PR complex.

In the present work, we studied the size and spectroscopic properties of Pchld- and Chld-PR complexes solubilized with

* Present address: Solar Energy Group, Riken, Wako-shi, Saitama 351, JAPAN

C. Sironval and M. Brouers (eds.); Protochlorophyllide Reduction and Greening. ISBN 90 247 2954 8
© 1984, Martinus Nijhoff/Dr W. Junk Publishers, The Hague/Boston/Lancaster.

32

Triton X-100 and the activity of PR in the early phase of greening of squash etioplasts. Light-induced changes in the structure of the prolamellar body were also investigated to understand
the role of PR in the prolamellar body on a structural basis.

MATERIAL AND METHODS

Plant material. Etiolated seedlings of squash(Cucurbita
moschata Durch. var. melonaeformis Makino cv. Tokyo) were grown
in the dark for 7 days at 25 C. For greening, they were exposed
to light from fluorescent lamp(3,000 lux) for various periods.

Preparation of plastid membranes. Intact plastids were
isolated from cotyledons by Percoll density gradient centrifugation as described elsewhere(14). They were ruptured hypoto-
nically by suspending in 40 mM Tris-HCl(pH 7.7), 1 mM Na_2EDTA,
1 mM $MgCl_2$ and 5 mM 5-mercaptoethanol and the plastid membranes
were pelleted by centrifugation at 4,500 xg for 5 min.

Fractionation of PR. The Plastid membranes(2 mg protein/ml)
were treated with 2 mM Triton X-100 in 20 mM HEPES-KOH(pH 7.0),
5% glycerol, 1 mM NADPH, 1 mM Na_2EDTA and 1 mM $MgCl_2$ and centrifuged at 30,000 xg for 30 min. The supernatant containing
the solubilized pigment-PR complexes was fractionated by 10-30%
glycerol density gradient centrifugation in the presence of 1 mM
Triton X-100 according to Ikeuchi and Murakami(6).

Regeneration of photoactive Pchld. Etiolated cotyldons
were illuminated for various periods to convert Pchld to Chld,
then incubated for 10 min in the dark at 25 C for regeneration
of photoactive Pchld. The plastid membranes prepared as
described above were incubated with 1 mM NADPH and irradiated
with a single flash of sufficient intensity. Newly formed
Chld showed an absorption peak at 684 nm irrespective of the
period of illumination.

Other procedures. The assays of pigments and the activity
of PR were carried out as described previously(6). Absorption
and fluorescence spectra were recorded, respectively, with
UV 3,000 spectrophotometer(Shimadzu) and RF 502 spectrofluorome-
ter(Shimadzu) at liquid nitrogen temperature. The pigments were
excited at 436 nm. The data were not corrected for variation

in the sensitivity of photomultiplier with wavelength. The method for electron microscopy was described previously(15) except that Spurr's resin was used instead of Luft's Epon mixture.

RESULTS

The distribution pattern of the activity of PR in the fractions prepared in a glycerol density gradient containing 1 mM Triton X-100(Fig. 1A) was essentially the same as that reported previously(6). The same distribution of Pchld in the fractions was obtained, revealing that the Pchld-PR complex was solubilized and fractionated. Our previous study(6) showed that the Pchld-PR complex-enriched fraction was located between the coupling factor-enriched fraction and a fraction of free pigments probably bound to Triton X-100 micelles[mol wt 90,000(16)]; the Pchld-PR complexes seemed to be in their aggregated form. Flash illumination of the solubilized Pchld-PR complexes induced full photoconversion of Pchld to Chld, but did not affect at all the sedimentation profile of the aggregate of Pchld-PR complex(Fig. 1B). The distribution of Chld was very similar to that of the PR activity in the glycerol density gradient. It is evident that the aggregates of pigment-PR complexes do not change their

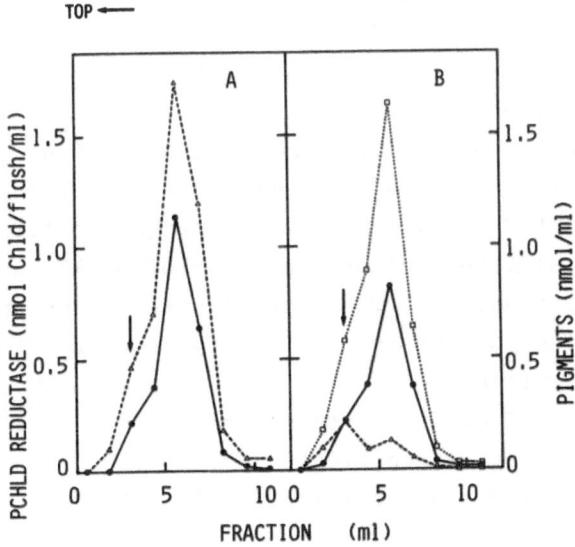

FIGURE 1. Sedimentation profile of solubilized PR complex in glycerol density gradient. PR solubilized with 2 mM Triton X-100 was fractionated before(A) and after(B) flash illumination to convert all photoactive Pchld. (–●–), PR; (–▵–), Pchld; (···□···), Chld. Arrows indicate carotenoid-enriched fraction.

34

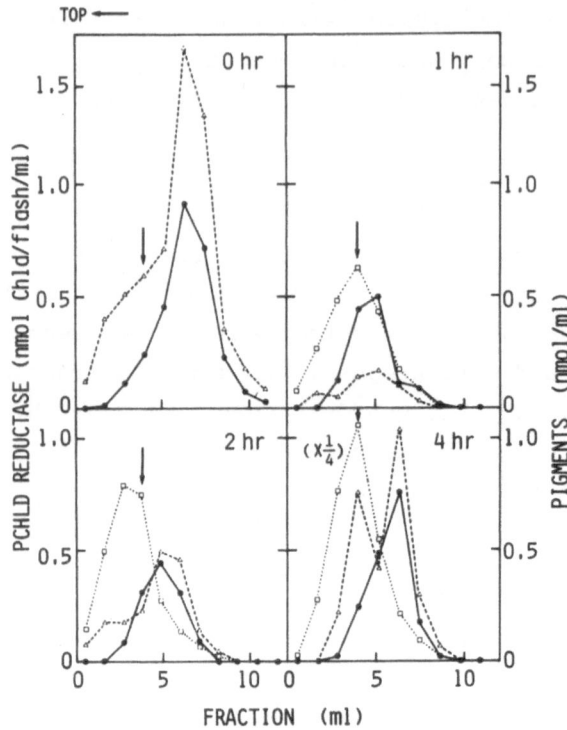

FIGURE 2. Sedimentation profile of solubilized PR complex from etioplasts(0 hr) and etiochloroplasts continously illuminated for various periods(1 hr, 2 hr, 4 hr). (—●—), PR; (-▵-), Pchld; (···□···), Chld. Arrows indicate carotenoid-enriched fraction.

size upon illumination and that most of Chlds are still associated with the PR immediately after illumination.

Sedimentation profiles of the aggregates of pigment-PR complexes solubilized with 2 mM Triton X-100 from etioplasts and greening etiochloroplasts are shown in Fig. 2. The sedimentation of the aggregates which was not altered immediately after flash illumination, retarded significantly after 1-2 hr of illumination and reverted to the original position after 4 hr of illumination. When analyzed by SDS-polyacrylamide gel electrophoresis, the PR polypeptide from 1 hr-illuminated etiochloroplasts had the same size(36,000 daltons) as that of unilluminated etioplasts(data not shown). These results show that the aggregate of pigment-PR complexes is partially dissociated to a smaller one during the first 1 hr of illumination and restored its original size after 4 hr with no change in the molecular weight of the monomer polypeptide.

The distribution of Chl(d) in the fraction was different

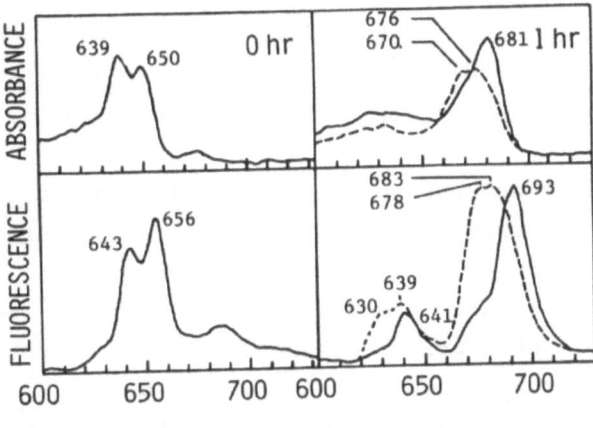

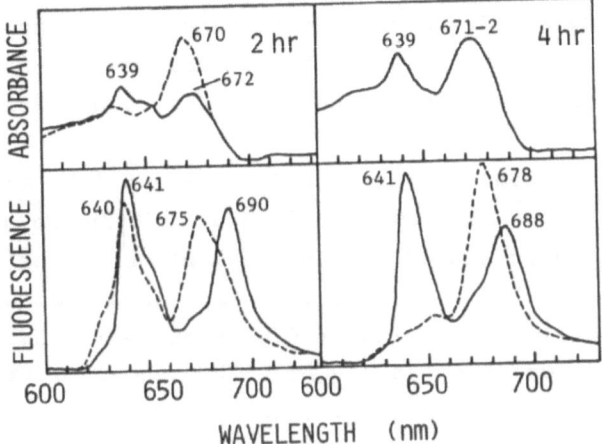

FIGURE 3. Absorption and relative fluorescence spectra of PR-enriched and carotenoid-enriched fractions at -196 C. Fractions were obtained from the glycerol density gradient shown in Fig. 2. (——), PR-enriched fraction; (--), carotenoid-enriched fraction.

from that of the activity of PR and rather similar to that of carotenoids. This seems to show that Chl(d)s are contained in detergent micelles together with carotenoids. On the other hand most of regenerated Pchl(d)s in 2-4 hr-illuminated etiochloroplasts were found in the fraction with high PR activity, although some Pchl(d)s were in the carotenoid-enriched fraction. This indicates that considerable amounts of regenerated Pchl(d) are associated with the PR.

Absorption and fluorescence spectra at low temperature of the PR-enriched fractions in glycerol density gradients from etioplasts and etiochloroplasts are compared with those of the carotenoid-enriched fractions(Fig. 3). The Pchld-PR complex extracted from etioplast membranes showed absorption peaks at 639 nm and 650 nm and fluorescence peaks at 643 nm and 656 nm. This indicates that there are two forms of the photoactive Pchld, Pchld 640 and Pchld 650, in the Pchld-PR complex prepared in the presence of Triton X-100.

The PR-enriched fraction prepared from 1 hr-illuminated etiochloroplasts showed an absorption peak at 681 nm and fluorescence peaks at 693 nm and 641 nm. This indicates that Chl(d) and Pchl(d) are associated with the PR. The carotenoid-enriched fraction showed a broad absorption peak of Chl(d), in which two small peaks could be recognized at 670 nm and 676 nm. Two fluorescence peaks of Chl(d) were also recognized at 678 nm and 683 nm. Pchl(d)s emitting at 630 nm and 639 nm were also found in this fraction. This indicates that both free and protein-associated pigments are contained in the carotenoid-enriched fraction.

The PR-enriched fractions isolated from 2 hr- and 4 hr-illuminated etiochloroplasts similarly contained Chl(d) absorbing at 671-672 nm and emitting at 688-690 nm and Pchl(d) absorbing at 639 nm and emitting at 641 nm, whereas the carotenoid-enriched fractions contained Chl(d) absorbing at 670 nm and emitting at 675-678 nm and Pchl(d) emitting at 640 nm with a shoulder around 630 nm. From the observations presented in Fig. 3 we favor the view that Pchl(d) and Chl(d) in the PR-enriched fraction are associated with PR, whereas those in the carotenoid-enriched fraction are mostly free pigments, although they may be associated with detergent micelles.

When etioplasts were illuminated, the activity of PR drastically decreased within 1 hr(Fig. 4). Subsequent illumination for 3 hr hardly exerted further decrease of the activity, but after 4 hr the activity started to decrease again. It has been previously reported that a 36,000-dalton polypeptide of the PR dcreased gradually during greening of the etiochloroplasts under continous illumination(17, 18). The drastic change in the activity of PR in the first 1 hr of greening does not match the change in the amount of PR protein. The slow decrease of the activity of PR after 4 hr of illumination seemed to be coupled

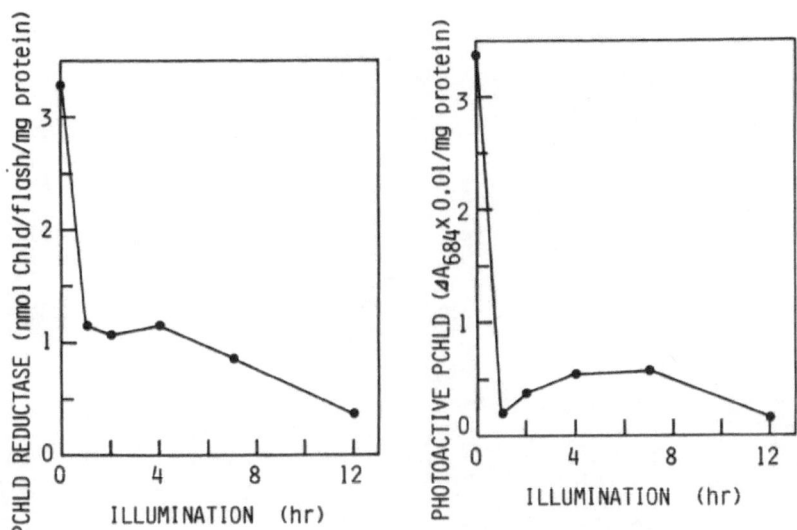

FIGURE 4. Change in the activity of PR of plastid inner membranes during greening.

FIGURE 5. Regeneration of photoactive Pchld during greening. Etioplasts were illuminated for various periods, then incubated for 10 min in the dark. Regenerated Pchld was photoconverted to Chld and estimated.

with a gradual decrease of PR protein. In addition Chl synthesis started after 2 to 4 hr lag and reached the maximal rate after 6 hr(15). This change in Chl synthesis also does not match that in the activity of PR.

It is reasonable to assume that the regeneration of Pchld, substrate of the PR, might regulate the rate of Chl synthesis. The regeneration of the photoactive Pchld was extremely low after 1 hr of illumination but increased during subsequent illumination of up to 7 hr(Fig. 5). Further illumination caused a decrease of the regeneration of Pchld. This change in Pchld regeneration seems to correspond with the change in Chl synthesis during the early phase of greening.

In squash etioplasts Pchld and PR are located mostly in a paracrystalline prolamellar body(14). Structural changes of the prolamellar bodies during greening are shown in Fig. 6. After 1 hr of illumination, the paracrystalline configuration of the prolamellar body was completely disordered, loosened

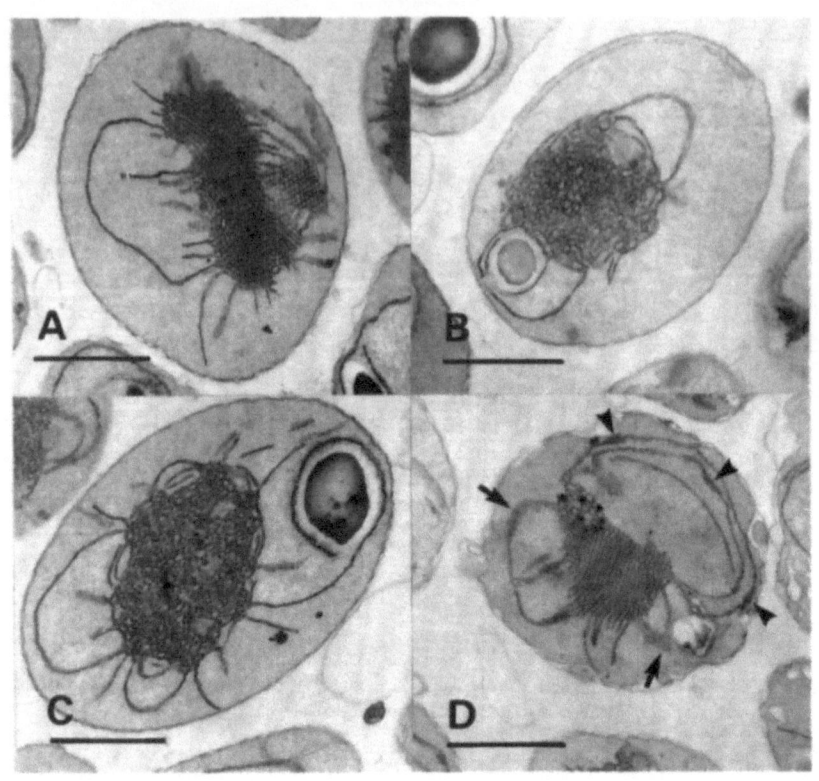

FIGURE 6. Electron micrographs of representative etioplasts(A) and greening etiochloroplasts(B, 1 hr; C, 2 hr; D, 4 hr illumination). Arrow, perforated prothylakoid; arrow head, grana stack. Bar indicates 1 μm.

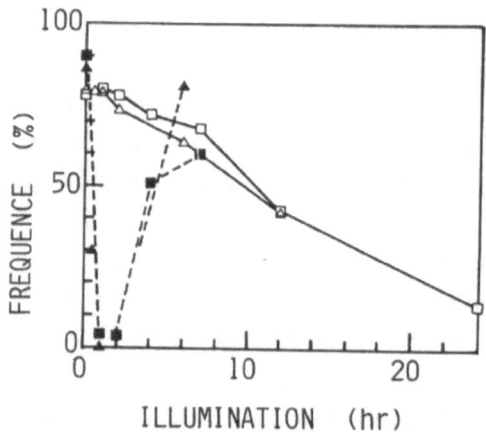

FIGURE 7. Frequency of the para-crystalline prolamellar bodies and the plastids with prolamellar body in electron micrographs of plastid preparations during greening. (--■--), (--▲--), para-crystalline prolamellar body; (--□--), (--△--), plastid with pro-lamellar body. Experiment I, (■, □); Experiment II; (▲, △).

and rearranged, resulting in a cluster of randomly arranged
vesicles or wavy tubules which are connected to each other.
Such configuration of the illuminated prolamellar body remained
unchanged during the next hour. However, an illumination of
more than 3 hr caused reformation of the paracrystalline prolamellar body similar to that found in the unilluminated etioplasts. These structural changes can be seen more evidently
in Fig. 7 which shows the time course of the light-dependent
disordering and reformation of the paracrystalline structure
of the prolamellar body. Frequency of the plastids with prolamellar bodies during greening process is also shown in Fig. 7.
It decreased gradually after 2 hr of illumination. The average
size of the prolamellar body seemed to decrease in a similar
way, since the size of the plastids did not change up to 12 hr
of illumination.

DISCUSSION

Squash PR forms a ternary complex associated with Pchld and
NADPH, and has a single polypeptide with a molecular weight of
36,000. The Pchld-PR complex solubilized with Triton X-100
sedimented between the coupling factor ATPase(mol wt, ca.
350,000) and the carotenoid-enriched fraction in glycerol density gradient(6). Carotenoids are likely associated with
Triton X-100 micelles of molecular weight of about 90,000(16).
Evidently the Pchld-PR complex solubilized with Triton X-100
is not a monomer, but an aggregate, or polymeric form of the
complexes.

Sedimentation profiles of the solubilized pigment-PR complexes presented in this work have shown a reversible change
in the size of the aggregate of the complexes in the early
phase of plastid development. Light-induced decrease of size,
or disaggregation of Pchld holochrome in vitro has been reported by other investigators(2-4). Cannani and Sauer(3) observed with bean that the Pchld holochrome(mol wt 600,000) was
disaggregated to produce a Chld-protein complex with a molecular weight smaller than 70,000, concomitantly with disintegration of the subunit of the holochrome(mol wt 45,000) to a small

Chld-binding polypeptides(mol wt 16,000), when crude Pchld holochrome was illuminated then incubated in the dark. This is
contradictory to our observation with purified Pchld-PR complex
of squash. Molecular weight of the PR polypeptide(mol wt
36,000) remained unchanged before and after illumination of the
etioplasts at least up to 4 hr, and neither fragments nor dis-
integrated products from the 36,000-dalton PR polypeptide were
detected in squash(15) and barley plastids(17) after illumination. This indicates that changes in the size of the aggregate
of the pigment-PR complex are not caused by restricted proteolysis of the PR polypeptide.

It was shown that during continuous illumination of etioplasts up to 4 hr, the physical state of the pigments changed
as seen in their sedimentation profiles(Fig. 2) and spectral
properties(Fig. 3). These light-induced changes in pigments
bound to the PR leading to changes in configuration of PR protein or pigment-PR interaction, may be responsible for changes
in the size of aggregate of the pigment-PR complexes.

Immediately after photoconversion most, if not all, of Chlds
were associated with the PR whereas considerable amounts of
Chl(d) were released from the PR after more than 1 hr of illumination. Within 1 hr of illumination, newly formed Chld is
phytylated to produce Chl(19) which has probably less affinity
to the PR than Chld since Pchl has less affinity to the PR than
Pchld(6). This resulted in a release of Chl. In exchange for
released Chl(d), regenerated Pchld can bind to the PR molecule.
In addition, photosynthetic Chl-binding proteins seemingly at-
tract Chl released from the PR(13). A small amount of Chl(d)
which was still associated with the PR is probably nascent Chld
newly formed during illumination.

The activity of PR in squash plastids decreased rapidly during the first 1 hr of illumination, remained nearly constant up
to 4 hr, then decreased gradually under continued illumination.
Similar drastic decrease of the PR activity was reported in
barley(17, 18). These changes do not follow the time course of
degradation of 36,000-dalton PR polypeptide during greening of
squash cotyledons(15). The activity of PR in squash etiochloro-

plasts illuminated for 1 hr was one third of that in the unil-luminated etioplasts, whereas the amount of the enzyme polypep-tide decreased by only 10 per cent in 1 hr-illuminated etiochlo-roplasts. Therefore, degradation of the PR polypeptide can be excluded for the cause of the rapid activity decrease. It has been established that not only Pchld but also NADPH are required to stabilize a high PR activity(6, 20). Pchld is exhausted upon illumination and is not supplied until Pchld is regenerated. Illumination also induced a decrease in NADPH content(21). Con-sequently the shortage of substrate may be responsible for the low activity of the PR in etiochloroplasts during the early phase of greening.

In etiochloroplasts illuminated for 6 to 12 hr, Chl was syn-thesized at maximal rate, although the activity of PR was rather low. Therefore, in these plastids photoconversion of Pchld to Chld which is catalyzed by PR seems unlikely to limit the rate of Chl synthesis in vivo when an appropriate amount of Pchld is sup-plied. During continuous illumination Chl synthesis started after a 2 hr lag, then proceeded progressively with the increase of regeneration of Pchld. In support to this, Nadler and Granick(21) reported that supply of δ-aminolevulinic acid limited the rate of Chl synthesis in vivo in bean etiochloro-plasts.

Pchld accumulated in the dark was converted to Chld immedi-ately upon illumination; a significant amount of Pchld was, however, regenerated during the subsequent 4 hr of illumination. In a similar time course the size of the aggregate of Pchld-PR complex decreased and was restored as seen in the sedimentation profiles, indicating reversible changes in physicochemical prop-erties of the Pchld-PR complex. Light-induced structural changes of prolamellar body also followed a similar time course; para-crystalline configuration of the prolamellar body was disordered by a brief illumination and reformed after 4 hr in the light. These findings suggest that rebinding of Pchld to the PR leads to a restoration of photoactive Pchld-PR complex and concomitantly recrystallization of tubular elements in the prolamellar body. If this is the case we can conclude that Pchld-PR complex is

responsible for the paracrystalline structure of the prolamellar
body in etioplasts.

ACKNOWLEDGMENT
    We thank Prof. Y. Inoue of Riken for his valuable comments
and technical advise.

REFERENCES
1.   Schopfer P, Siegelman HW. 1968. Plant Physiol. 43, 990.
2.   Henningsen KW, Kahn A. 1971. Plant Physiol. 47, 685.
3.   Canaani OD, Sauer K. 1977. Plant Physiol. 60, 422.
4.   Stummann BM. 1979. Physiol. Plant. 45, 122.
5.   Apel K, Santel HJ, Redlinger TE, Falk H. 1980. Eur. J.
     Biochem. 111, 251.
6.   Ikeuchi M, Murakami S. 1982. Plant Cell Physiol. 23, 1089.
7.   Krasnovskii AA, Kosobutskaya LM. 1952 Dokl. Akad. Nauk. SSSR
     85, 177.
8.   Smiths JHC. 1952. Carnegie Inst. Wash. Year B. 51, 151.
9.   Boardman NK. 1966. In Vernon LP, Seely GR. eds., The
     Chlorophylls. Academic Press, New York and London. pp. 437.
10. Virgin HI. 1981. Ann. Rev. Plant Physiol. 32, 451.
11. Boardman NK. 1967. In San Pietro A, Greer FA, Army TJ. eds.,
     Harvesting the Sun. Academic Press, New York. pp. 211.
12. Bogorad L, Laber L, Gassman M. 1968. In Shibata K, Takamiya
     A, Jagendorf AT, Fuller RC. eds., Comparative Biochemistry
     and Biophysics of Photosynthesis. University Park Press,
     State College. pp. 299.
13. Henningsen KW, Thorne SW, Boardman NK. 1974. Plant Physiol.
     53, 419.
14. Ikeuchi M, Murakami S. 1983. Plant Cell Physiol. 24, 71.
15. Ikeuchi M, Murakami S. 1982. Plant Cell Physiol. 23, 575.
16. Helenius A, Simons K. 1975. Biochim. Biophys. Acta 415, 29.
17. Santel HJ, Apel K. 1981. Eur. J. Biochem. 120, 95.
18. Mapleston RE, Griffiths WT. 1980. Biochem. J. 189, 125.
19. Henningsen KW, Thorne SW. 1974. Physiol. Plant. 30, 82.
20. Steve AK, Griffiths WT. 1983. Plant Physiol. 72, 229.
21. Mapleston RE, Griffiths WT. 1978. FEBS Lett. 92, 168.
22. Nadler K, Granick S. 1970. Plant Physiol. 46, 240.

PROTEOLYTIC EFFECTS ON THE PROTOCHLOROPHYLLIDE-OXIDOREDUCTASE
IN OAT

URSULA RÖPER and CORNELIUS LÜTZ
Botanisches Institut der Univ. zu Köln,Gyrhofstr.15, 5 Köln
41,W. Germany.

The protochlorophyllide oxidoreductase (PCR) is one of the major
intrinsic membrane proteins in etioplasts (Griffiths 1978, Apel
et al. 1980, Röper et al. 1983). It catalyzes the first light-
dependent step in chlorophyll biosynthesis. The enzyme activity
is mainly found in PT-membranes of etioplasts, in comparison to
PLBs (Lütz et al. 1981, Gerday et al. 1982),which holds also
for PCR-protein (Ryberg and Sundqvist 1982, Lütz, in preparation). In case of etioplast membrane preparations from oat or
bean plants, this enzyme is described as a 37/35 KD doublet in
SDS-PAGE. It occurs as a single polypeptide (37 KD) in most
other plants. However, Röper et al. (1983) could show, that for
the oat membrane fractions the doublet is an isolation artefact,
possibly due to internal protease activity in younger etioplasts.
Bergweiler et al. (1983) demonstrated the appearance of the
35 KD band of the PCR during senescence of etiolated oat plants.

Recent work has shown, that enzyme activity as well as the PCR-
protein are easily broken down in vivo and in vitro during
illumination (Mapleston and Griffiths 1980, Kay and Griffiths
1983, Dehesh and Apel 1983). These experiments were performed
with the proteolytic activities already present in the membranes,
which were also described by Hampp and De Fillipis (1980).

Here we present studies on the destruction of the PCR using
avaible proteases with known cleavage specificities. The
resulting polypeptide pattern is described. Further the influence
of proteolytic digestion under different conditions on the
activity of the enzyme to form Chlide is followed.

Abbreviations: BE: broken etioplasts, PCR: protochlorophyllide-
oxidoreductase, PLB: prolamellar body, PT: prothylakoid

C. Sironval and M. Brouers (eds.); Protochlorophyllide Reduction and Greening. ISBN 90 247 2954 8
© 1984, Martinus Nijhoff/Dr W. Junk Publishers, The Hague/Boston/Lancaster.

Materials and Methods

Oat plants (Avena sativa var. "Flämingskrone") were grown in complete darkness for 6-7 days. Isolation of etioplasts was performed according to the methods described by Röper et al. (1983). Broken etioplasts membranes were prepared after osmotic treatment of isolated plastids (Lütz et al. 1981) and used as suspensions containing 1-2 mg protein/ ml.

Membrane proteins were digested with a) trypsin (E.C.3.4.21.4.) (Serva) b) thermolysine (E.C.3.4.24.4.)(Merck) and pronase P (Sigma).Digestions were performed in shock buffer -minus EDTA-, with 5 μg Ca^{++}/ 100 μg added for thermolysin (mild digestion), and in grinding medium minus EDTA + 15 μg Ca^{++}/ 100 μg pronase P for all other studies. In the case of the mild protease treatment an enzyme/ total protein ratio of 1:150-200 was used, and samples were transferred into 80% acetone after an incubation at 15°C for 2,15 and 60 minutes.

To study the influence of pronase P treatment on the PCR-activity and on polypeptide break-down, the enzyme was used in concentrations of 20 to 100 μg/ml suspension. Details are given in the respective figure legends. The ability of photoconversion of PChlide during or after proteolytic digestion of the PCR was determined with BEs or Triton X-100- solubilized membranes (Apel et al.1980). Cofactors were added at different stages of the experiments (1mg/ml $NADPH_2$, 3μM PChlide). Samples were illuminated for 30 sec with 1500 lux white light and extracted with 80 % acetone, or aliquots were used with Triton X-100 to run spectra of light-dark cycles (2 min darkness, 30 sec. light) during proteolysis.

Protein determination and SDS-PAGE on 11%-16% Gels are described in Bergweiler et al. (1983).

Results

1. Mild digestion of membrane proteins

A selective digestion is obtained if proteolysis takes place with low enzyme concentrations,used up to 60 min. The changes in the polypeptide pattern are shown in Fig. 1. The main interest is focused on the PCR-Doublet (35/37 KD)

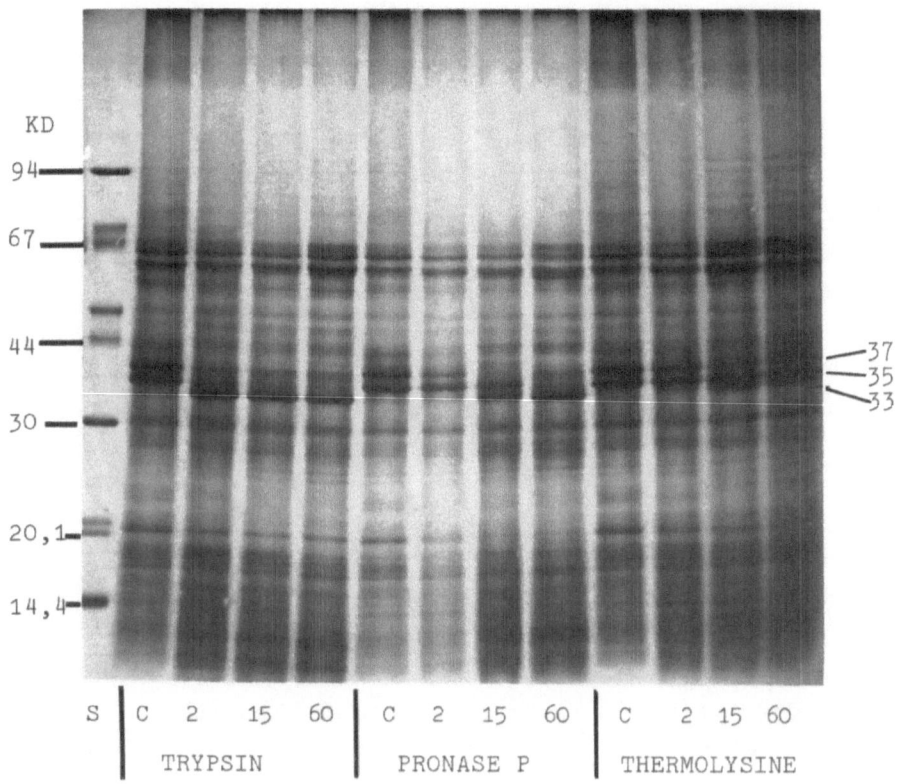

FIGURE 1. PAGE analysis after a mild treatment of BE-membranes with proteases. Digestion times: 2 min to 60 min. Enzymes were added in amounts of 1:150 (thermolysin) or 1:200 (others) to the BE-samples. S = Standard

Already after 2 min of trypsin treatment, most of the enzyme seems to run as a polypeptide of 34,2 KD. Longer treatment only slightly enhances this formation. Additionally, two other bands are formed running faster than the 14 KD reference.

Whether they are related to the PCR, has to be examined. Despite the high
cleavage capability, pronase p does not attack as fast as trypsin
does. While trypsin action affected the 35/ 37 kD peptides
simultaneously, pronase P digests the 37 kD peptide first, leading
to 35/ 34,2 kD peptides (2 min to 15 min), but after one hour
the third band appears to be the only stable band. Thermolysin
splits less effectively : after 15 min of incubation some of the
37 kD polypeptides survive and a third peptide of about 34,5 kD
is simultaneously formed. This pattern does not significantly
change within 60 min.

It is remarkable that a mild proteolytic digestion mainly affects the 35/37
kD polypeptides of the PCR, ending up in a relatively stable
peptide of 34,2 - 34,5 kD. Other proteins of the whole BE protein
pattern are not significantly digested under these conditions, but
remain stable. This is valid expecially for a 31 kD- and a 17 kD
polypeptide.

2. Complete digestion of membrane proteins

If higher concentrations of pronase P are used (1:100-1:50), a
complete change of the polypeptide pattern from BE is observed
(Fig. 2): after complete digestion of the 35/37 kD polypeptide
only degradation products in a range of molecular weights lower
than 12 kD are visible.

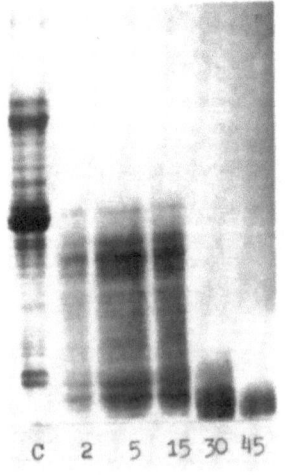

FIGURE 2. PAGE after digestion
of BE-membranes. 25 µg pronaseP/ml
and 1 mg NADPH$_2$/ml were added to
the sample and aliquots taken
after 2, 5 and 15 min. To the
30 min aliquot additional 25 µg
pronaseP/ml and to 45 min aliquot
50 µg/pronaseP/ml sample were added.
C = control.

If the experiment is performed with membranes, solubilized in
Triton X-100 the same degradation occurs, but slightly faster.

After the addition of $NADPH_2$ during pronase P digestion and subsequent illumination for 30 sec at 1500 Lux the photoconversion
of the endogenous PChlide-pool was measured. The enzyme treated
samples showed the same high Chlide formation as was observed
in the controls without enzyme: all values varied between 69%
to 75% photoconversion. Because the proteins are completely
degraded after 45 min of incubation, the photoconversion can
only be related to the small peptides being unresolved in the
lower part of the gel. In the Triton treated sample the rate of
Chlide formation is slightly lower (c. 65%, data not shown).

3. Protection of the PCR by its substrates
The influence of the substrates PChlide and $NADPH_2$ on the activity
of the PCR with or without proteolysis is presented in Fig. 3. In
this experiment the PCR, present in the isolated BE-membranes, was
digested, until a residue of 10 - 20% of the original content
remained (estimated after densitometry of the PAGE-separation).
Nevertheless, the spectra show that the degradation of the 35/37
kD PCR does not   drastically influence the Chlide formation
from exogenous PChlide. The relative increase of Chlide due to
added substrate was about 36% in the pronase-treated sample and
about 41 % in the control. The increase is referred to the Chlide
from the endogenous source, as 100%.
This experiment shows that the break-down products of the PCR are
also able to transform exogenous PChlide if $NADPH_2$ is added.

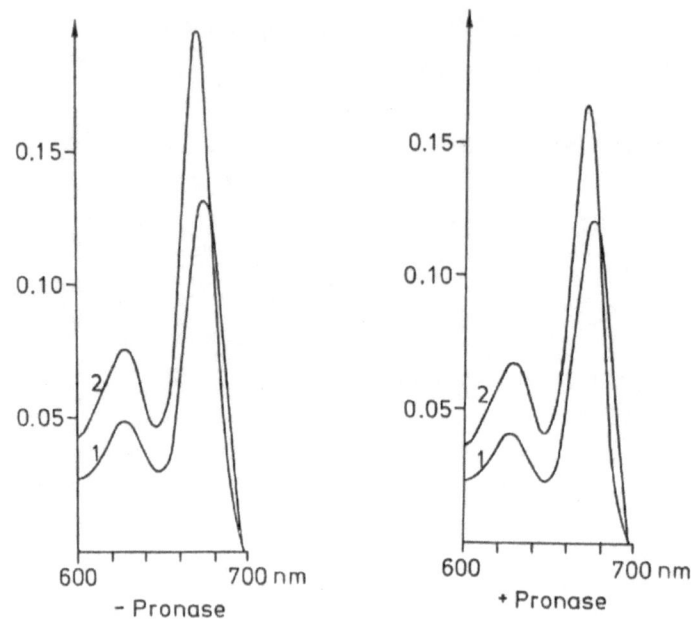

FIGURE 3: Test of protection of PCR by the substrates (in vivo spectra). Aliquots of the samples were illuminated first to empty the internal pools of PChlide and NADPH$_2$(Curves (1)). Then substrates were added, and after a dark incubation of 5 min an illumination of 5 min (1500 Lux) followed (Curves(2)).

4. PCR activity during several light treatments

After digestion of more than 85% of Triton solubilized PCR with pronase P, the cofactor (NADPH$_2$) was added to facilitate formation of Chlide. As substrate the internal PChlide pool was used. Four consecutive light treatments (each 30 sec., 1500 lux) were given to observe a continuous PCR activity. Figure 4 shows the spectra, recorded after each light treatment. The comparison indicates that the activity of the enzyme is more or less the same, either after digestion of the 35/37 kD-polypeptides or in the controls. Calculations for pigment turnover and photoconversion are given in table 1:

Table 1: Assay of pigments after several light treatments(cf.Fig4)

		PChlide µg/ml	Chlide µg/ml	Photo-conversion%	µg Chlide formed
− PRONASE	0	2,512	-	-	
	1	1,741	0,483	21,7	0,483
	2	1,475	0,703	32,2	0,220
	3	1,288	0,874	40,4	0,171
	4	1,194	0,959	44,5	0,085
+ PRONASE	0	2,101	-	-	
	1	1,791	0,434	19,5	0,434
	2	1,515	0,605	27,7	0,171
	3	1,455	0,752	34,0	0,147
	4	1,327	0,849	39,0	0,097

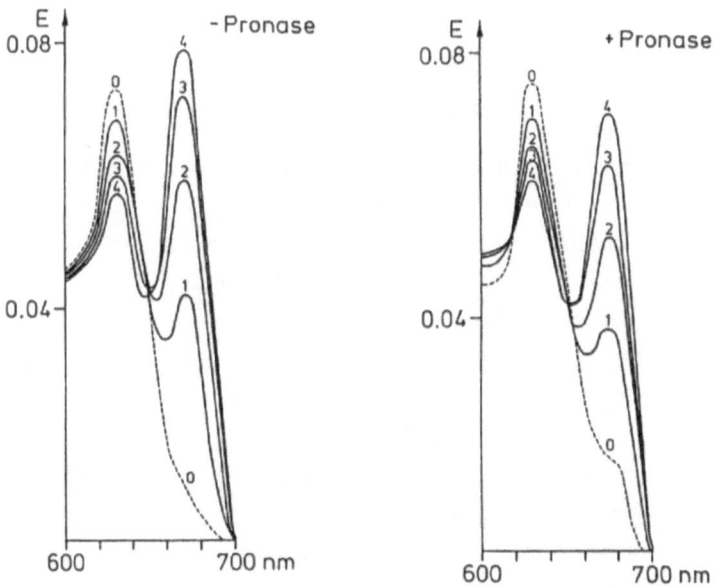

FIGURE 4: Assay of PCR activity ± Pronase in the presence of
Triton. After degradation of more than 85% of the PCR with pronase
P, the samples were incubated for 5 min in the dark, followed by
4 illuminations (30 sec. each). After each illumination, immediately a spectrum was recorded. Similar to the control, the digested
sample showed high activity of Chlide formation, originating from
the endogenous PChlide pool.

Pronase digestion of PCR in the presence of Triton X-100 .
obviously does not drastically change the activity of the enzyme
under the conditions used. A slight decrease is observed in %
photoconversion and in total Chlide formed after each illumination.

Discussion

All experiments were performed with etioplasts, isolated according
to conventional methods, and therefore showing in the PAGE the
35/37 kD doublet of the PCR (cf. Röper et al. 1983). If etioplast
membranes from oat plants are prepared so that they contain the
37 kD-PCR moiety, the same experimental results are obtained.For
a better comparison we chose the BE-membranes containing the
doublet, because this is the well-known and often described PCR-
pattern in oat. Further, it is possible to follow also changes of
the 35 kD-polypeptide, which is also photoactive (Oliver and
Griffiths, 1981).

The digestion pattern of the doublet under mild proteolytic conditions is surprising, when the cleavage specificities of the three
proteases are considered. Thermolysine, which splits mainly after
several hydrophobic amino acids (Matsubara 1970), is not efficient.
Pronase P even if used in low concentrations, would be expected to
act more rapidly than observed. However, trypsin turned out to split
off parts of the enzyme very rapidly: 2 min are efficient to form a
stable band of about 33kD. Therefore one may conclude that at
least a small part of the PCR molecule (less than 4 kD) contains
Arginine and/ or Lysine, the specific sites of action for trypsin.
Higher amounts of these amino acids in the PCR-molecule would
also explain its basic pI (Ikeuchi and Murakami 1982, Bergweiler et al. 1983).

All three enzyme digestions end after 60 min in formation of a
c. 33kD polypeptide. It seems that exogenous proteases cannot
digest further as long as the membrane structure is maintained
to protect the remaining part of the PCR. However, with higher
concentrations of pronase P (and not of trypsin or thermolysin)
the PTs are digested, while PLBs are only slightly affected
(Lütz and Tönissen, in preparation). Under these conditions,

PAGE shows only small peptides, and it is not known at the
moment which low molecular weight peptides are formed from
PCR.

Several experiments are described here to figure out possible
effects of protease attack under in vitro conditions on the activity of the PCR. Membrane structure was destroyed either by
strong pronase P action alone, or together with Triton X-100.
Though the reductase was digested and no protection by intact
membranes was possible, no significant enzyme inactivation was
observed. Endogenous as well as exogenous PChlide could be redu--
ced to Chlide -and this activity should be still present in the
smaller break-down products, as the 37/35 kD polypeptides can no
more be seen in PAGE. It is concluded. that the PCR needs only a
very small peptide to remain active and not the intact 37 kD
polypeptide.

In vivo, the PCR seems to be proteolytically degraded on illumination together with a loss of activity (Dehesh and Apel 1983,
Kay and Griffiths 1983). Protection of enzyme degradation by the
substrates could be demonstrated for the in vivo conditions; this
is different from our experiments. The membrane-bound etioplast
protease may therefore act in a different way than the proteases
used in this study, either because of other cleavage properties,
or because of a possible penetration into a more hydrophobic
environment around the active site of the PCR.

Our study also indicates that there might be no close correlation
between PCR-activity measured and the amount of the 37 kD
polypeptide present in etioplasts.

Summary

The Chlide forming enzyme protochlorophyllide oxidoreductase was
subjected to digestion with different proteases in vitro. As long
as the membrane structure is maintained, only a small peptide of
max 4 kD is removed, possibly enriched in basic amino acids. A
complete digestion of the 37/35 kD doublet of the enzyme in oat
did not significantly change the photoreduction properties.
Degradation of the enzyme by pronase P is not inbibited by the

substrates $NADPH_2$ or PChlide.It is concluded, that PCR. activity
and enzyme structure are not necesseraly related.

Acknowledment:

This work was supported by the Deutsche Forschungsgemeinschaft to
C. Lütz.

REFERENCES

1. Apel,K., Santel,H.J., Redlinger,T.E., Falk, H. (1980). The
   protochlorophyllide holochrome of barley (Hordeum vulgare L.).
   Isolation and characterisation of the NADPH: protochlorophyllide
   oxidoreductase. Eur. J. Biochem. 111, 251-258
2. Bergweiler,P., U.Röper, C. Lütz. (1983). The development and
   ageing of membrane proteins from etioplasts of Avena sativa L.
   Physiol. Plantarum, in press.
3. Dehesh,K., Apel,K. (1983). The function of proteases during
   light dependent transformation of etioplasts to chloroplasts
   in barley (Hordeum vulgare L.). Planta 157, 381-383
4. Gerday,C.,Michel-Wolwertz,M.R.,Brouers,M. (1982). Some properties of purified fractions from bean etioplast membranes. In:
   Cell Function and Differentiation, Part B, p. 25-32, Alan Liss
   Inc. New York.
5. Griffiths, W.T. (1978). Reconstitution of Chlorophyllide Formation by isolated Etioplast Membranes. Biochem. J. 174, 681-692
6. Hampp,R.,DeFilippis,L.F. (1980). Plastid protease activity
   and prolamellar body transformation during greening.
   Plant Physiol. 65, 663-668
7. Ikeuchi,M.,Murakami,S. (1982). Behavior of the 36000-dalton
   protein in the internal membranes of squash etioplasts during
   greening. Plant Cell Physiol. 23, 575-583
8. Kay,S.A., Griffiths,W.T. (1983). Light-induced breakdown of
   NADPH-protochlorophyllide oxidoreductase in vitro. Plant Physiol.
   72, 229-236
9. Lütz,C.,Röper,U.,Beer,N.S.,Griffiths,W.T. (1981). Sub-etioplast
   localization of the enzyme NADPH: protochlorophyllide oxidoreductase. Eur. J. Biochem. 118, 347-353
10.Mapleston,R.E.,Griffiths,W.T. (1980). Light modulation of the
   activity of protochlorophyllide reductase. Biochem. J. 189,
   125-133
11. Matsubara,H. (1970). Purification and assay of Thermolysin. In:
   Meth. Enzymol. XIX, 642-651, Academic Press, New York.
12.Oliver,R.P.,Griffiths,W.T. (1981). Covalent labelling of the
   NADPH:protochlorophyllide oxidoreductase from etioplast membranes with (3H)N-phenylmaleimide. Biochem. J. 195, 93-101
13. Röper,U.,Bergweiler,P.,Lütz,C. (1983). The occurence of the
    protochlorophyllide reductase from oat etioplasts as a single
    37 kD polypeptide. Z. Pflanzenphysiol. 112, 89-93
14. Ryberg,M.,Sundqvist,C. (1982). Characterisation of prolamellar
    bodies and prothylakoids fractionated from wheat etioplasts.
    Physiol. Plantarum 56, 125-132

CHARACTERIZATION OF PCHLIDE PROTEIN COMPLEXES IN PT AND PLB
ENRICHED FRACTIONS OF BEAN ETIOPLASTS. EFFECT OF NADPH.

C. GERDAY[*], M-R. MICHEL-WOLWERTZ[**], M. BROUERS[*]

[*] Laboratory of Photobiology
[**] Laboratory of Biochemical Systematics. Department of Botany
B.22, University of Liege, 4000 Liege, Belgium.

1. INTRODUCTION

Etioplasts are developed from proplastids in seedlings
grown in the dark. They are characterized by a paracrystalline
membrane structure, the prolamellar body (PLB) with attached
prothylakoïds (PT).

Etiolated bean leaves accumulate at least three PChlide-
protein complexes differing by their spectral properties.
Two of these complexes, $P_{645-637}$ and $P_{657-650}$, are reduced
by light to Chlide·proteins while the third complex $P_{633-628}$
is not photoreducible. Griffiths (1974) reported evidence
suggesting that NADPH is the hydrogen donor for PChlide photoreduction. The illumination of an etiolated leaf produces
first two forms of Chlide complexes : $P_{688-678}$ and a low
amount of $P_{675-670}$. $P_{688-678}$ is transformed into $P_{696-684}$
within a few min at room temperature (Sironval and Michel,
1967). Then $P_{696-684}$ undergoes the Shibata shift giving $P_{685-673}$
within 30 min (Shibata, 1957).

Studying the Chlide forms which accumulate when etiolated
leaves are illuminated by a series of brief flashes, Michel

Abbreviations

PT = prothylakoïd; PLB = prolamellar body; PChlide = protochlorophyllide or/and protochlorophyll; Chlide = chlorophyllide;
Px-y = pigment-protein complex, x refers to the fluorescence
emission maximum and y to the red absorption maximum at 77°K;
TES = (N-tris [hydroxymethyl]methyl 2 - aminomethane) sulphonic
acid; HEPES = 4-(2-hydroxyethyl)-1-piperazine ethane sulphonic
acid; NADPH = Nicotinamide adenine dinucleotide phosphate
(reduced form); CF1 ATPase = chloroplast factor-1-adenosine
5'-triphosphatase; SDS = sodium dodecylsulphate; mol.wt =
molecular weight; δALA =δaminolevulinic acid.

C. Sironval and M. Brouers (eds.); Protochlorophyllide Reduction and Greening. ISBN 90 247 2954 8
© 1984, Martinus Nijhoff/Dr W. Junk Publishers, The Hague/Boston/Lancaster.

and Sironval (1973) have shown that this illumination produces preferentially $P_{675-670}$ or $P_{696-684}$, depending on the frequency of the flashes. They have suggested that there were two forms of photoreducible complexes yielding $P_{675-670}$ or $P_{696-684}$ after illumination. Belyaeva and Litvin (1981) have shown further that the spectral characteristics of the Chlide complexes was dependent on the wavelength of the actinic light (blue or red). They have proposed that there were two photoreactions with different action spectra which involved two separate pools of PChlide complexes. Neither Belyaeva and Litvin, nor Michel and Sironval attached specific spectral characteristics to the photoreducible PChlide pools which their results suggested to be.

El Hamouri and Sironval (1983) obtained fractions enriched in photoreducible PChlide-protein complexes which differed by the spectral properties of the Chlide-protein complexes formed on illumination. They found that the constitutive polypeptides of these fractions were different.

The separation procedure followed by El Hamouri and Sironval -based on sucrose density gradient centrifugation of lysed and sonicated etioplast preparations- is often used to prepare PTs and PLBs enriched fractions. Recently, the separation of PTs from PLBs has been the subject of several researches which have rised questions about the preferential localization of photoreducible PChlide complexes :

1. Preparing PT and PLB fractions from oat etioplasts, Lütz (1978), Lütz et al (1981) have claimed that the amounts of total PChlide, of the photoreducible part of this pigment and of the NADPH-photooxidoreductase were higher in the PT fraction than in the PLB fraction.

2. The results of Ryberg and Sundqvist (1982a) on wheat leaves indicated "the presence of phototransformable PChlide in the prolamellar bodies proper" while the bulk of the proteins was found located in the prothylakoïds.

3. Ikeuchi and Murakami (1983) working on squash cotyledons claimed that "the photoactive PChlide-NADPH reductase complex in etioplasts was concentrated in the prolamellar body and

that the physical state of PChlide in the prolamellar body
differed from that of PChlide in prothylakoïds".

As bean has been often used in studies on the early events
occuring after illumination, we have previously applied the
separation process of El Hamouri and Sironval to bean leaves
(Gerday et al, 1982). We have separated two PChlide containing
fractions which differed by the Chlide complexes appearing
on illumination. Electron microscopy analysis indicated,
however, that the PTs and PLBs were not properly separated
by the procedure used. In the present paper, we apply a some-
what different procedure which tends to purify the fractions.
The separation is performed with and without supply of exogenous NADPH. Our purpose is to examine whether one can separate photoreducible PChlide-protein complexes differing by
the Chlide complexes they yield, and thus to answer the question on whether different PChlide/Chlide complexes are located
within different etioplast membrane structures.

2. MATERIALS AND METHODS

2.1. Plant Material

Bean (*Phaseolus vulgaris* L. cv. Commodore) seedlings were
grown in the dark, at $20 \pm 1°C$, for 15 to 17 days, on a moist
mixture of vermiculite-perlite. The etiolated leaves were
harvested under a green safe light. Further operations were
performed either in the dark or under green safe light, at $4°C$.

2.2. Etioplast isolation and PT-PLB separation

The method described by Lütz (1978) was used with some
modifications, according to Gerday et al (1983). The media
are described elsewhere (Gerday et al, 1983).

C.a. 70g of leaf material were ground in a homogenizer (Braun,
type Mx322; 1x5 sec, speed 2) in the isolation medium. Filtration was performed through a layer of 30 µm mesh nylon tissue.
The residue of the first grinding was collected, ground again
and filtered in the same conditions. This procedure was performed 4 times in 300, 200, 150, 150 ml of medium. The successive supernatants were pooled and centrifuged at 270 g for
2 min (pellet P_I discarded). The supernatant S_I was centrifuged

at 2000 g for 10 min (supernatant S_{II} discarded).

The pellet P_{II} was resuspended in 10 ml of isolation medium (step 1) and layered on the top of a continuous sucrose density gradient 10-60% (w/w) (step 2) which was then centrifuged at 23000 g for 30 min in a MSE swing-out rotor. Intact etioplats layered at density c.a. 1.21 were removed using a Pasteur pipette.

They were ruptured osmotically by dilution with 5 volumes of sucrose free medium (step 3) and then centrifuged at 39100 g for 15 min (supernatant S_{III} discarded). The pellet P_{III} was resuspended in 3-5 ml of sucrose free medium (step 4), homogenized in a glass-homogenizer and sonicated 3 x 5 sec, low level, setting 5 (MSE sonicator). The membrane suspension (medium density 1.00) was layered on the top of a discontinuous sucrose density gradient 30-46 % (w/w) (step 5) (density 1.12 and 1.20 respectively, fig. 1) and centrifuged at 75000 g for 30 min.

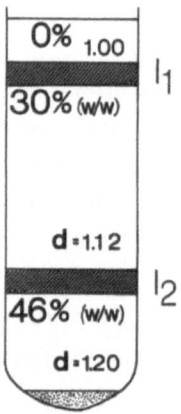

FIGURE 1. Discontinuous sucrose density gradient used for fractionation of purified etioplast membranes.

The material concentrated at both interfaces I_1 (separating media 0-30% (w/w) sucrose of densities 1.00-1.12) and I_2 (separating densities 1.12-1.20) were collected using a Pasteur pipette.

When the PT-PLB separation is performed in the presence of exogenous NADPH, 1 mM NADPH was added at the steps 1 to 5 (except in step 4 where 0.5 mM NADPH was added).

2.3. Electron microscopy

Thin sectioning of resin-embedded samples were prepared
and examined in a Siemens Elmiskop 101 electron microscope
as described in Lütz (1978).

2.4. Absorption and fluorescence emission spectra

Absorption spectra were recorded at 77°K using a Cary 17
spectrophotometer. Fluorescence emission spectra were recorded
at 77°K using the home-built apparatus described by Sironval
et al (1968). The spectra were not corrected for wavelength
variations of the photomultiplier response.

2.5. Illumination of the samples

The samples were irradiated with an electronic flash
(Multiblitz 50 Report Porba 125 J 5800 K). The duration of
each flash was 1 msec. At each light treatment, two successive flashes (separated by 3 sec intervals) were given to
the samples placed 20 cm from the light source.

2.6. Pigment analysis

The amount of initial photoreducible and non photoreducible PChlide complexes were estimated from measurements of
absorbance at 678nm and at 628nm (corrected for contribution
of Chlide, spectra recorded at 77°K) of illuminated fractions.
The percentage of photoreducible PChlide complexes was calculated from the ratio :

$$R = \frac{[Chlide] \times 100}{[Chlide] + [PChlide\ 628]}$$

where [Chlide] and [PChlide 628] are the concentration of
Chlide formed after illumination and of remaining non photoreducible PChlide respectively.

Total pigment content was estimated after extraction in
acetone-water (80%-20%). Acetone extracts were centrifuged
for clarification and absorption spectra were recorded at
room temperature. Pigments were quantified from measurements
of the absorbance at 663 and 626 nm, using the equations of
Brouers and Michel-Wolwertz (1983).

2.7. Protein and polypeptide analysis

Proteins were quantified according to Bramhall et al (1969).
SDS-polyacrylamide gel electrophoresis was performed as des-

58

cribed by Laemmli (1970). The separating gel and the stacking
gel were 12% and 6% acrylamide-bisacrylamide respectively.

3. RESULTS

After centrifugation, the etioplast membranes were concentrated at the two interfaces of the discontinuous sucrose
density gradient (see materials and methods, sect.2.2., fig.1).
At the upper interface I_1, the material was green and non
diffusive while at the I_2 interface, it appeared milky.

3.1. Ultrastructure

Electron micrographs of these materials are shown in
fig. 2. The I_1 fraction contained mainly vesicles and broken
membranes. Some contaminant mitochondria and few PLB fragments
were also seen in this fraction. Intact PLBs were mainly found
in the I_2 fraction. In this fraction, attached vesicles, presumably PT contaminations, remained. The I_1 and I_2 fractions
will be called below, the PT and the PLB fractions respectively.

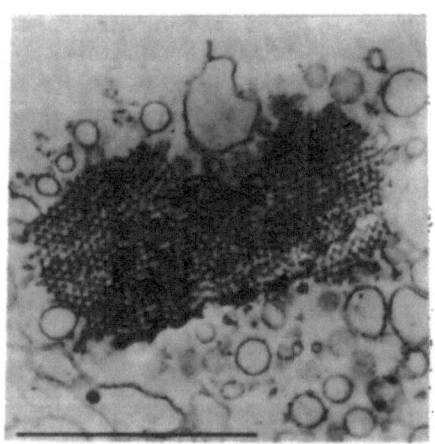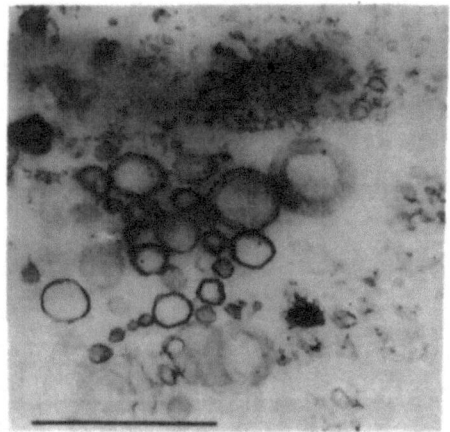

FIGURE 2. Electron micrograph of the materials most often
observed in the I_1 fraction (PT fraction, on the left) and in the
I_2 fraction (PLB fraction, on the right). The bar indicates
1 μm.

3.2. Properties of control fractions prepared without addition of NADPH

3.2.1. Spectral properties of the etiolated materials.

Fig. 3 shows the red absorption and the fluorescence emission spectra of control PT and PLB fractions.

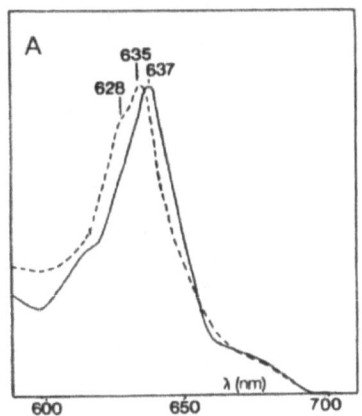
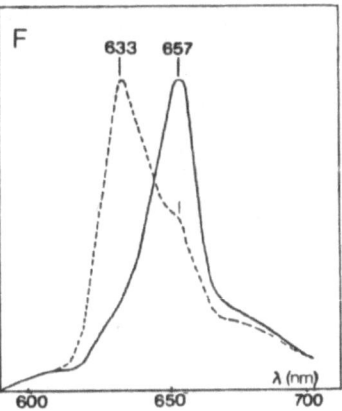

FIGURE 3. Red absorption and fluorescence emission spectra recorded at 77°K of etiolated PT (——) and PLB (---) fractions.

The PT fraction contained nearly exclusively the photoreducible complex $P_{645-637}$ (absorption maximum at 637nm). The position of the fluorescence maximum at 657nm, instead of 645nm, is indicative of an energy transfer from $P_{645-637}$ to a residue of $P_{657-650}$.

The absorption maximum of the etiolated PLB fraction was located at 635nm. This results from the presence of a high amount of $P_{633-628}$ which leads to a shift of the 637nm band towards shorter wavelengths. A shoulder was seen at 657nm in the fluorescence spectrum, showing that the photoreducible complex $P_{645-637}$ transfered energy to a trace of $P_{657-650}$.

3.2.2. Spectral properties of illuminated materials. The absorption and fluorescence emission spectra of the control PT and PLB fractions recorded immediately after illumination, are shown in fig. 4. In both fractions, the Chlide complex $P_{688-678}$ was formed. The percentages of photoreducible PChlide (parameter R, material and methods, sec. 2.6.) were however different : 53% in PT fraction, 25% in the PLB fraction.

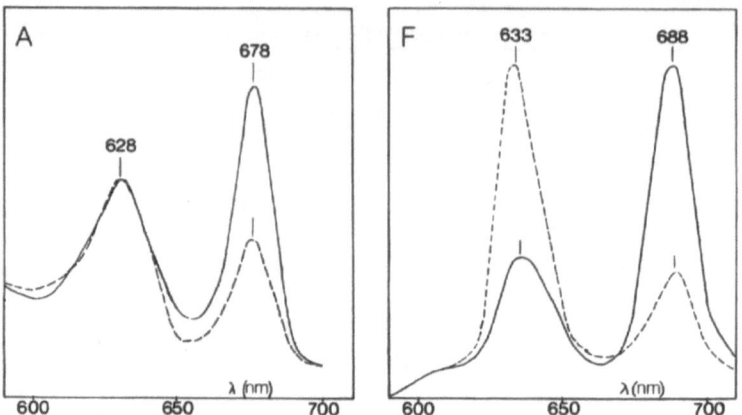

FIGURE 4. Red absorption and fluorescence emission spectra of PT (———) and PLB (---) fractions recorded at 77°K imme-diately after light treatment.

The fluorescence emission maximum of the Chlide complexes shifted towards 683nm in both control fractions during a subsequent 30 min dark period (see fig. 8).

3.2.3. Distribution of pigment and protein between PT and PLB control fractions. Table 1 shows the percentage of PChlide and protein recovered in the fractions. PChlide was more abundant in the PT fraction although PT and PLB fractions contained approximately the same amount of proteins.

	PChlide (%)	Proteins (%)	PChlide/protein (nmole/mg)
PT	77	49	86
PLB	23	51	24

Table 1 Distribution of PChlide and total proteins between control PT and PLB fractions.

3.2.4. Polypeptide content. A doublet of mol. wts 35000-36000 was characteristic of the PT fraction besides a poly-peptide of mol. wt 55000 (fig. 5). The PLB fraction was characterized by a large amount of a doublet of mol. wts 54000-56000, accompanied by a few minor components (the doublet of mol. wts 35000-36000; a polypeptide of mol. wt 30000).

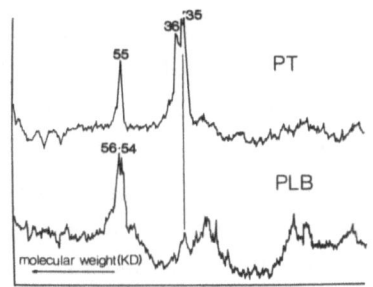

FIGURE 5. Densitometric tracings of SDS-polyacrylamide gel after electrophoresis of PT and PLB fractions.

3.3. Effects of the addition of NADPH

When NADPH was supplied after the separation of the PTs and the PLBs, there was no effect on the spectral properties of the PChlide/Chlide complexes. However the result was different when NADPH was added at the steps of the separation procedure indicated in materials and methods, (chap. 2.2.). In this case, the presence of NADPH modified the characteristics and the properties of the two fractions, as compared with the control without NADPH.

3.3.1. Effect of NADPH on the photoreducible PChlide complexes recovered in the fractions
In the presence of NADPH, a higher amount of photoreducible PChlide complexes was recovered in both fractions, especially in the PLB fraction (fig. 6). The percentage of the photoreducible PChlide complexes was 62.9% in the NADPH-treated PT fraction and 43% in

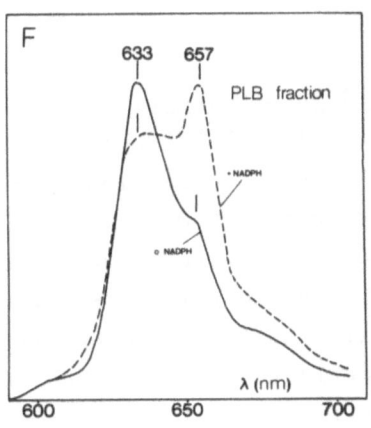

FIGURE 6. Fluorescence emission spectra recorded at 77°K of etiolated PLB fraction prepared with (---) and without (——) supply of NADPH.

the PLB fraction against 53% and 25% respectively in the control PT and PLB fractions. The ratios were estimated from absorption spectra (Materials and Methods, sect. 2.6.) except the second (43%), estimated from fluorescence emission spectrum.

The fluorescence spectrum of the PLB fraction depended strongly on whether this fraction was prepared with or without the supply of NADPH. The emission band at 657nm which was reduced to a shoulder in the absence of NADPH was the main fluorescence peak of the NADPH treated fraction.

The absorption spectra proved that the supply of NADPH led to a higher amount of the PChlide complex $P_{657-650}$ in the PT fraction (fig.7). The low pigment concentration did not allow to analyse this effect in the PLB fraction.

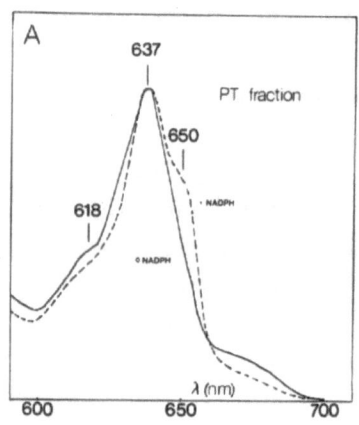

FIGURE 7. Red absorption spectra recorded at 77°K of etiolated PT fraction prepared with (---) and without (——) supply of NADPH.

3.3.2. <u>Effect of NADPH on the Chlide complexes</u>. The illumination yielded first $P_{688-678}$ in both NADPH supplemented fractions (as it was the case in the control fractions). But the fate of this Chlide complex during the 30 min dark period following the illumination was dependent on the fraction when NADPH had been supplemented (fig.8). During this dark period, the fluorescence maximum of the Chlide shifted to 696nm in the PT fraction while it shifted partly to 678nm in the PLB fraction (a shoulder at 688nm was still observed after 30 min).

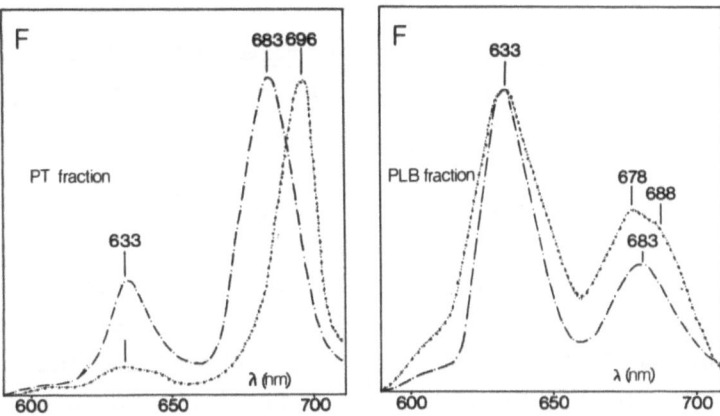

FIGURE 8. Fluorescence emission spectra of PT fraction (on the left) and of PLB fraction (on the right) recorded at 77°K after a 30 min dark period following the illumination. (···) fractions prepared with supply of NADPH. (—.—) fractions prepared without supply of NADPH.

3.3.5. Effect of NADPH on the polypeptide content. Electrophoresis pattern of the NADPH-treated fractions (fig. 9) showed that the relative amount of polypeptides of mol. wts 35000-36000 was higher in both NADPH-treated fractions. This effect was more obvious in the PLB fraction however (compare . fig 9 with fig 5). Moreover, in this fraction a polypeptide of mol. wt 30000 was observed in a larger quantity than in the control fraction.

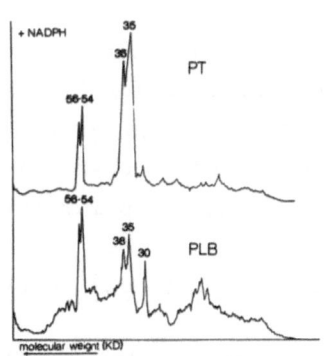

FIGURE 9. Densitometric tracings of SDS-polyacrylamide gel after electrophoresis of PT and PLB fractions prepared with supply of NADPH.

4. DISCUSSION

4.1. On the localization of the photoreducible PChlide complexes

We have prepared two etioplast membrane fractions by density gradient centrifugation. The low density fraction was found enriched in membrane vesicles, presumably PTs (PT fraction). It contained the main part of the total PChlide and most of the photoreducible PChlide-protein complexes. This is in agreement with the results of Lütz et al (1981). The amount of the polypeptides of mol. wts 35000-36000 attributed by Oliver and Griffiths (1980) to PChlide-photooxidoreductase, was found to be the highest in this fraction. The polypeptide of mol. wt 55000 which this fraction contains also could be a subunit of the chloroplast enzyme CF_1-ATPase. •

The high density fraction was characterized by a low amount of PChlide-photooxidoreductase. This is in contrast with the results of Ikeuchi and Murakami (1983) who found the main part of the photoenzyme in the prolamellar body fraction prepared from squash in the presence of NADPH. The PLB fraction was rich in non photoreducible complexes $P_{633-628}$ in contrast with the data of Ryberg and Sundqvist (1982a) who observed a high amount of photoreducible PChlide in the PLBs (in accordance with Ikeuchi and Murakami, 1983). It seems unlikely that the high amount of $P_{633-628}$ in our PLB fraction was due to desactivation, owing to the protective effect of the high concentration of sucrose which we have used (Ryberg and Sundqvist, 1982b).

It cannot be excluded that the photoreducible complexes are to be ascribed to a contamination of our PLB fraction by PTs. In this hypothesis, the PTs would possess alone the apparatus for starting greening. This assumption has already been put forward by Lütz (1981). The lack of specific markers for PTs and PLBs does not permit to conclude.

4.2. On the separation of distinct PChlide-protein complexes yielding distinct Chlide complexes.

When NADPH was not supplied, the photoreducible complexes had the same spectral properties in both PT and PLB fractions. This illumination produced Chlide-protein complexes with the

same spectral properties ($P_{688-678}$). In the dark, the emission
maximum shifted towards 683nm in both fractions. Thus, in the
absence of NADPH, the separation of two PChlide complexes,
differing by the Chlide complexes they yield when illuminated
(Gerday et al, 1982) was not achieved.

In contrast, when the PT-PLB separation was performed in
the presence of NADPH, the Chlide complexes changed in the
dark in a way which depended on the fractions : the emission
maximum shifted towards longer wavelengths in the PT fraction,
while it shifted towards shorter wavelengths in the PLB fraction . This can be due to a difference between fractions in
the accessibility of NADPH at some protein site. However, the
fact that NADPH acts, in both fractions, -although differen-
tly-, on the PChlide complexes before illumination suggests
that the accessibility is not concerned in the fate of the
Chlide complexes. It is however possible that the protein
site of the fixation of NADPH is not the same before illumination than after illumination.

One of the effects of the supply of NADPH is the obvious
increase in the amount of $P_{657-650}$ in the PT fraction, whereas
only traces of $P_{657-650}$ were observed in this fraction in
the absence of NADPH. This fraction produced $P_{696-684}$ after
illumination. This result has connections with what we have
observed previously with crude fractions (Gerday et al, 1982).
In this case, both fractions collected at the interfaces I_1
and I_2 contained very different amounts of $P_{657-650}$, and the
fates of the Chlide complexes were dependent on the fractions.
It may be suggested that these circumstances are linked one
to another : Ikeuchi and Murakami (1983) have observed that
only their fraction enriched in $P_{657-650}$ produced a
long wavelength absorbing Chlide on illumination. This is
also what we have observed.

4.3. On the regulation of the amount of photoreducible complexes
     by NADPH.

Several modes of actions of NADPH can be suggested to explain the strong increase in photoreducible PChlide complexes
which results from NADPH addition to the PLB fraction.

1. This increase could be due to the high amount of non photoreducible complex $P_{633-628}$ in the PLB fraction. $P_{633-628}$ is known to be changed into a photoreducible PChlide complex in the presence of NADPH (Wolwertz and Brouers, 1981; El Hamouri et al, 1981).

2. NADPH could also have a protective effect. It might be that the desactivation of PChlide photooxidoreductase is due to the release of NADPH from the protein site. Would the naked protein become sensitive to proteolysis ?

3. The addition of NADPH could also modify the distribution of PTs and PLBs between fractions. In this case, the effect would be for instance, at the sonication step. In the presence of NADPH, the contamination of the PLB fraction with PTs would increase as a result of a less pronounced release of the PT membranes.

If the NADPH content is able to modify the distribution of PTs and PLBs between fractions, we have to be carefull when comparing the results obtained by the authors, as only part of them add NADPH during fractionation.

ACKNOWLEDGMENTS

The authors wish to thank Dr C. Lütz for electron microscopy studies and for many stimulating discussions. This work was supported by the Institut pour l'Encouragement de la Recherche Scientifique dans l'Industrie et l'Agriculture (IRSIA) and by the Fonds National de la Recherche Scientifique (FNRS) - Belgium.

REFERENCES

1. Belyaeva OB, Litvin FF. 1981. Photosynthetica, 15, 210-215.
2. Bramhall S, Noack N, Loewenberg JR. 1969. Analytical Biochemistry, 31, 146-148.
3. Brouers M and Michel-Wolwertz M-R. 1981. in Chloroplast Development. Akoyunoglou ed. Balaban International Science Services p. 185-196.
4. Brouers M and Michel-Wolwertz. 1983. Photosynthesis Research 4, 265-270.
5. El Hamouri B and Sironval C. 1983. Photobiochem. and Photobiophysics 5, 263-272.
6. El Hamouri B, Brouers M, Sironval C. 1981. Plant Science Letters 21, 375-379.
7. Gerday C, Michel-Wolwertz M-R and Brouers M. 1982. in Cell function and Differentiation, part B, 25-32 Alan R Liss Inc ed. NY 10011.

8. Gerday G, Lütz C, Michel-Wolwertz M-R and Brouers M. 1983. Proceedings of VIth International Congress on Photosynthesis, Août 1983, Bruxelles (sous presse).
9. Griffiths WT. 1983. FEBS letters 46 (1), 301-304.
10. Ikeuchi M and Murakami S. 1982. Plant and Cell Physiol. 24 (1) 71-80.
11. Laemmli UK. 1970. Nature 227, 680-685.
12. Lütz C. 1978. in Chloroplast Development G. Akoyunoglou et al eds p. 481-488, Amsterdam Elsevier.
13. Lütz C. 1981. Protoplasma 108, 99-115.
14. Lütz C, Roper U, Beer NS, Griffiths WT. 1981. Eur. J. Biochem. 118, 347-353.
15. Michel J-M and Sironval C. 1973. Physiol. Vég. 11, 291-300.
16. Oliver RP and Griffiths WT. 1980. Biochem. J. 191, 277-280.
17. Ryberg M and Sundqvist C. 1982a. Physiol. Plant. 56, 125-132.
18. Ryberg M and Sundqvist C. 1982b. Physiol. Plant. 56, 133-138.
19. Shibata K. 1957. J. of Biochem. 44, 147-173.
20. Sironval C, Michel J-M. 1967. Books of Abstracts, European Photobiology Symposium, Hvar, Yougoslavia p. 105-108.
21. Sironval C, Brouers M, Michel J-M and Kuiper Y. 1968. Photosynthetica. 2, 268-287.

PROTOCHLOROPHYLL(IDE) AND CHLOROPHYLL(IDE) IN ISOLATED ORIGINAL
AND REFORMED PROLAMELLAR BODIES

MARGARETA RYBERG AND CHRISTER SUNDQVIST
University of Göteborg,Depart.of Plant Physiol., Carl Skottsbergs Gata 22, S-413 19 Göteborg, Sweden.

INTRODUCTION

Etioplasts in dark-grown leaves contain two structurally different systems, the prolamellar bodies (PLBs) and the prothylakoids
(PTs). During the early stages of greening the PLBs are transformed and the thylakoidal system is built up during conversion of
the PTs to mature thylakoids and final formation of grana. There
are strong indications that PLBs are reorganized into thylakoids
(1, 2). The initial stage in this process is characterized by the
reduction of protochlorophyllide (PChlide) to chlorophyllide
(Chlide). Whether this reduction is the real trigger for the
structural changes has not been conclusively shown. Several circumstances indicate however, that this is the fact. The first
step in the ultrastructural changes of the PLBs is the tube transformation. The wavelength dependence of this process is similar
to that of the reduction of PChlide to Chlide (3). However, the
reduction can take place without any following tube transformation if the temperature is kept close to 0 $^{\circ}$C (4) or if the seedlings are aged (5). The early studies by Virgin and coworkers (3)
together with the observation of strong red fluorescence from the
PLBs (6) made it plausible to place the PChlide in the PLBs. A
close correlation between accumulation of PChlide$_{650-657}$ (see
Abbreviations) and formation and enlargement of PLBs was shown
by Klein and Schiff (7). Further indications for a correlation
between PLBs and PChlide is that mutants incapable of synthesizing
PChlide also lack PLBs (8). Occasionally there have been reports
on plant tissues which have PLBs but lack PChlide (9, 10).

More penetrating investigations with the use of modern biochemical methods to reveal the composition of PLBs and PTs were
started by Lütz (11) and Wellburn and Hampp (12). The use of oat

C. Sironval and M. Brouers (eds.); Protochlorophyllide Reduction and Greening. ISBN 90 247 2954 8
© 1984, Martinus Nijhoff/Dr W. Junk Publishers, The Hague/Boston/Lancaster.

as an experimental material complicated the interpretation of the
results as large amounts of saponins from the vacuoles were found
to bind to the PLBs (13). Nevertheless, the main conclusion from
these investigations changed the prevailing theory that PChlide
was located mainly in the PLBs and it was concluded that most of
the PChlide was localized to the PTs (14).

The reduction of PChlide to Chlide is a photoenzymatic process.
The enzyme was found to function in a buffered water solution
(15). Smith used the term protochlorophyll-holochrome for this
pigment-protein complex. The enzymatic nature was, however, not
shown in vitro. The accumulation of excess amounts of PChlide by
treatment with δ-aminolevulinic acid (ALA) made it possible to
suggest some kind of enzymatic behaviour in vivo (16-18). The
development by Griffiths (19) of an enzymatic assay in vitro
opened up a more straightforward way to follow what occurs during
the transformation of etioplasts to chloroplasts. The molecular
weight of the holochrome was found to be approximately 600 kD.
SDS-polyacrylamide gelectrophoresis showed the holochrome to be
composed of polypeptides of 45 kD for Phaseolus (20) and 21 and
29 kD for Zea (21). The enzyme PChlide-oxidoreductase purified
to homogenity was found to have a molecular weight of 35-37 kD
(22-24).

In vivo the PChlide-protein complex has an absorbance maximum
at ca 650 nm. During isolation of the holochrome or the PChlide-
-protein complex with methods typical for holochrome preparation
or PChlide-oxidoreductase preparation the maximum is shifted
towards shorter wavelengths (630-636 nm). The addition of NADPH
increased the 650/630 nm absorbance ratio in membrane preparations containing the enzyme (25, 26). At present the more detailed
distribution of PChlide between PLBs and PTs is under debate. In
the result and discussion section of this paper we will present
our view on this question, based on experiments on saponin-free
wheat seedlings (27-32).

Abbreviations

ALA, δ-aminolevulinic acid; Chlide$_x$, chlorophyll and/or chlorophyllide, x denotes absorbance maximum; PChlide$_{x-y}$, protochloro-

phyll and/or protochlorophyllide, x denotes absorbance maximum, y denotes fluorescence emission maximum; PLB, prolamellar body; PT, prothylakoid; SDS-PAGE, sodiumdodecylsulphate polyacrylamide gelelectrophoresis.

MATERIAL AND METHODS

Six- to seven-day-old dark-grown seedlings of wheat (Triticum aestivum L. cv. Starke II, Weibull) were grown and harvested as described earlier (27). Only 3 cm parts 1.5-2 cm from the tip of the leaves were used. All experiments were performed under green safelight at 5 $^{\circ}$C unless otherwise stated.

The procedure for isolation of plastids and separation of PLBs and PTs (27) is based on the method described by Lütz (11). The major change is the use of bottom-loaded sucrose density gradients instead of top-loaded gradients for separation of PLBs and PTs which includes sonication of the inner plastid membranes in 50% sucrose medium instead of shock medium prior to separation in a 10-50% (w/w) sucrose gradient. This alteration resulted in a better preservation of $PChlide_{650-657}$, especially in the PT-fraction.

Pigment and protein analyses as well as preparation of samples for transmission electron microscopy are described earlier (27). Samples for scanning electron microscopy were dehydrated in ethanol and freon and taken to dryness in a critical point chamber. Reformed PLBs were isolated from dark-grown wheat seedlings after illumination for 4 h with white light (500 lux) and a further 16 h in darkness (31). Treatment of leaf pieces with 8-hydroxyquinoline and ALA was performed as described by Ryberg (30).

RESULTS AND DISCUSSION
Separation of PLBs and PTs

To minimize the contamination of membrane fragments from other organelles the preparation of PLBs and PTs was performed from isolated etioplasts, mainly according to the method developed by Lütz (11). However, several modifications have been made to increase the purity and to keep the spectral forms of the pigments in their natural state (27, 28). As it was known that a glycerol

solution with low water potential stabilized PChlide$_{650-657}$ a
method was developed which included sonication in a sucrose medium with low water potential and loading of the sonicated solution in the bottom of the gradient. In this way it was possible
to avoid long time exposure to high water potential which is the
case when isolation is performed using top-loaded gradients. This
method increased the 650/630 nm absorbance ratio for the PT-fraction (Fig. 1). The absorbance and fluorescence properties of the
PLB-fraction were very similar after separation in top- and in
bottom-loaded gradients. This insensitivity of PLBs against different sucrose concentrations is one of the most evident differences found between PLBs and PTs.

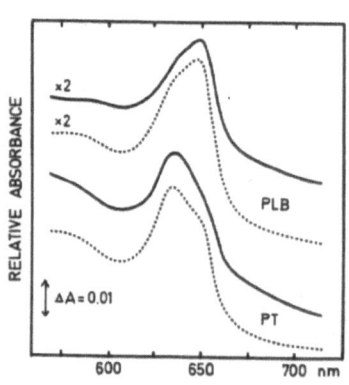

FIGURE 1. Absorbance spectra of
prolamellar body and prothylakoid fractions after isolation
in top-loaded (————) and in
bottom-loaded (·····) gradients.
Redrawn from (27).

Increased sonication also improved the separation, probably
due to higher degree of breakage between PLBs and PTs at the
attachment points. One interesting result was the reversed location of PLBs and PTs in the gradients after centrifugation. With
bottom-loaded gradients PLBs were found mainly at a density of
1.17 g ml^{-1} and PTs at 1.21 g ml^{-1}. In top-loaded gradients PLBs
were found at the same density but PTs stayed in the upper part
of the gradient as a broad diffuse band. In most experiments
each band was removed from the gradient, diluted and pelleted.
This procedure was necessary especially when top-loaded gradients
were used since the PT-enriched band needed to be concentrated.
In bottom-loaded gradients the bands were more narrow and distinct and concentration was not necessary. This also contributed

to a better preservation of PChlide$_{650-657}$. However, when a
centrifugation step was omitted the purification steps were
fewer. Still, as judged from electron micrographs, the purity
of the fractions was high.

Addition of NADPH to the solutions used during the isolation
was also very effective in preventing any change in absorbance
at 650 nm. If NADPH is present the negative effect of low sucrose
concentration in the medium is more or less overcome (Fig. 2).

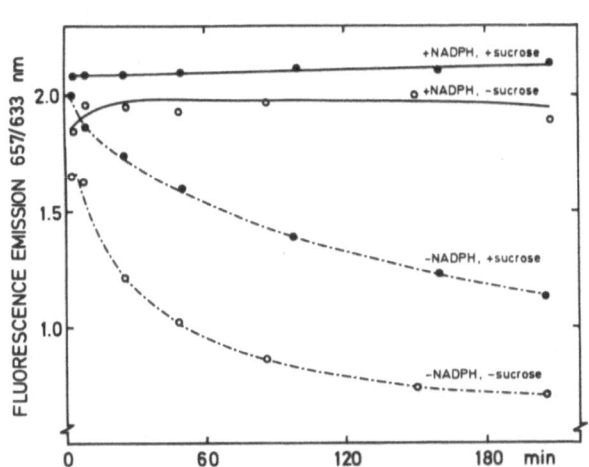

FIGURE 2. Effects of
NADPH and sucrose on
the 657/633 nm fluor-
escence emission ratio
at -196 °C of unsepara-
ted prolamellar body-
-prothylakoid membra-
nes from wheat etio-
plasts. Excitation at
440 nm. The membra-
nes were suspended in
isolation medium
(+ sucrose) or in
sucrose-free medium
(- sucrose). From (28).

Composition of PLBs and PTs

The PLB- and PT-fractions when separated in bottom-loaded gradients, differed strongly in protein composition (27). The PLB-
-fraction was dominated by one single protein regarded to be the
PChlide-oxidoreductase (cf. 22-24). SDS-PAGE showed that this
enzyme was present in high amounts also in the PT-fraction (Fig.
3). This fraction contained several other proteins, including
high amounts of CF$_1$ chloroplast ATPase. PLBs and PTs separated
from squash cotyledons (33) were shown to have very similar protein patterns as the PLB- and PT-fractions described in this
paper. It is important to have this protein distribution in mind
when comparisons are made of other constituents on a protein basis. As no specific markers for PLBs or PTs have been found,
other than the ultrastructural appearance in the electron microscope of the PLBs, it is still difficult to make conclusive

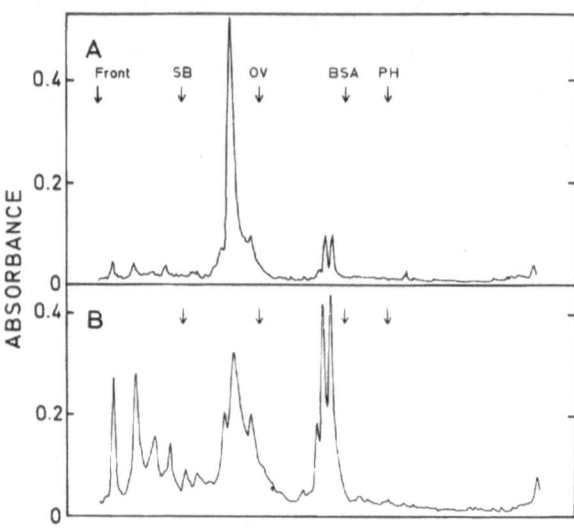

FIGURE 3. Polypeptide pattern of prolamellar body and prothylakoid fractions after SDS--PAGE. The molecular weight for the standard proteins are SB 21500, OV 43000, BSA 68000, PH 96000. From (27).

comparison. No real efforts have been made to study contamination of other membrane fragments in the PLB- and PT-fractions. Even if the preparation proceeds via the purification of etioplasts, one cannot exclude presence of other membrane fragments and certainly not of etioplast envelops. The presence of only one polypeptide dominating in the PLB-fraction shows, however, that this fraction is comparatively pure. The problem is therefore mainly confined to the PT-fraction.

About 40% of the recovered PChlide was found in the PLB-fraction and 60% in the PT-fraction. The PChlide content per protein was four times higher in the PLBs than in the PTs. We have shown that there are prominent differences in spectral properties between the isolated PLB- and PT-fractions (27, 28). The PLB-enriched fraction showed a higher ratio phototransformable PChlide ($PChlide_{650-657}$ and/or $PChlide_{636-657}$) to non-phototransformable PChlide ($PChlide_{628-632}$) than did the PT-enriched fraction. The results lead to the conclusion that phototransformable PChlide is better protected from degradation in PLBs than in PTs and that the PLB may serve as a structure suitable for storing PChlide in complex with the PChlide-oxidoreductase and NADPH (34).

Separation of PLBs and PTs from ALA-treated leaves further elucidates the differences between these membrane systems. Two

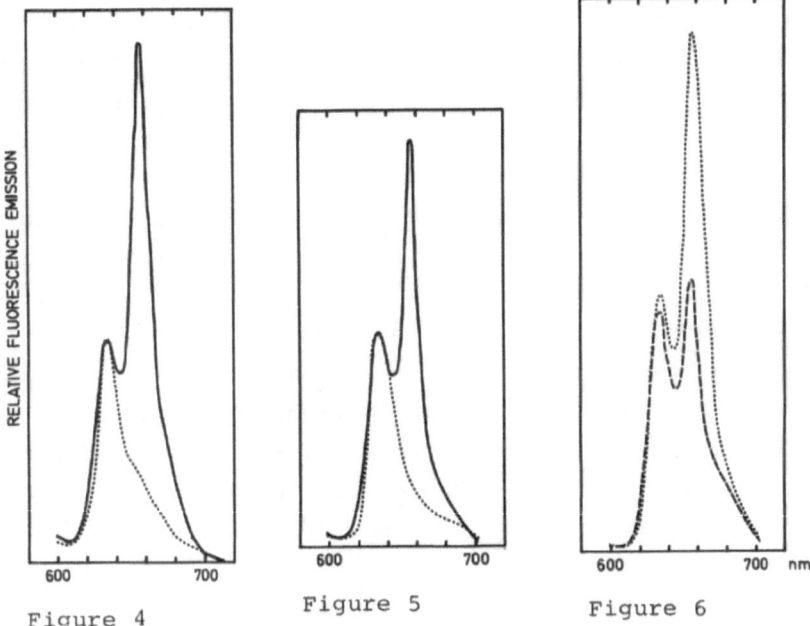

Figure 4

Figure 5

Figure 6

FIGURE 4. Fluorescence emission spectra at -196 °C of prolamellar body (———) and prothylakoid (·····) fractions isolated from ALA-treated dark-grown wheat leaves. Excitation at 440 nm. The spectra are equalized to unity at 633 nm.

FIGURE 5. Fluorescence emission spectra at -196 °C of saponin--treated prolamellar body (———) and prothylakoid (·····) fractions. Excitation at 440 nm. The spectra are equalized to unity at 633 nm.

FIGURE 6. Fluorescence emission spectra at -196 °C of etioplast inner membranes from oat (-----) and wheat (·····). Excitation at 440 nm. The spectra are equalized to unity at 633 nm.

fluorescence emission maxima were seen in the PLB-enriched fraction, at ca 633 nm due to non-phototransformable PChlide and at ca 657 nm due to phototransformable PChlide (Fig. 4). In the PT--enriched fraction only a peak at 633 nm appeared, probably due to a high amount of PChlide synthesized from exogenous ALA. One reason for an emission maximum only at 633 nm in the PT-fraction could be a degradation of $PChlide_{650-657}$, since this isolation was performed without addition of NADPH and using top-loaded gradients. However, it has previously been shown that ALA-treated leaves show a higher fluorescence emission yield at 632-643 nm than do untreated leaves (35). In addition, fluorescence emission spectra of the etioplast suspension isolated from ALA-treated leaves showed a lower 657/633 nm fluorescence emission ratio than did etioplasts from untreated leaves (results not shown). Thus a more reasonable explanation for a high fluorescence emission at 633 nm in the PT-enriched fraction is that ALA-induced PChlide is synthesized and enriched in the PTs. To incorporate PChlide into the PLBs it has to be as PChlide$_{650-657}$. Similar differences between PLBs and PTs were seen when 1.3% saponin was added to the membrane suspension prior to sonication and separation in a top-loaded gradient (Fig. 5). No fluorescence emission maximum at 657 nm was seen in the PT-fraction, indicating a degradation of the PChlide-reductase-NADPH complex through the action of the detergent. The PLB-fraction did not differ significantly from untreated PLB-fractions (cf. 27). Isolation of etioplast membranes from saponin-containing oat leaves also resulted in a lower fluorescence emission at 657 nm than in the corresponding fraction from wheat (Fig. 6).

The experiments described above, ALA-feeding, saponin-treatment and isolation from saponin-containing oat further strengthen earlier conclusions that there are fundamental differences between PLBs and PTs with respect to the PChlide forms, and that PChlide$_{650-657}$ is better protected from degradation in the PLBs than in the PTs.

Results leading to the conclusion that PLBs are built up mainly by saponins (36-38) and the conclusion that glycolipids are bound exclusively to the PT-membranes (39), challenged us to investigate the lipid composition of the PLB- and PT-fractions. We found that glycolipids were the dominating lipids in both fractions (29). The molar ratio of monogalactosyl diacylglycerol to digalactosyl diacylglycerol was higher in the PLB-fraction (1.8) than in the PT-fraction (1.2). Calculated on a protein basis the PLB-fraction contained more than twice as much glycolipids as did the PT-fraction. Free sterols and sterol esters constituted less than 6 mol% of the total lipids. No saponins were detected. Recent results by Kesselmeier (13) and Lütz and Nordmann (40) support the opinion that saponins are not the main building

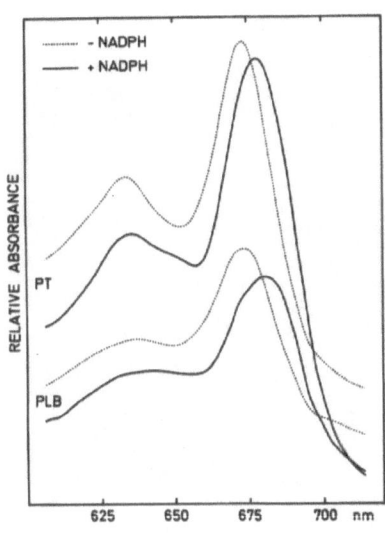

FIGURE 7. Absorbance spectra of prolamellar body and prothylakoid fractions isolated in the presence or absence of NADPH, recorded 24 h after irradiation.

units of PLBs (29, 41, 42).

Phototransformation of PChlide in PLBs and PTs

When PLBs and PTs are prepared in media with high sucrose concentration and in the presence of NADPH, $PChlide_{650-657}$ is well preserved (28). Phototransformation with three consecutive flashes of white light within a time interval of 15 s produced $Chlide_{684}$ in PLBs as well as in PTs (32). In the PLB-fraction almost all PChlide was transformed to Chlide, whereas in the PT-fraction small amounts of a short-wavelength PChlide form was present. At present we cannot conclude whether this is a true difference between PLBs and PTs in vivo or if it is a consequence of the more labile PChlide system found in the PTs. During a dark period following the flash sequence the absorbance maximum of the newly formed Chlide shifted towards shorter wavelengths (Fig. 7). This shift was not seen in the PLB-fraction in the presence of NADPH. Spectral analyses using fourth order derivatives of fluorescence emission spectra indicated the appeerance of a short-wavelength form in the PLB-fraction which was not found in the PT-fraction (32).

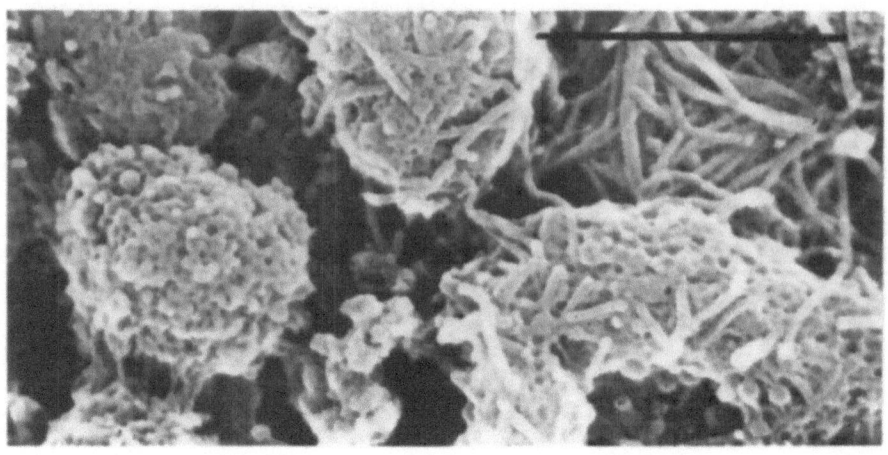

FIGURE 8. Prolamellar bodies as seen in the scanning electron microscope. The bar indicates 1 μm.

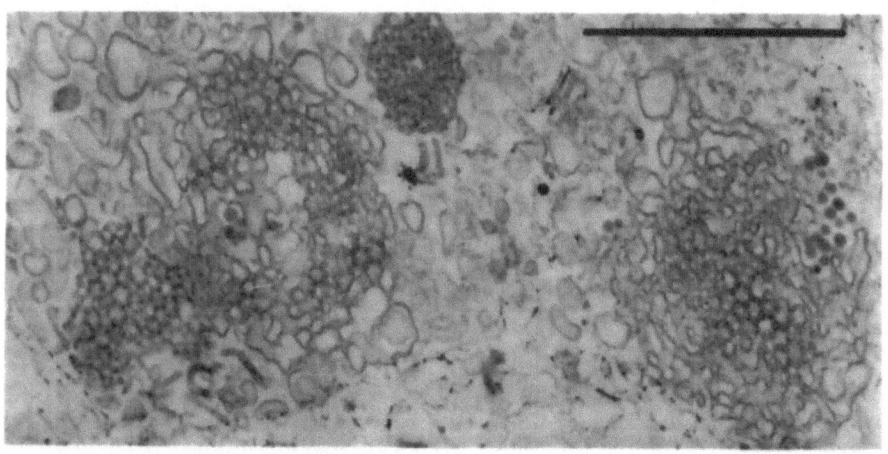

FIGURE 9. Prolamellar body fraction irradiated after isolation. The bar indicates 1 μm.

Structural alterations of isolated PLBs

Isolated PLBs appeared as cottonball-like structures in the scanning electron microscope (Fig. 8). Some of the PLBs showed a regular lattice as could be expected from the appearance in

the transmission electron microscope. Some PLBs showed a looser
tubular structure. After irradiation of a PLB-fraction isolated
from etiolated leaves a transformation of the PLBs might be expected. In the transmission electron microscope a transformation
of PLBs was seen (Fig. 9). In the scanning electron microscope
numerous vesicles were found on the surface of the PLBs (32). We
cannot tell whether this vesicle formation corresponds to the
transformation as seen in the transmission electron microscope,
or if it is a different type of reaction caused by having the
PLBs in an isolated state.

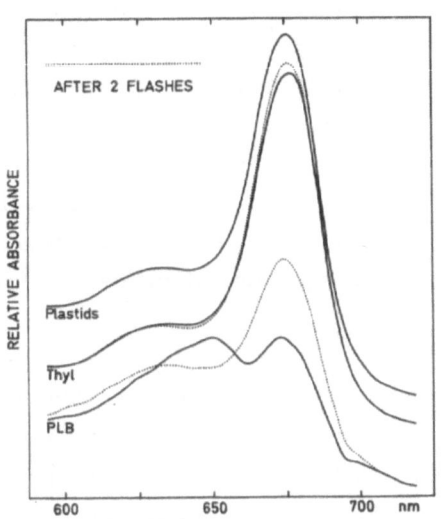

FIGURE 10. Absorbance spectra
of plastids, reformed prolamellar
bodies and thylakoids isolated
from partly greened wheat seedlings.

Pigment composition in reformed prolamellar bodies

When dark-grown leaves are irradiated with continuous white
light PLBs are transformed and PChlide is converted to Chlide.
When the seedlings are placed back into darkness PChlide starts to
reaccumulate and PLBs to reform. PLBs and thylakoids were isolated
from such partly greened material which contained well developed
reformed PLBs, mainly of the narrow type (31). It was found that
most of the protein in these PLBs were the PChlide-oxidoreductase.
The pigments were distributed between both PLBs and thylakoids.
The PLBs contained more PChlide in relation to Chlide than did
the thylakoids. The PChlide was difficult to recognize in absor-

bance spectra because of the high amount of Chlide found in the PTs. In spite of the presence of only one polypeptide in the PLBs both Chlide and PChlide were found in this fraction (Fig. 10). At present it is, however, difficult to conclude if the Chlide was artifically bound to the PLBs during preparation or if it is a true constituent. Treffry (9) showed reformation of PLBs without detectable amounts of PChlide and suggested that an unphytylated form of chlorophyll is associated with the development of PLBs. Reformation of PLBs in darkness did not take place in leaves where the resynthesis of PChlide was inhibited by photodecomposition of $Chlide_{684}$ (43). When the later appearing $Chlide_{672}$ was photodecomposed PLBs were reformed and PChlide accumulated. The conclusion was drawn that $Chlide_{684}$ is closely associated with the enzymes involved in PChlide synthesis. According to Oliver and Griffiths (44) $Chlide_{684}$ is a complex between the pigment molecule and the enzyme PChlide-oxidoreductase, while $Chlide_{672}$ is the free pigment. Thus, the results by Ryberg and coworkers (43) strongly indicate that reformation of PLBs is connected with presence of PChlide-oxidoreductase.

Localization of Mg-protoporphyrin

Treatment of dark-grown leaves with ALA increases the PChlide content. A pre-treatment with 8-hydroxyquinoline inhibits the synthesis of PChlide and causes the accumulation of Mg-protoporphyrin. This treatment also results in an altered ultrastructural appearance of the PTs indicating a correlation between PTs and PChlide synthesis (45). Analyses of isolated PLBs and PTs from leaves treated in this way showed that the distribution of Mg-protoporphyrin and PChlide between the PLB- and PT-fractions differed (30). Mg-protoporphyrin was found to an even higher extent in the PTs than was PChlide. The conclusion was drawn that the synthesis of membrane-bound chlorophyll precursors is localized to the PTs.

CONCLUSIONS

PChlide is present in both PLBs and PTs. Because of the lack of markers for the different membranes the quantitative distribution of PChlide between these systems is still uncertain. The biosynthesis of PChlide is probably connected to the PTs. The PChlide-oxidoreductase is the dominating protein in the PLBs and seems to be more stable in these membranes than in the PTs. The PLB might function as a structure suitable for protecting and storing phototransformable PChlide (Pchlide$_{650-657}$). The reason for such a storage compartment has to be elucidated by future work.

Acknowledgement. The authors are grateful to Ms Lena Töyrä for skilful technical assistance with the electron microscopy work.

REFERENCES
1. Henningsen KW, Boynton JE. 1974. IX. Development of plastid membranes during greening of dark grown barley seedlings. J. Cell Sci. 15, 31-55.
2. Bradbeer JW, Gyldenholm AO, Ireland HMM, Smith JW, Rest J, Edge HJW. 1974. Plastid development in primary leaves of Phaseolus vulgaris. VIII. The effects of the transfer of dark--grown plants to continuous illumination. New Phytol. 73, 271-279.
3. Virgin HI, Kahn A, von Wettstein D. 1963. The physiology of chlorophyll formation in relation to structural changes in chloroplasts. Photochem. Photobiol. 2, 83-91.
4. Treffry T. 1970. Phytylation of chlorophyllide and prolamellar-body transformation in etiolated peas. Planta 91, 279-284.
5. Henningsen KW, Boynton JE. 1969. Macromolecular physiology of plastids. VII. The effect of a brief illumination on plastids of dark-grown barley leaves. J. Cell Sci. 5, 757-793.
6. Boardman NK, Wildman SG. 1962. Identification of proplastids by fluorescence microscopy and their isolation and purification. Biochim. Biophys. Acta 59, 222-224.
7. Klein S, Schiff JA. 1972. The correlated appearance of prolamellar bodies, protochlorophyll(ide) species, and the Shibata shift during development of bean etioplasts in the dark. Plant Physiol. 49, 619-626.
8. von Wettstein D, Henningsen KW, Boynton JE, Kannangara GC, Nielsen OF. 1971. The genic control of chloroplast development in barley. In Autonomy and biogenesis of mitochondria and chloroplasts (NK Boardman, ed.) pp. 205-223. North Holland Publishing Company, Amsterdam. ISBN 0-7204-4087-4.
9. Treffry T. 1973. Chloroplast development in etiolated peas: Reformation of prolamellar bodies in red light without accumulation of protochlorophyllide. J. Exp. Bot. 24, 185-195.

10. Michel-Wolwertz MR, Bronchart R. 1974. Formation of prolamellar bodies without correlative accumulation of protochlorophyllide or chlorophyllide in pine cotyledons. Plant Sci. Lett. 2, 45-54.

11. Lütz C. 1978. Separation and comparison of prolamellar bodies and prothylakoids of etioplasts from Avena sativa L. In Chloroplast Development (G Akoyunoglou and JH Argyroudi-Akoyunoglou eds) pp. 481-488. Elsevier/North Holland, Amsterdam, The Netherlands. ISBN 0-444-80084-0.

12. Wellburn AR, Hampp R. 1979. Appearance of photochemical function in prothylakoids during plastid development. Biochim. Biophys. Acta 547, 380-397.

13. Kesselmeier J. 1982. Steroidal saponins in etiolated, greening and green leaves and in isolated etioplasts and chloroplasts of Avena sativa. Protoplasma 112, 127-132.

14. Lütz C, Klein S. 1979. Biochemical arfd cytological observations on chloroplast development. VI. Chlorophylls and saponins in prolamellar bodies and prothylakoids separated from etioplasts of etiolated Avena sativa L. leaves. Z. Pflanzenphysiol. 95, 227-237.

15. Smith JHC, Kupke DW. 1956. Some properties of extracted protochlorophyll holochrome. Nature 178, 751-752.

16. Granick S, Gassman M. 1970. Rapid regeneration of protochlorophyllide$_{650}$. Plant Physiol. 45, 201-205.

17. Sundqvist C. 1969. Transformation of protochlorophyllide, formed from exogenous δ-aminolevulinic acid, in continuous light and in flashlight. Physiol. Plant. 22, 147-156.

18. Sundqvist C. 1970. The conversion of protochlorophyllide$_{636}$ to protochlorophyllide$_{650}$ in leaves treated with δ-aminolevulinic acid. Physiol. Plant. 23, 412-424.

19. Griffiths WT. 1978. Reconstitution of chlorophyllide formation by isolated etioplast membranes. Biochem. J. 174, 681-692.

20. Canaani OD, Sauer K. 1977. Analysis of the subunit structure of protochlorophyllide holochrome by sodium dodecyl sulfate--polyacrylamide gel electrophoresis. Plant Physiol. 60, 422--429.

21. Guignery G, Luzzati A, Duranton J. 1974. On the specific binding of protochlorophyllide and chlorophyll to different peptide chains. Planta 115, 227-243.

22. Apel K, Santel HJ, Redlinger TE, Falk H. 1980. The protochlorophyllide holochrome of barley (Hordeum vulgare L.). Isolation and characterization of the NADPH:protochlorophyllide oxidoreductase. Eur. J. Biochem. 111, 251-258.

23. Oliver RP, Griffiths WT. 1980. Identification of the polypeptides of NADPH-protochlorophyllide oxidoreductase. Biochem. J. 191, 277-280.

24. Beer NS, Griffiths WT. 1981. Purification of the enzyme NADPH:protochlorophyllide oxidoreductase. Biochem. J. 195, 83-92.

25. Griffiths WT. 1974. Source of reducing equivalents for the in vitro synthesis of chlorophyll from protochlorophyll. FEBS Letters 46, 301-304.

26. Brodersen P. 1976. Factors affecting the photoconversion of protochlorophyllide to chlorophyllide in etioplast membranes isolated from barley. Photosynthetica 10, 33-39.

27. Ryberg M, Sundqvist C. 1982a. Characterization of prolamellar bodies and prothylakoids fractionated from wheat etioplasts. Physiol. Plant. 56, 125-132.
28. Ryberg M, Sundqvist C. 1982b. Spectral forms of protochlorophyllide in prolamellar bodies and prothylakoids fractionated from wheat etioplasts. Physiol. Plant. 56, 133-138.
29. Ryberg M, Sandelius AS, Selstam E. 1983. Lipid composition of prolamellar bodies and prothylakoids of wheat etioplasts. Physiol. Plant. 57, 555-560.
30. Ryberg M. 1983, in press. The localization of magnesium-protoporphyrin and protochlorophyllide in separated prolamellar bodies and prothylakoids of wheat treated with 8-hydroxyquinoline and δ-aminolevulinic acid. Physiol. Plant.
31. Ryberg M, Minkov I. 1983, in press. Characteristics of in-vivo reformed prolamellar bodies after isolation. In Proceedings of the Sixth International Congress on Photosynthesis (C. Sybesma, ed.).
32. Ryberg M, Sundqvist C. 1983, in press. Spectroscopical and ultrastructural characteristics of irradiated isolated prolamellar bodies and prothylakoids in the presence and in the absence of NADPH. In Proceedings of the Sixth International Congress on Photosynthesis (C. Sybesma, ed.).
33. Ikeuchi M, Murakami S. 1983. Separation and characterization of prolamellar bodies and prothylakoids from squash etioplasts. Plant & Cell Physiol. 24, 71-80.
34. Ryberg M. 1982. Chlorophyll formation in dark-grown wheat. A study of the localization of membrane-bound chlorophyll precursors. PhD thesis. University of Göteborg, Göteborg, Sweden. ISBN 91-86022-11-3.
35. Sundqvist C, Klockare B. 1975. Fluorescence properties of protochlorophyllide in flash irradiated dark grown wheat leaves, treated with δ-aminolevulinic acid. Photosynthetica 9, 62-71.
36. Kesselmeier J, Budzikiewicz H. 1979. Identification of saponins as structural building units in isolated prolamellar bodies from etioplasts of Avena sativa L. Z. Pflanzenphysiol. 91, 333-344.
37. Kesselmeier J, Ruppel HG. 1979. Relations between saponin concentration and prolamellar body structure in etioplasts of Avena sativa during greening and re-etiolating and in etioplasts of Hordeum vulgare and Pisum sativum. Z. Pflanzenphysiol. 93, 171-184.
38. Lütz C. 1981a. On the significance of prolamellar bodies in membrane development of etioplasts. Protoplasma 108, 99-115.
39. Lütz C. 1981b. Distribution of carotenoids and lipids in separated prolamellar bodies and prothylakoids of etioplasts from Avena sativa L. Z. Pflanzenphysiol. 104, 43-52.
40. Lütz C, Nordmann U. 1983. The localization of saponins in prolamellar bodies mainly depends on the isolation of etioplasts. Z. Pflanzenphysiol. 110, 201-210.
41. Lichtenthaler HK, Bach TJ, Wellburn AR. 1982. Cytoplasmic and plastidic isoprenoid compounds of oat seedlings and their distinct labelling from ^{14}C-mevalonate. In Biochemistry and Metabolism of Plant Lipids (JFGM Wintermans, PJC Kuiper, eds) Elsevier Biomedical Press B.V.

42. Murakami S, Ikeuchi M, Miyao M. 1983. Steroidal saponins are not main building units of the prolamellar body in etioplasts. Plant & Cell Physiol. 24, 581-586.
43. Ryberg H, Axelsson L, Klockare B, Sandelius AS. 1981. The function of carotenoids during chloroplast development. III. Protection of the prolamellar body and the enzymes for chlorophyll synthesis from photodestruction, sensitized by early forms of chlorophyll. In Photosynthesis. V. Chloroplast Development (G Akoyunoglou, ed.) pp. 295-304, Balaban International Science Services, Philadelphia, Pa. ISBN 0-86689-010-6.
44. Oliver RP, Griffiths WT. 1982. Pigment-protein complexes of illuminated etiolated leaves. Plant Physiol. 70, 1019-1025.
45. Ryberg M, Ryberg H. 1981. Chlorophyll precursors and plastid ultrastructure in dark-grown wheat leaves treated with 8--hydroxyquinoline and δ-aminolevulinic acid. In Photosynthesis. V. Chloroplast Development (G Akoyunoglou, ed.) pp. 177-183, Balaban International Science Services, Philadelphia, Pa. ISBN 0-86689-010-6.

INTERMEDIATES IN
PROTOCHLOROPHYLLIDE
PHOTOREDUCTION.

THE LONG-WAVELENGTH-ABSORBING QUENCHERS FORMED DURING ILLUMI-
NATION OF PROTOCHLOROPHYLLIDE-PROTEINS.

E.DUJARDIN
Photobiology Laboratory B22 Liège University, B4000,Sart-Tilman
Belgium.

In etiolated leaves of angiosperms, protochlorophyllide
(P) is converted into chlorophyllide(C) within a very short
time by a light flash. The reaction has been shown to be a photoenzymatic reduction where the electron donor is NADPH. The
formation of a ternary complex between protochlorophyllide,
the apoprotein (protochlorophyllide photoreductase) and NADPH
seems to be a prerequisite for synthesizing chlorophyllide in
the light (Griffiths, 1978).

There are two photoactive species of the ternary complex.
They differ by their spectroscopic properties. The first species has a main absorption band at 650 nm and emits at 657 nm.
The second species absorbs at 640 nm and emits at 645 nm. Both
species are converted by light into a chlorophyllide which e-
mits a fluorescence at 688 nm and has a red absorption band at
677 nm.

$$\left\{ \begin{array}{l} P_{657-650} \\ \\ P_{645-640} \end{array} \right. \quad \xrightarrow{\quad h\nu \quad} \quad C_{688-677}$$

This reaction is slown down when the temperature is lowered: Smith (1954), and later Goedheer and Verhulsdonck (1970)
observed the formation of chlorophyllide at 193 K, but not at
77 K.

Intermediate pigments.

Rubin et al.(1962) were the first to show that, when illuminating etiolated leaves at 153 K, the intensity of the fluorescence band of the protochlorophyllide was decreased although
there was no chlorophyllide fluorescence appearing. The fluorescence of chlorophyllide at 688 nm appeared later when the

C. Sironval and M. Brouers (eds.); Protochlorophyllide Reduction and Greening. ISBN 90 247 2954 8
©1984, Martinus Nijhoff/Dr W. Junk Publishers, The Hague/Boston/Lancaster.

leaf was warmed up to 183 K in darkness. Sironval and Kuiper
(1972) were able to prove that the illumination of protochlorophyllide at 178 K provoked the accumulation of an unknown
state "X" of the protochlorophyllide complex, characterized by
a low fluorescence yield. All these authors did not investigate
the absorption spectra.

Using fresh or freezed-dried etiolated bean leaves in which
it is known that the protochlorophyllides remain photoactive
(Dujardin,1973), Dujardin and Sironval(1977) succeeded in de-
monstrating that long-wavelength-absorbing pigments were accumulated during an illumination at 178 K. At about the same time,
Raskin discovered similar pigments in fresh leaves(1976,1977).
When trapped in liquid nitrogen these intermediate pigments have a broad absorption band extending from 680 to 750 nm, with
a maximum around 690-695 nm(spectrum 2 of fig.1a). They do not
emit any detectable 77 K fluorescence (spectrum 2 of fig.1b).
They disappear by warming up in the dark to 223 K.

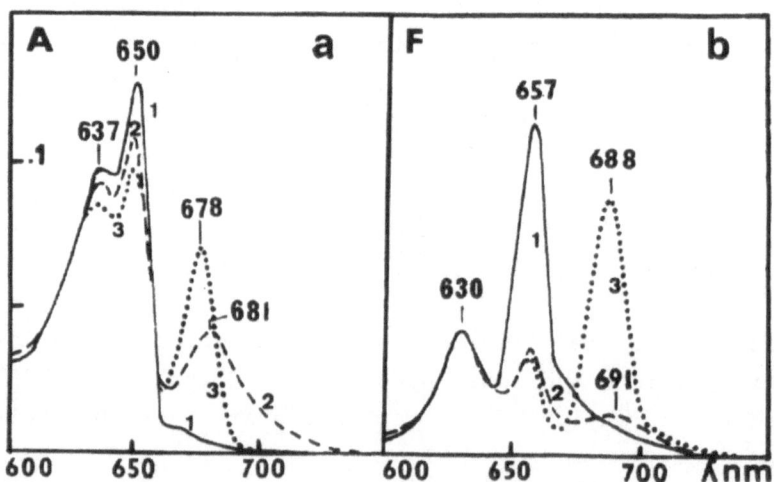

FIGURE 1. Spectra 1 (full lines): red absorption (a) and fluorescence (b) spectra recorded at 77 K, of an etiolated fresh
leaf. Spectra 2 (dashed lines): the leaf has been warmed up
to 178 K, illuminated during 30 sec with a polychromatic 250
Watt Sylvania projector and cooled again at 77 K. Spectra 3
(dotted lines): the same leaf has been warmed up to 250 K during 1 minute and cooled again at 77 K.

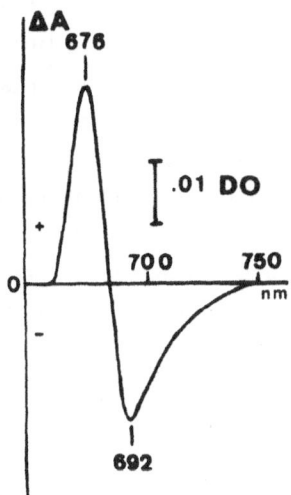

FIGURE 2. Difference absorption spectrum between spectum 3 and spectrum 2 of fig. 1 a showing the transformation of the long-wavelength-absorbing pigments into chlorophyllide $C_{688-678}$.

Then, the 678 nm band characteristic for chlorophyllide (spectrum 3 of figure 1a) and the 688 nm fluorescence emission of chlorophyllide can be recorded (spectrum 3 of figure 1b). The difference spectrum (3a-2a) of figure 2 shows that the intermediates absorb around 692 nm.

Thus, it appears that the photoreduction of protochlorophyllide into chlorophyllide comprises at least 2 steps, as follows :

$$\left\{\begin{array}{l} P_{657-650} \\ P_{645-640} \end{array}\right. \xrightarrow[(1)]{h\nu \quad (178 \text{ K})} P_{-,680 \to 750} \xrightarrow[\text{darkness}]{\text{warming}} C_{688-678}$$

This scheme applies also for extracts from etiolated leaves (Dujardin, 1978; Franck, Mathis, 1980).

Reversibility of light reaction (1).

It is possible to reduce all the photoactive protochlorophyllides of an etiolated leaf using a single 5-10 picosecond intense laser flash given at room temperature. The reduction occurs through a non-fluorescent state of the system (Dobek et al., 1981). However, when decreasing the temperature, more light is needed to reduce completely the protochlorophyllide into chlorophyllide. At 173 K, 30 s of an intense illumination (0.02 W.cm^{-2}) are required to convert only a part of the precursors into long-wavelength-absorbing intermediate pigments (Dujardin, Sironval,1977); at 143 K,60 s are required (Raskin, 1977). At liquid nitrogen temperature, only a small fraction (2 to 5%) of the photoactive protochlorophyllides yields long-wavelengths absorbing pigments after a prolonged illumination at high intensity (Dujardin, Correia,1979). These results are interpreted by assuming that light reaction (1) is reversible and temperature-dependent (Dujardin , Correia, 1979):

$$\begin{cases} P_{657-650} \\ P_{645-640} \end{cases} \xrightarrow[\text{77 K(1b)}]{\text{h}\nu \ \ (1a)} P_{-,680} \rightarrow 750$$

The same conclusion has been reached by Litvin et al.(1981) who performed the following experiment: they have accumulated a certain amount of long-wavelength-absorbing pigments at a low temperature and they have illuminated them with 694.3 nm flashes from a powerful ruby laser observing the reverse reaction:

$$P_{-,680} \rightarrow 750 \xrightarrow[\text{77 K}]{\text{694.3 nm light}} \begin{cases} P_{657-650} \\ P_{645-640} \end{cases}$$

We have been looking for this reversion in Liège. Until now, we did not succeed to observe it, possibly because our ruby laser is 100 times less powerful than that of the Russian authors.

'As they involve two light reactions with opposite directions, Litvin's result suggests a cycle which should work at low temperature:

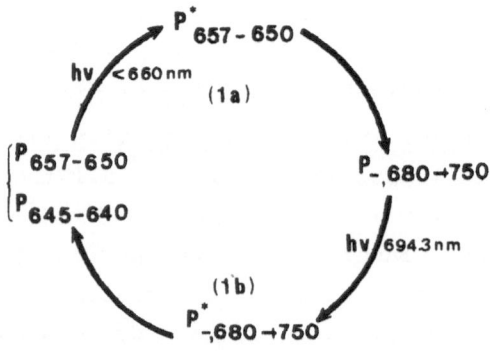

Such a cycle has not been demonstrated to work at room temperature.

Several intermediate pigments.

The formation of the intermediate pigments comprises several steps implying several short-lived species. Using a raw extract from etiolated leaves, Franck and Mathis (1980) had observed a species absorbing at 695 nm and another species absorbing around 685 nm 0.5 μs after a 2 ns flash. Inoue et al. (1981) reached similar results with a solution of protochlorophyllide holochrome prepared according to the method of Boardman.

Further data on the intermediate steps are given in this volume. They point out to the formation of:

1.- a very short-lived species arising during the first 100 picoseconds after a flash, with an absorption band around 690-700 nm (Iwai et al,1984)
2.- a species|occuring within 1 ns after the flash (van Bockhove et al., 1984)
3.- a third species arising during 250 ns after the flash (absorption band at 680 ->725 nm (Iwai et al,1984).

Dark reaction (2).

Raskin (1981) has shown that intermediates are spontaneously transformed in the dark into $P_{688-678}$ at temperature higher than 123 k. This reaction limits the life-time of the intermediates; it is temperature-dependent and it is certainly complicated as demonstrated by Franck (1982). Franck detected three phases by investigating the kinetics of the rise of the fluorescence of chlorophyllide after a brief actinic illumination of etiolated leaves at various temperatures. Part of the intermediates pigments have a much longer life-time than the bulk of them (for example, $t_{1/2}$ > 5ms at 254 K).

Below 123 K, the inhibition of the transformation of the intermediates into chlorophyllide should favor back reactions, especially back reactions triggered by light (like reaction (1b)).

Dark reduction of protochlorophyllide.

When etiolated leaves containing long-wavelength-absorbing pigments are warmed up in darkness, the absorption of the remaining photoreducible protochlorophyllides decreases (fig.1a, compare spectrum 3 to spectrum 2). This phenomenon has been investigated by Sironval and Kuiper (1972). This result suggests that there is a dark reduction of protochlorophyllide.

$$\left\{ \begin{array}{l} P_{657-650} \\ P_{645-640} \end{array} \right. \xrightarrow{\text{dark; } T> 123 \text{ K}} C_{688-678}$$

The occurence of this reduction might be due to the formation by light of a high-energy state which would be stable at a low temperature (below 123 k) and which would result in the reduction when the temperature is rised up. It is possible that a strong reductant is produced as follows:

(a) $\left\{ \begin{array}{l} P_{657-650} \\ P_{645-640} \end{array} \right. \xrightarrow{\text{light}} \left[\begin{array}{c} \text{high-energy} \\ \text{state} \end{array} \right] \longrightarrow \left[\begin{array}{c} \text{strong} \\ \text{reductant} \\ \text{stable at} \\ T<123 \text{ K} \end{array} \right]$

(b) $\begin{array}{c} \text{strong} \\ \text{reductant} \end{array} + \left\{ \begin{array}{l} P_{657-650} \\ P_{645-640} \end{array} \right. \xrightarrow[\text{dark}]{T> 123 \text{ K}} C_{688-678}$

We are totally ignorant of the mechanism of such a phenomenon. We do not know either if (and how) it could involve the intermediate pigments.

$P_{680 \to 750}$ intermediates as fluorescence quenchers.

The most striking characteristic of the $P_{-,680\to750}$ intermediates is their ability to quench the fluorescence emitted by other pigments. This property is observed at room temperature (Dobek et al.,1981) as well as at a low temperature (Raskin, 1977; Dujardin,Correia, 1979; Dujardin,1982; Franck, 1982). Only the fluorescence of the photoinactive protochlorophyllide $P_{633-628}$ is poorly quenched, or is not quenched at all (Dujardin, Correia,1979). For this reason, $P_{630-628}$ may be considered as a reference for studying the quenching of the fluorescence of other pigments.

Fig. 1 shows how the fluorescence of the photoactive protochlorophyllide $P_{657-650}$ is quenched by the $P_{-,680->750}$ pigments at a low temperature. A stated above this phenomenon is observed in leaves as in extracts. In the later case, however, a quenching of the fluorescence emitted at 645 nm by the photoactive protochlorophyllide is the most often observed (Dujardin, unpublished). On the other hand, etiolated leaves fed whith δ-aminolevulinic acid (δ-ALA) accumulate a photoinactive protochlorophyllide emitting at about 637-640 nm and absorbing at about 632-634 nm i.e. at longer wavelengths than the photoinactive $P_{630-628}$ endogenous protochlorophyllide. In these δ-ALA-treated leaves illumination at 173 K provokes the quenching of the 657 nm emission, but not of the emission at 635-6 nm (Dujardin, unpublished).

The intermediates as probes for the study of pigment organization.

    a) The etioplast membrane.

The $P_{-,680->750}$ quenchers may be used for studying the organization of the pigments in the membranes. For instance, the fact that the photoactive protochlorophyllides $P_{657-650}$ and $P_{645-640}$ are quenched indicates that they are closely connected with the quenchers inside the etioplast membrane, while the major part of the photoinactive $P_{630-628}$ protochlorophyllide should not be connected closely (Dujardin and Correia, 1979; Dujardin, Correia and Sironval, 1980; Correia and Dujardin, 1983). The protochlorophyllide accumulated by feeding etiolated leaves with δ-ALA behaves like $P_{630-628}$, and could also be located rather far from the photoactive protochlorophyllides.

The quenching of the fluorescence of the photoactive protochlorophyllides is very efficient in etiolated leaves. As an average, the fluorescence of 20 to 50 protochlorophyllides is quenched at 77 K by the $P_{-,680->750}$ pigments formed from only 5 to 7% of the photoactive protochlorophyllides (Dujardin, Correia, 1979).

For interpreting this quenching, we must remained that in the etioplast membrane, the pigments are organized into units where energy transfers occur from $P_{645-640}$ to $P_{657-650}$, and from $P_{657-650}$ to $C_{688-678}$. These energy transfer units include 15 to 20 pigments in the leaf, or less in various extracts (Kahn et al., 1970; Brouers and Sironval, 1974; Dujardin, 1973). A theoretical description of these units has been made by Brouers and Sironval(1980; see also this volume: Strasser; Sironval et al.)

We should thus consider that $P_{-,680 \to 750}$ quenchers are formed from photoactive protochlorophyllides inside transfer units and therefore that they are included in the organization of these units. Several experiments show that the photoactive protochlorophyllides and the $C_{688-678}$ chlorophyllides (see fig.1) transfer their energy to the $P_{-,680 \to 750}$ quenchers, according to figure 3.

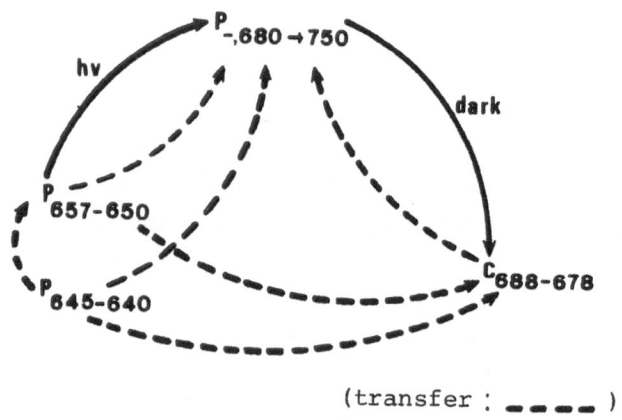

(transfer : ____)

FIGURE 3. The transfer unit in the etiolated leaf.

b) The greening membrane of the developing chloroplast.

It is easy to produce definite states of greening by submitting etiolated leaves to various light/dark treatments. The sequence of the very early greening steps in etiolated leaves is described by the P-C cycle (Sironval, 1981).

The fluorescence of all the chlorophyllides ($C_{688-677}$ and $C_{696-684}$) and of chlorophyll a ($C_{685-672}$) is quenched when leaves containing these pigments plus photoactive protochlorophyllide are illuminated at a low temperature. The quenching

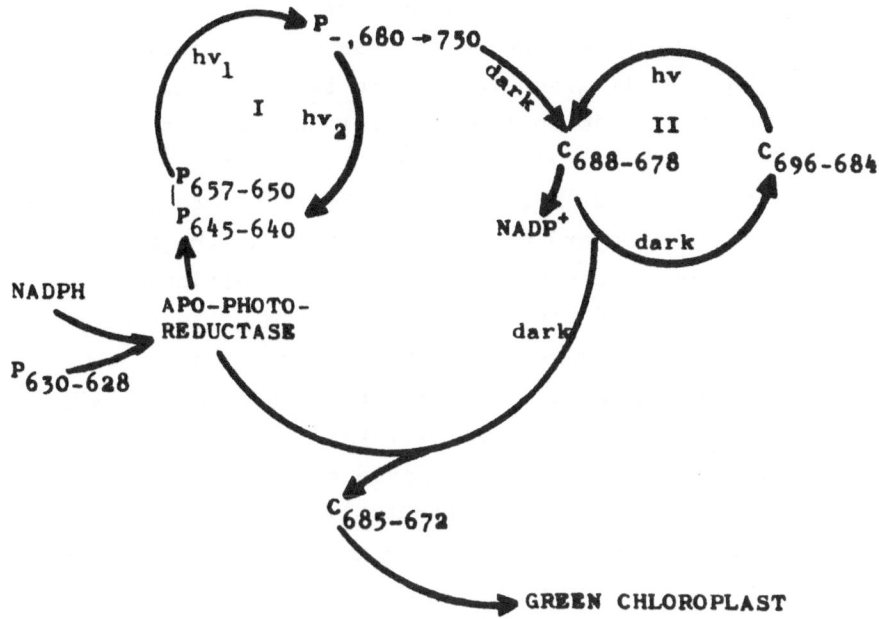

becomes more efficient as the amount of the protochlorophyllides becomes higher (Dujardin, Correia, 1979; Dujardin et al., 1981; Correia, Dujardin, 1983). Etiolated leaves illuminated by shadow day-light during 24 h, which still contain some photoactive protochlorophyllide, behave in the same way (Dujardin et al., 1981).

The long wavelength-absorbing quenchers can apparently be connected with a large amount of chlorophyllides and chlorophyll a in the greening chloroplast. The energy transfer from the chlorophyll(ide)s to the quenchers can be explained by the fact that the quenchers absorb at longer wavelengths and by the overlap of the absorption bands. This situation is similar to that of the active centres of photosynthesis which are also surrounded by pigments from which they collect energy.

c) The fully green membrane.

When leaves from green bean plants frozen in liquid nitrogen are illuminated with an intense actinic light, the intensities of the fluorescence at 686, 695 and 735 nm drop

dramatically (Correia and Dujardin, 1983). This drop may reach
40% of the intensity at 686 nm and 695 nm; it is always less
marked at 735 nm. If green leaves are submitted to post-etiolation (i.e. to a sojourn in the dark) before being frozen
in liquid nitrogen and illuminated, the drop of the intensity of the emissions at 686,695 and 735 nm becomes more important (Correia and Dujardin, 1983). Correia and Dujardin have
interpreted these facts by admitting the formation of the
$P_{-,680 \to 750}$ quenchers from the photoactive protochlorophyllides formed in the dark during post-etiolation.

As shown by the extensive work of Shlyk et al. (1969;
Shlyk, 1970; see also Shlyk et al., this volume) photoactive
protochlorophyllides are always present in fully green leaves
even when they have not been post-etiolated.We have thus to take
into account the possibility of the formation of $P_{-,680 \to 750}$
quenchers from these protochlorophyllides in the fully green
leaf also. Correia and Dujardin did not exclude, however, that
the pheophytin anion formed during the intense actinic illumination at 77 K of fully green leaves, could also act as a
quencher of the chlorophyll fluorescence (Breton, 1982; Klimov et al., 1977).

CONCLUSION

Several short-lived pigment-proteins are formed when the
photoactive protochlorophyllides-proteins of etiolated leaves
are illuminated.

These pigments absorb at longer wavelength than all the
pigments found in the leaf. They do not emit any detectable
fluorescence at a low temperature but a feeble fluoresence
seems to be emitted at room temperature (Dobek et al., 1981).
We call them $P_{-,680 \to 750}$.

They are transformed into chlorophyllide-proteins within
10 to 20 us, but part of them have a longer life-time.

They act as efficient quenchers of the fluorescence of
the other pigments, being part of the transfer units in the
etioplast membrane. Their spectral properties and the fact
that they are included in the organization of the transfer

units suggest that the $P_{-,680 \to 750}$ pigment-proteins could behave as do the active centres in green photosynthetic membranes. They might play a role similar as PS I centres. This suggestion is supported by the fact that in etiolated leaves, an actinic 694 nm light absorbed by the $P_{-,680 \to 750}$ pigments and not by the protochlorophyllides- causes the reduction of protochlorophyllide into chlorophyllide at room temperature (Dujardin, 1978).

In greening and fully green leaves, the possibility of the formation of the $P_{-,680 \to 750}$ quenchers from protochlorophyllides has also been demonstrated.

We are brought to ask the following questions:
1.- Are the $P_{-,680 \to 750}$ pigments merely intermediates in the synthesis of chlorophyllide ?
2.- Is the role of these pigments limited to fluorescence quenching inside the membrane organization ?
3.- Is the energy which they trap utilized for photosynthesis ?

ACKNOWLEDGMENTS

The author thanks the F.N.R.S. (Fonds national de la Recherche scientifique), the Ministère de la Politique Scientifique (Belgique), and the "Actions Concertées nr 80/85-18/1983, Belgique" for financial support.

REFERENCES.

1. Breton A.1982, FEBS Lett.,147, 116-201.
2. Brouers M, Sironval C. 1974, Plant Sc. Letters, 2, 67-72.
3. Brouers M, Sironval C. 1980, Photosynthetica, 14 ,213-221.
4. Correia M, Dujardin E. 1983, Photobiochem. Photobiophys. 5, 281-292.
5. Dobek A, Dujardin E, Franck F, Sironval C, Breton A, Roux E. 1981, Photobiochem. Photobiophys. 2, 35-44.
6. Dujardin E. 1978,"Photosynthetic Oxygen Evolution", Metzner H ed. Academic Press N.Y. and London, 83-98.
7. Dujardin E, Sironval C.1977, Plant Sc. Letters, 10,347.
8. Dujardin E. 1973, Photosynthetica, 7, 121-131.
9. Dujardin E, Correia M. 1979, Photobiochem. Photobiophys. 1, 25-32.
10. Dujardin E.1981, in"Photosynthesis V Chloroplast Development" Akoyunoglou G. ed. Balaban International Science Services . Philadelphia Pa , pp 31-38.
11. Dujardin E, Correia M, Sironval C. 1981, in "Photosynthesis V Chloroplast Development" Akoyunoglou G ed. Balaban International Science Services.Philadelphia Pa.pp 21-29.

12. Dujardin E. 1982, in "Cell Function and Differentiation" part B. Akoyunoglou G ed. Alan R. Liss Inc. 150 Fifth Av. New-York, 43-52.
13. Franck F. 1982, Dissertation. Liège University, Belgium.
14. Franck F, Dujardin E, Sironval C. 1980, Plant Science Letters. 18, 375-380.
15. Franck F, Mathis P. 1980, Photochem. Photobiol. 32, 799-803.
16. Goedheer J C, Verhülsdonck C A H. 1970, Biochem. Biophys. Res. Comm. 39, 260-266.
17. Griffiths W T. 1978, Biochem. J. 174, 681-692.
18. Inoue Y, Kobayashi T, Ogawa T and Shibata K. 1981, Plant Cell Physiol. 22, 197-206.
19. Iwai J, Ikeuchi M, Inoue Y, Kobayashi T. 1984, this volume.
20. Klimov V V, Klenavik A V, Shuvalov V A, Krasnovsky A A. 1977, FEBS Lett. 82, 183-186.
21. Litvin F F, Ignatov N V, Belyaeva O B. 1981, Photobiochem. Photobiophys. 2, 233-237.
22. Raskin V I. 1976, Vestsi Akad. Nauk BSSR ser Biol., 5, 43-46. (russian).
23. Raskin V I. 1977, Dokl. Akad. Nauk SSSR ser Bioph.XXI, 272-275.
24. Raskin V I. 1981, Photoreduction of Protochlorophyllide. Minsk "Nauka i Technika".
25. Shlyk A A. 1970, "Chlorophyll metabolism in green plants" Godnev T ed. Progr. Sci. Transl, Jerusalem,(Engl. transl.) Israël.
26. Shlyk A A, Savchenko G Y, Averina N G. 1969, Biofizika 14 , 119-129 (Engl. transl).
27. Sironval C. 1981, in "Photosynthesis V Chloroplast Development" Akoyunoglou G ed. Balaban International Science Services. Philadelphia Pa , pp 3-14.
28. Sironval C, Strasser R, Brouers M. 1984, this volume.
29. Smith J H C, Benitez A. 1954, Plant Physiol. 29, 135-143.
30. Strasser R. 1984, this volume.
31. van Bochove A C, Griffiths W T, van Grondelle R. 1984, this volume.

EARLY PROCESSES OF PROTOCHLOROPHYLLIDE PHOTOREDUCTION AS MEASURED BY NANOSECOND AND PICOSECOND SPECTROPHOTOMETRY

J. IWAI[1], M. IKEUCHI[2], Y. INOUE[2] and T. KOBAYASHI[1]

[1] Department of Physics, University of Tokyo, Hongo, Tokyo 113, Japan.
[2] The Institute of Physical and Chemical Research, Wako, Saitama 351, Japan.

SUMMARY

The process of Protochlorophyllide (PChlide) photoreduction was investigated by means of nsec and psec time resolved spectrophotometry, and three intermediates (X_1, X_2 and X_3) have been detected. Of these three, X_1 and X_2 are newly detected intermediates, while X_3 is identical with X-690, the previously reported intermediate. The photoreduction process is summarized as in the following scheme:

$$\text{PChlide} \xrightarrow[\text{1-2 nsec}]{h\nu} X_1 \xrightarrow[\text{35-250 nsec}]{} X_2 \xrightarrow[\text{1-2 } \mu\text{sec}]{} \substack{X_3 \\ (\text{X-690})}$$

$$\xrightarrow[\text{12 } \mu\text{sec}]{} \text{Chlide}$$

1. INTRODUCTION

Many spectroscopic studies have been conducted on the photoreduction process in etiolated leaves. It is now well established that the photoactive pigment is protochlorophyllide [PChlide], which is protochlorophyll [PChl] lacking phytol group (1-3). It was found that there are two forms of protochlorophyllide by early spectroscopic studies on the leaves of dark-grown bean seedlings (1). A main component of the two forms has a red absorption maximum at 650 nm which is called protochlorophyllide-650 [PChlide-650] while a minor component has a maximum at 636 nm (1). Only one of the two forms, PChlide-650, is converted to chlorophyllide a [Chlide a] on the illumination of the plants. The PChlide in etiolated leaves is bound to a protein to form a pigment protein complex called PChlide holochrome. PChlide holochrome is found to be localized in membranes called prolamellar bodies, which exist in proplastids in the leaves of dark-grown bean plants (3,4).

An important step in the development process of chloroplasts in higher

C. Sironval and M. Brouers (eds.); Protochlorophyllide Reduction and Greening. ISBN 90 247 2954 8

plants is the production of Chlide by reduction of the precursor PChlide. Chlide is then esterified to yield chlorophyll *a*. The reduction is an enzymatic reaction triggered by light and it involves the attachment of two hydrogen atoms to one of the C=C double bonds in magnesium porphyrin ring (4). One mole of PChlide is reduced by 1 mole of hydrogen. The two hydrogen atoms needed for the reduction are provided by the protein which construct the PChlide holochrome. Thus, the photoreduction is restricted to the PChlide molecule on PChlide holochrome.

Etiolated leaves of Plant seedlings which have been grown in darkness supply a good starting material since their etioplasts do not contain any chlorophyll or Chlide, but instead a large amount of PChlide is accumulated. A large fraction of PChlide can be photoreduced into Chlide by a single short illumination of light.

Shibata (1) found that a form of Chlide having absorption maximum at 684 nm (C 684) is produced after 1 min of illumination of PChlide-650 at physiological temperature. Subsequently, Gassman et al. (5) found that another form of Chlide having absorption maximum at 678 nm (C 678) preceeds the appearance of C 684. Subsequently it was found that Chlide is formed in less than 1 msec following illumination (6). Recent studies has shown that illumination at low temperature (below -80°C) of several materials permits the accumulation of species which seem to be stabilized intermediates in the photoreduction of PChlide (7-11). These intermediates are characterized by the absorption at long wavelength (around 700 nm, whereas PChlide absorbs at 630-650 nm and Chlide usually at 670-680 nm) and by the absence of any detectable fluorescence from room temperature down to 77K, whereas both PChlide and Chlide are highly fluorescent at low temperature. These intermediates quench the fluorescence of neighbour Chlide and PChlide. When these species are formed, rewarming of the material leads to their disappearance and to the formation of Chlide. In a more recent study on frozen etiolated leaves at -35°C, it has been shown that the light-induced decrease in PChlide-650 concentration and the increase in C 678 concentration in frozen etiolated leaves with fluorescence as the probe, and found that the two kinetics do not agree (12). It was observed that the increase of 690 nm fluorescence due to C 678 formation proceeds more slowly than the decay of the 661 nm fluorescence which disappears within 10 nsec due to the decrease in PChlide concentration (12). It was thus predicted that an unknown non-fluorescent intermediate exists before the formation of C 678 (12).

Recent studies have shown that PChlide photoreduction involves inter-
mediated steps with short lifetimes (13). Unstable species have been shown
to be formed within a time period of the order of 0.5 μsec in a non-purified,
etiolated leaf extract illuminated by a single 20 nsec Nd:YAG laser flash
at room temperature (13); these species are intermediary between PChlide
and Chlide and absorb at 695 nm. Chlide with absorption at 675 nm is formed
in a time of several μsec after a 20 nsec flash.

In a previous paper (14) we studied the transient absorption change
for the photoreduction of PChlide in a solution of extracted PChlide holochrome and in etiolated leaves. The actinic light source was the second
harmonic (532 nm, 5 nsec) of a Q-switched Nd:YAG laser. We obtained the
following results: (1) A short-lived intermediate (X-690) with an absorption maximum at 690 nm appears before the formation of C 678. (2) The
intermediate, X-690, appears instantaneously at the onset of the actinic
laser and decays with a rate constant of 3.4×10^5 sec^{-1}. (3) This decay
constant for the intermediate, X-690, agreed well with the rate constants
for Chlide formation, 3.1×10^5 sec^{-1} for the extract and 4.6×10^5 sec^{-1}
for the etiolated leaves.

In the present paper we utilized the pulses of 5 nsec and 30 psec of
the stimulated Raman light shifted to 630 nm from the pumping pulse of a
Q-switched Nd:YAG laser to investigate the earlier phenomena in PChlide
photoreduction.

2. EXPERIMENTAL

2.1. Samples

The sample was prepared as follows: squash plants (*Cucurbita moschata*
Durch.var., *melonaeformis* Makino cv. Tokyo) were grown in vermiculite in
the dark at 25°C for 9 days before their cotyledons were harvested. During
all preparative procedures, leaves and extracts received no light except
from a dim green safe light. These etiolated leaves were homogenized with
0.4 M tricine-NaOH, 1 mM MgCl$_2$, and 1 mM EDTA (pH 7.7) and centrifuged
at 2000 x g for 5 minutes. The precipitate was resuspended with 40 mM
tricine-NaOH and 1 mM MgCl$_2$ (pH 7.7) and centrifuged at 9200 x g for 10
minutes. Then the precipitate was suspended with 1% digitonin and 1 mM
NADPH for 10 minutes. The suspension was centrifuged at 9200 x g for
10 minutes and then at 100,000 x g for one hour. The supernatant was
condensed at a photoactive pigment concentration of about 0.1 mM.

2.2. <u>Nanosecond spectroscopy</u> (15)

All the spectroscopic measurements (nanosecond and picosecond spectroscopy and ordinary stationary spectrophotometry) were performed with the
sample solution in a cell with 5 mm optical pathlength. *Figure 1* shows

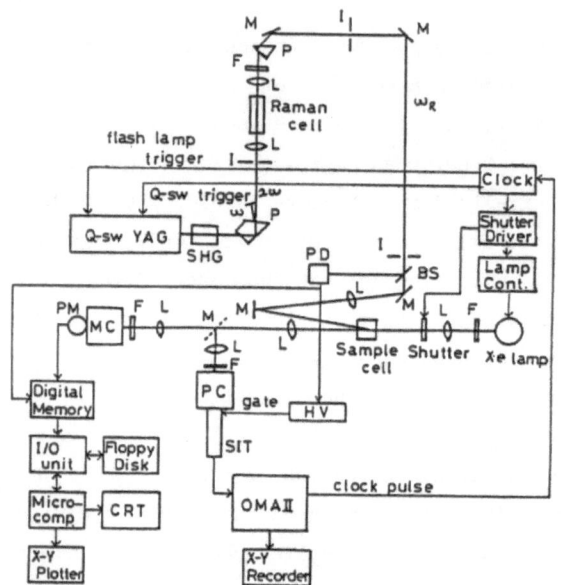

*Figure 1. Block diagram
of nanosecond laser photolysis apparatus used in
the present study.*

F : *filter*
I : *iris*
L : *lens*
M : *mirror*
P : *prism*
BS : *beam splitter*
MC : *monochromator*
PC : *polychromator*
PD : *photodiode*
PM : *photomultiplier*

ω : *fundamental 1064 nm*
2ω : *2nd harmonic 532 nm*
ω_R : *Stokes Raman 630 nm*

the block diagram of nanosecond time-resolved absorption spectroscopy
apparatus used in the present study. The experiment of nanosecond time-
resolved spectroscopy was performed in two different modes. In one mode,
the time dependence of absorbance change (TDAC) at a fixed wavelength was
measured. In the other mode, the time-resolved difference spectrum (TRDS)
at a fixed delay-time after excitation was measured. In both modes of
experiments, the excitation light source was the first Stokes (630 nm)
light generated by the stimulated Raman scattering of the second harmonic
(532 nm) of Nd:YAG laser (Quanta-Ray, DCR-2A) in acetone. The excitation
pulsed light at 630 nm with 5 nsec width was focused on the sample with
30 cm focal length lens. The second harmonic of laser was eliminated by
a cut-off filter (Hoya 056). The light emitted from the Xe lamp of Nd:YAG
laser was removed using a prism. The probe light source was a pulsed xenon

lamp (300 W, Varian, Xenon illuminator VIX 300 F). The angle between
excitation laser beam and probe light beam from the Xe flash lamp was 10
degree. The light intensity of the xenon flash lamp was kept constant for
500 µsec by use of light intensity monitor which gave feedback to the power
supply of the lamp. A mechanical electromagnetic shutter (Copal, EMS #0)
was used to protect samples from being damaged and to avoid the overload
current in a photomultiplier. The probe light was selected with a monochromator (Shimadzu-Bausch & Lomb, f = 17 cm with 1,200 grooves/mm grating)
and was detected by a photomultiplier (Hamamatsu Photonics, R666S). The
output signal of the photomultiplier was digitized with a transient recorder
(Iwatsu, DM901, 9 bits, 10 nsec/point in the fastest writing mode). The
digitized signals (1024 points) were analyzed by a microcomputer (Nippon
Electric, PC8012). The detection system of TRDS was composed of a grating
polychromator (Shimadzu, 20 cm focal length, 500 nm blazed grating with
600 grooves/mm), a gated SIT (Princeton Applied Research, 1254), and an
optical multichannel analyzer (OMA 2: Princeton Applied Research, 1215,
1216).

In the measurement of TDAC both excited and non-excited data were
integrated over 16 shots. The sample solution was stirred before every
shot. In the measurement of TRDS, the probe beam was split into two beams
(reference and monitor) and both were detected simultaneously. The monitor
and reference beams passed through the excited and non-excited regions of
the same sample, respectively.

2.3. Picosecond spectroscopy (16)

Figure 2 shows the block diagram of picosecond time-resolved absorption
spectroscopy used in the present study. A mode-locked Nd:YAG laser
(Quantel, Model 472) was used for both excitation and probe light. A
single pulse was selected out of an output pulse train from the oscillator
of Nd:YAG laser with the use of a Pockel's cell. Xenon flash lamps for
pumping the laser were fired at the repetition rate of 10 Hz, while the
laser oscillation took place at 1 Hz which was determined by the opening
rate of the mechanical electromagnetic shutter. A selected single pulse
was amplified twice by a two-stage amplifier. The energy, pulse width
and power of the amplified single pulse at 1064 nm were 60 mJ, 30 psec,
and 20 GW, respectively. The generated second harmonic (532 nm) by KDP
crystal had 10 mJ energy, 30 psec pulse width, and 3 GW peak power. The
second harmonic was split into two beams with almost equal intensity.

One of the two beams was focused into 10 cm cell containing acetone to generate the stimulated Raman scattered light (first Stokes) at 630 nm which was used as the excitation light pulse. The other was focused into the 10 cm length cell containing water or heavy water to generate a picosecond continuum. The excitation pulse was focused on the sample after being reflected by a movable prism which varied the arrival time of the excitation pulse at the sample cell relative to the probe pulse. The

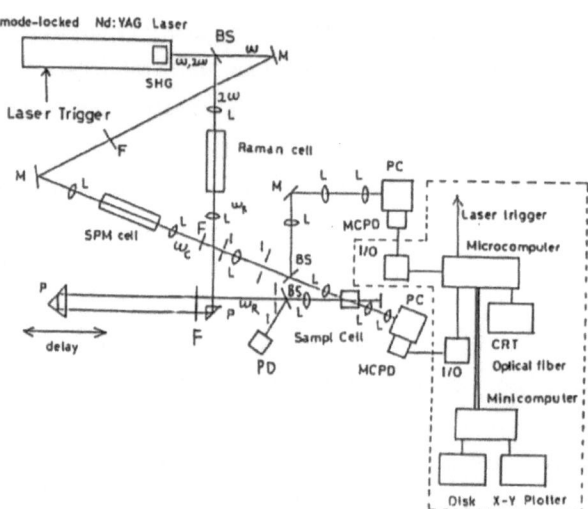

Figure 2. Block diagram of picosecond laser photolysis apparatus used in the present study.

M : *mirror*
F : *filter*
L : *lens*
P : *prism*
I : *iris*
PD : *photodiode*
PC : *polychromator*
BS : *beam splitter*

ω : *fundamental 1064 nm*
2ω : *2nd harmonic 532 nm*
ω_R : *Stokes Raman 630 nm*
ω_C : *continuum*

monitor beam of the picosecond continuum was split into two of nearly equal intensity. Two sets of polychromators and multichannel photodiodes were used as a detection system. One of the beams was used as the probe beam, which was focused on the sample and then focused again on the entrance slit of a polychromator coupled with a multichannel photodiode (Union Giken, MC-100). The other beam was used as a reference beam, which was focused on the other sample cell without excitation and then focused again on the entrance slit of the other polychromator coupled with the other multichannel photodiode. The analog data on the spectrum of the picosecond continuum was converted to digital values and stored in the memory of a microcomputer (Nippon Electric, PC8000). The stored memory data are processed in the microcomputer or transferred to a minicomputer (DEC, MINK 11) through light emitting diodes and optical fibers.

3. RESULTS AND DISCUSSION

3.1. <u>Nanosecond spectroscopy</u>

   3.1.1. <u>The time dependence of absorbance change (TDAC)</u>. The time dependence of absorbance change was measured at 585, 600, 620, 660, 675, 690,
705, 720, and 730 nm with two different resolution times at each wavelength.
The typical results are shown in *Figures 3, 4* and *5*. In *Figure 3*, the
time dependence of the absorbance change at 675 nm is shown. In all curves,
the negative-going spikes at time zero are artifacts due to the scatterd
laser light and/or fluorescence of the sample. *Figure 3(a)* and *(b)* show
the linear plot of absorbance change in the time domain of -1 to 9 μsec,
and -10 to 90 μsec, respectively. The time dependence of the absorbance
change at 675 nm in the time ranges of 0 to 90 μsec is triphasic: the
absorbance change first increased rapidly with a time constant shorter than
0.25 μsec, then increased with an intermediary time constant of 1 - 2 μsec
during the following 1 - 4 μsec, and then kept increasing slowly until about
50 μsec with a time constant of about 12 μsec. The time constants of the
above triphasically growing process were $\simeq$ 0.25 μsec, 2.3 μsec and
11.2 $\pm$ 0.3 μsec. In *Figure 4*, the time dependence of the absorbance change
at 690 nm is shown. At this wavelength the absorbance change increased also
rapidly, then increased at an intermediary rate during the first 3 - 5 μsec,
and then decreased at a slow rate until 50 μsec. The respective time
constants of the these processes were $\leq$ 0.25 μsec, 1.2 μsec and 13.0 $\pm$ 1.0 μsec.
The time dependence of the absorbance change at 705 nm is shown in *Figure 5*.
At this wavelength, the absorbance change increased rapidly with a time
constant shorter than 0.25 μsec, then decreased slowly during the subsequent
1 - 50 μsec. There was no clear phase with a time constant of 1 - 2 μsec.
The time constant for the slow absorbance decrease was 11.7 $\pm$ 0.3 μsec.

   The above data suggest that there are three processes in the time
ranges between 0 and 50 μsec: the first rapid process with a time constant
$\leq$ 0.25 μsec, the second intermediary process with a time constant of 1 - 2 μsec,
and the third slow process with a time constant of 12 μsec. Based on these,
the early processes of Chlide formation can be given as follows:

$$PChlide \xrightarrow{\leq 0.25 \text{ μsec}} X_2 \xrightarrow{1 - 2 \text{ μsec}} X_3 \xrightarrow{12 \text{ μsec}} Chlide$$

As will be discussed in the following section, another intermediate, X_1,
preceding X_2 was found by psec measurements.

106

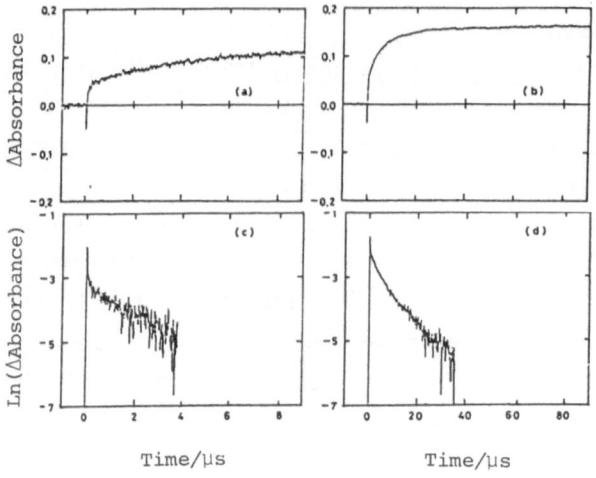

Figure 3. Time dependence of the absorbance change at 675 nm due to the generation of intermediates by the excitation of protochlorophyllide by 5 nsec laser pulse at 630 nm. The time resolution was 10 nsec in (a) and 100 nsec in (b).

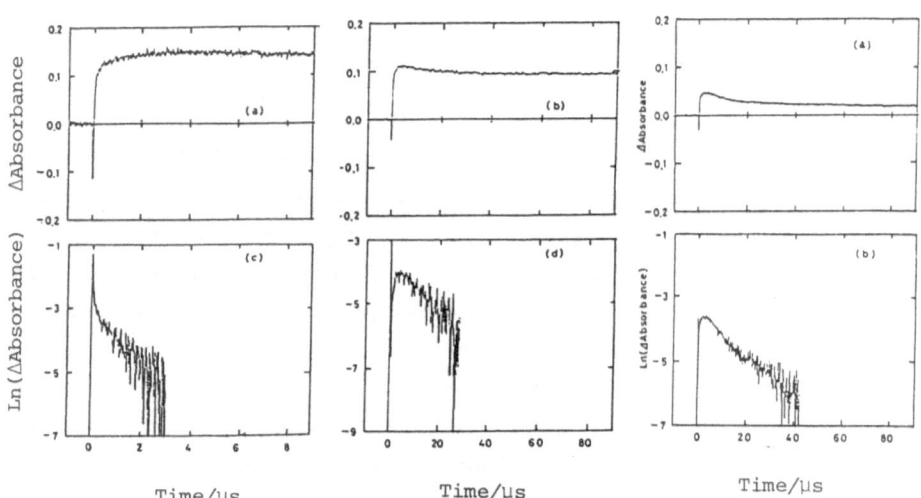

Figure 4 (left). Time dependence of the absorbance change at 690 nm due to the generation of intermediates by the excitation of protochlorophyllide by 5 nsec laser pulse at 630 nm. Time resolution: (a), 10 nsec; (b), 100 nsec.

Figure 5 (right). Time dependence of the absorbance change at 705 nm due to the generation of intermediates by the excitation of protochlorophyllide by 5 nsec laser pulse at 630 nm. Time resolution: (a), 10 nsec; (b), 100 nsec.

3.1.2. <u>Nanosecond time-resolved difference spectra (TRDS)</u>. Time-resolved absorption difference spectra of PChlide holochrome solution are shown in *Figure 6*. The difference spectrum measured at 0.25 µsec after excitation showed a positive peak at 688 nm. The peak position shifted from 688 nm to 680 nm when the delay times was increased from 0.25 to 15.5 µsec: at 687 nm (delay time t_d = 0.7 µsec), 685 nm (t_d = 2 µsec), 684 nm (t_d = 6.5 µsec), and finally at 680 nm (t_d = 15.5 µsec). The spectral width decreased with increasing the delay time; 700 cm^{-1} (t_d = 0.25 µsec), 675 cm^{-1} (t_d = 0.7 µsec), 635 cm^{-1} (t_d = 2.0 µsec), 605 cm^{-1} (t_d = 3.5 µsec), 560 cm^{-1} (t_d = 6.5 µsec), and 530 cm^{-1} (t_d = 15.5 µsec). Two isosbestic points were observed. One was located at 660 nm on the shorter-wavelength side of the absorption band and appeared at time ranges of 0 - 3.5 µsec, and the other was located at

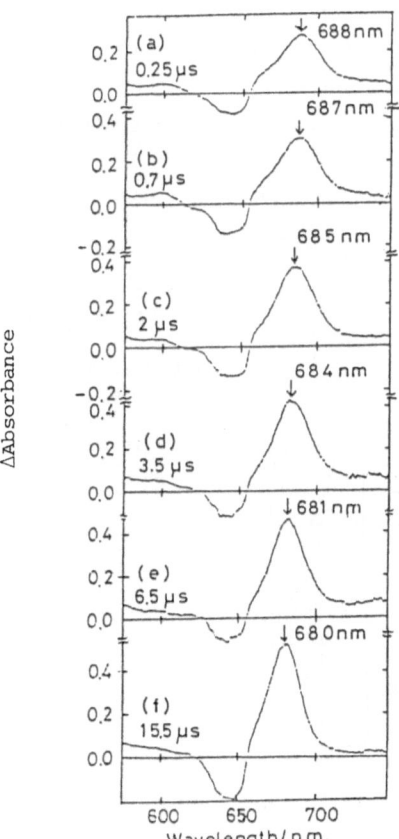

Figure 6. Transient difference absorption spectra of intermediates generated by the excitation of protochlorophyllide by 5 nsec actinic laser pulse at 630 nm.

Delay time of the gate of OMA vidicon was set at 0.25 (a), 0.7 (b), 2.0 (c), 3.5 (d), 6.5 (e), and 15.5 µsec (f). The half gate width for 0.25 µsec delay was 0.15 µsec, so that the difference spectrum shown in (a) is the spectrum averaged over 0.1 - 0.4 µsec. The half gate width of the vidicon detector in the cases (b), (c), (d), (e), and (f), was 1.0 µsec. The experimental error in absorbance change was less than 0.01 in all the measured spectral region except the region between 630 nm and 655 nm.

687 nm on the longer-wavelength side of the absorption band and appeared at time ranges of 3.5 - 20 μsec. The isosbestic point at 660 nm is due to the formation of X_2, and that at 687 nm is due to the transformation of X_3 to C 678. The fact that these two isosbestic points appear at different time ranges implies that the spectral changes observed during 0.25 to 15.5 μsec is composed to more than two different processes. This view is consistent with the scheme assumed in the previous section.

The difference spectrum at 0.25 μsec and that at 3.5 μsec may correspond to two intermediates, X_2 and X_3, respectively: X_2 has an absorption maximum at 688 nm and X_3 at 684 nm. It is noteworthy that both X_2 and X_3 have broad absorption in 680 and 690 nm region. Probably, the intermediate with absorption maximum at 690 nm (X-690) previously reported by us (14) and by Franck and Mathis (13) corresponds to X_3. Although there was an indication of the presence of an earlier intermediate, we were not convinced in the previous paper (14). In the present study, however, the two intermediates were clearly separated in kinetics with the two different time constants in *Figures 3 - 5* experiments and with the two different isosbestic points appearing at different time ranges in *Figure 6* experiment. We note that neither X_2 nor X_3 can be an excited state in PChlide, since their lifetimes

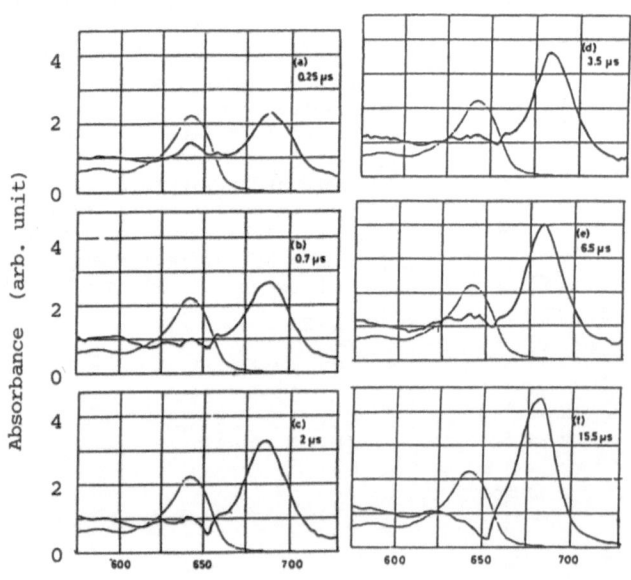

Figure 7. Reconstructed absorption spectra of intermediates generated by the excitation of protochlorophyllide by 5 nsec actinic laser pulse at 630 nm.

Wavelength/nm

are too long to be assigned to singlet excited states.

The time-resolved absorption spectra of these intermediates are shown in *Figure 7*, in which the time-resolved difference spectra in *Figure 6* were corrected by using the absorption spectra before and after full photo-conversion of PChlide holochrome. The spectrum at the delay time of 0.25 μsec may be the absorption spectrum of X_2, and that at 3.5 μsec may be that of X_3.

It is of note that the absorption spectra of both X_2 and X_3 are similar in shape to that of the final product C 678.

3.2. Picosecond spectroscopy.

Picosecond time-resolved difference spectra at the delay time of 100 psec *(a)*, 700 nsec *(b)*, 1.5 nsec *(c)*, and 4.4 nsec *(d)* are shown in *Figure 8,* and those at 12 nsec *(a)*, 17 nsec *(b)*, and 35 nsec *(c)* are shown in *Figure 9*. The sharp spikes at 630 nm in the difference spectra are the artificial signal due to the excitation light scattered by the sample. The typical feature of the difference spectra at the delay time ranges between 100 psec and 4.4 nsec are as follows: (1) No characteristic absorption peak was found in 680 - 690 nm region where X_2, X_3 or Chlide has absorption

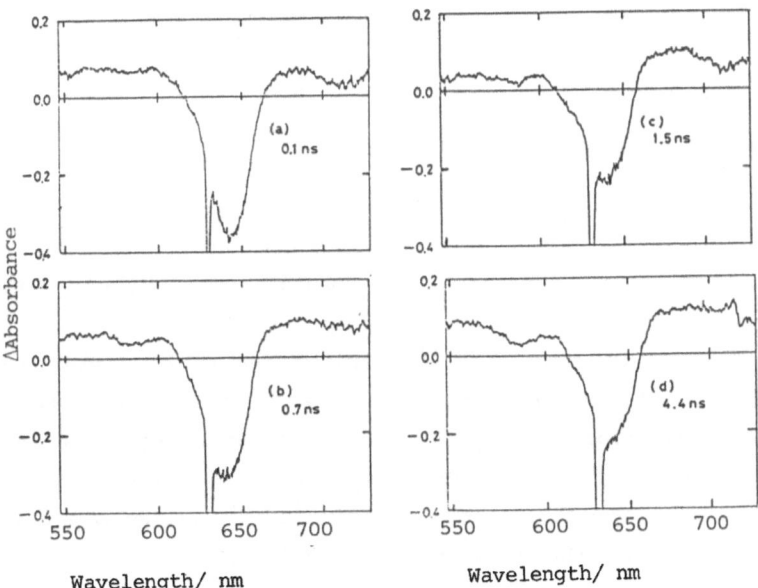

Figure 8. *Time-resolved difference spectra after the excitation of protochlorophyllide by 30 psec laser pulse at 630 nm.*

110

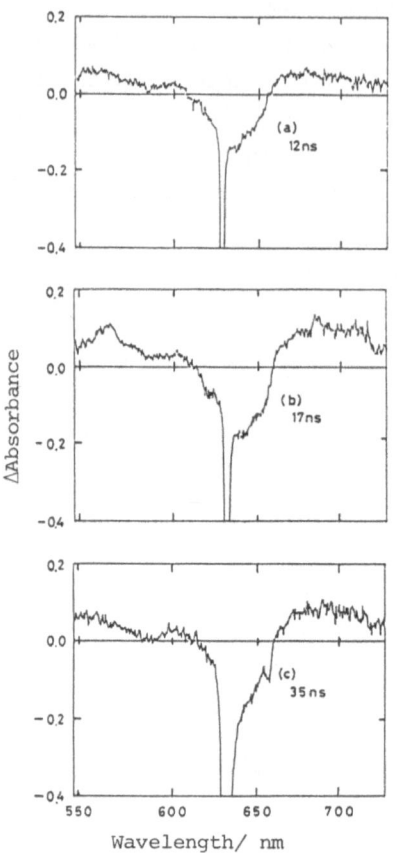

Figure 9. *Time-resolved dif-*
ference spectra after the ex-
citation of protochlorophyllide
by 30 psec laser pulse at 630 nm.

maximum. (2) The negative absorbance change between 620 and 660 nm showed a maximum at 644 nm. This wavelength is slightly longer than that of the absorption maximum (639 - 642 nm) of PChlide. (The possibility of artifacts due to fluorescence was precluded by blank experiments.) (3) The wavelength for the negative maximum shifted toward shorter wavelengths with lapse of delay time. (4) The two wavelengths for zero absorbance change initially observed at 620 nm and at 660 nm shifted toward shorter wavelengths with increasing delay time with a time constant of about 1 - 2 nsec. (5) The amplitude of the negative absorbance change decreased with increasing delay time with a time constant of about 1- 2 nsec.

The above experimental results lead us to the following view: there are two transient species which precede X_2. We denote one of them as X_0, which has a lifetime of about 1 - 2 nsec and an absorption maximum at $\leq$ 640 nm, and the other as X_1, which has a longer lifetime and an absorption maximum at $\geq$640 nm. The lifetime of the latter species is between 35 nsec and 250 nsec.

The time dependence of the absorbance changes at several wavelengths in the delay time ranges between 0 psec and 3.2 nsec are shown in *Figure 10,* which are the plots obtained from the experiments similar to *Figures 8* and *9* experiments. The data of *Figure 10,* support the previous view of the involvement of two transient species: the first species (X_0) has a lifetime of 1 - 2 nsec and the second one (X_1) has a lifetime much longer than 2 nsec. The formation time constant of X_0 is about 50 psec, which is only slightly longer than the width of both excitation and probe pulses (full width of

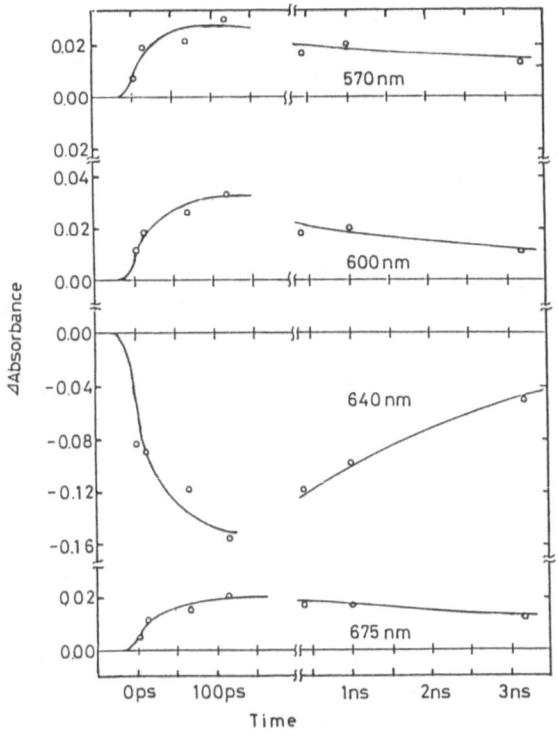

*Figure 10. Time depend-
ence of absorbance change
due to the formation of
photolysis intermediates
by the excitation of protochlorophyllide by 30 psec
laser pulse at 630 nm.*

about 30 psec). Although it is difficult to determine the accurate time
constant by the ordinary convolution-simulation method, it can be concluded
that the initial change is not instantaneous, but takes place with the time
constant of $\leq$ 50 psec. This time constant corresponds to the relaxation
from the Franck-Condon state (S_1^*) to the equilibrium state (S_1) in PChlide,
so that X_0 may not be an intermediate in chemical terms.

Summarizing these, the features of absorption changes in the time
ranges between 0 psec and 35 nsec can be described as follows: (1) There are
at least two species X_0 and X_1, (2) The time constant of X_0 formation is
$\leq$ 50 psec, (3) The time constant of X_2 formation is 1 - 2 nsec, (4) The
absorption maxima of X_0 ($\leq$ 640 nm) and X_1 ($\geq$ 640 nm) are located close to
that of parent PChlide, (5) The spectral widths of X_0 and X_1 are broader
than that of PChlide.

Based on the above results, the features of the initial process of
Chlide formation can be described as follows:

$$\text{PChlide}(S_0) \xrightarrow{\ h\nu\ } \text{PChlide}(S_1^*) \xrightarrow{\ \leq 50\ \text{psec}\ } \text{PChlide}(S_1) \text{ ---}$$
$$(= X_0)$$

$$\xrightarrow{\ 1\text{-}2\ \text{nsec}\ } X_1 \xrightarrow{\ 35\text{-}250\ \text{nsec}\ } X_2 \xrightarrow{\ 1\text{-}2\ \mu\text{sec}\ } X_3 \xrightarrow{\ 12\ \mu\text{sec}\ } \text{Chlide}(C\ 678)$$
$$(= X\text{-}690)$$

S_0, S_1 and S_1^* denote the ground state and the equilibrium and Franck-Condon states in the lowest excited singlet state S_1 in PChlide, respectively. In the present study, three intermediates involved in the process of PChlide photoreduction were kinetically detected. Among these three intermediates, X_3 is identical to X-690, the previously reported intermediate. Discussions on chemical identification of these intermediates will be published elsewhere.

ACKNOWLEDGMENTS

The nanosecond and picosecond spectroscopy apparatus have been constructed with the help by Dr. Hiroyuki Ohtani, Dr. Hisao Uchiki and Mr. Masayuki Yoshizawa at the University of Tokyo. This study was supported in part by a Grant-in-Aid from the Ministry of Education, Science and Culture and also by a grant for "Solar Energy Conversion by Means of Photosynthesis" given by the Science and Technology Agency of Japan to The Institute of Physical and Chemical Research (RIKEN).

REFERENCES

1. Shibata K. 1957. J. Biochem. 44, 147.
2. Wolff JB, Price L. 1957. Arch. Biochem. Biophys. 72, 293.
3. Boardman NK. 1966. In The Chlorophylls (ed. by Vernon LP, Seely GR), p.437. New York & London: Academic Press.
4. Kahn L. 1969. Plant Physiol. 43, 1769.
5. Gassman M, Granick S, Mauzerall D. 1968. Biochem. Biophys. Res. Commun. 30, 295.
6. Madsen A. 1963. Carnegie Inst. Wash. Yearbook 62, 371.
7. Raskin VI. 1977. Dokl. Akad. Nauk. SSSR (Biofizika), 21, 272.
8. Raskin VI. 1979. Dokl. Akad. Nauk. SSSR (Biofizika), 245, 1487.
9. Dujardin E, Sironval C. 1977. Plant. Sci. Lett. 10, 347.
10. Losev AP, Lyal'kova ND. 1979. Mol. Biol. (Moscow), 13, 837.
11. Dujardin E, Correia M. 1979. Photobiochem. Photobiophys. 1, 25.
12. Franck F, Dujardin E, Sironval C. 1980. Plant Sci. Lett. 18, 375.
13. Franck F, Mathis P. 1980. Photochem. Photobiol. 32, 799.
14. Inoue Y, Kobayashi Y, Ogawa T, Shibata K. 1981. Plant & Cell Physiol. 22, 197.
15. Kobayashi T, Iwai J, Uchiki H, Ohtani H. to be published.
16. Kobayashi T, Ohtani H, Iwai J, Yoshizawa M. to be published.

THE PRIMARY REACTION IN THE PHOTOREDUCTION OF PROTOCHLORO-
PHYLLIDE. A NANOSECOND FLUORESCENCE STUDY

A.C. van BOCHOVE[1], W.T. GRIFFITHS[2] and R.van GRONDELLE[3]
1) Depart. of Biophys.,Huygens Lab.of the State Univ.,P.O.Box
9504, 2300 RA Leiden, The Netherlands.
2) Depart.of Biochem.,The Medical School,Univ.of Bristol,Bris-
tol BS8 ITD, United Kingdom.
3) Depart.of Biophys.,Physics Lab. of the Free Univ.,De Boele-
laan 1081, HV Amsterdam, The Netherlands.

1. INTRODUCTION

The biosynthesis of chlorophyll by higher plants is a process which has
been the subject of many investigations during recent years. Especially the
use of advanced spectroscopic methods has provided useful new information
about the light requiring step of this process. Many of these spectroscopic
experiments have benefitted from the increase of knowledge about the biochemical definition of the photoconversion system.

From biochemical experiments it is now known that the photoreduction of
protochlorophyllide takes place in a photoactive enzyme-substrates complex.
The enzyme, protochlorophyllide reductase, has already been purified and
its molecular weight established. After illumination, hydrogen transfer
occurs from NADPH to protochlorophyllide, resulting in chlorophyllide formation.

Several recent experiments demonstrated the existence of intermediates
in the photoconversion reaction of protochlorophyllide. Dujardin and Correia
(1) discovered a non-fluorescent transient intermediate in a study of the
fluorescence spectra of etiolated leaves. More recently, rapid absorption
measurements on crude extracts from etiolated leaves have confirmed the
existence of long wavelength absorbing (690 nm) transients as intermediates
in the photoreduction of protochlorophyllide (2). The intermediate appeared
at room temperature within 0.5 μs after illumination by a laser flash, and
its decay (with a half-time of 8 μs) coincided with the appearance of
chlorophyllide. Transient emission spectra during the photoconversion
process were recently measured with a gated OMA detector, and assigned to
several new intermediates (3). Very recent experiments have been done on
etiolated bean leaves using a phase fluorimeter (4,5). With this method it
was possible to determine the average fluorescence decay time, which showed
a pronounced wavelength dependence. However, for a quantitative interpretation of the lifetime spectra, more kinetic information about the fluo-

C. Sironval and M. Brouers (eds.); Protochlorophyllide Reduction and Greening. ISBN 90 247 2954 8
©1984, Martinus Nijhoff/Dr W. Junk Publishers, The Hague/Boston/Lancaster.

rescence decay is desired.

In the present paper, time-resolved fluorescence measurements will be discussed (see also ref. 6). This study shows that the phototransformation in isolated membrane preparations is accompanied by rapid fluorescence transients. The connection between these transients and the photoconversion process will be demonstrated by the use of inhibitors of the photoreaction and by manipulation of the level of the hydrogen donor NADPH. The dependence of the fluorescence kinetics on temperature and emission wavelength has been studied, and a possible explanation of the results is presented.

2. MATERIALS AND METHODS

2.1. Biological material

Etiolated plant material, oats (Avena sativa var Pennel), was cultivated and the total etioplast membranes were isolated from the 7 day old dark grown plants by previously published procedures (7). The total etioplast membranes were also some times further purified by centrifugation on discontinuous sucrose gradients to yield purified membrane fractions enriched in protochlorophyllide reductase (8). The preparations were made in Bristol and transported to Leiden in a vacuum flask packed with solid CO_2.

2.2. Chemicals

Protochlorophyllide was prepared from etiolated plants as previously described (9). NADPH was a product of Boehringer (Mannheim). All other reagents were of Analar or the highest purity grade available.

2.3. Experimental methods

A frequency-doubled mode-locked Nd-YAG laser was used to excite fluorescence. The width of the laser pulse was 35 ps at a wavelength of 532 nm. During the emission measurements the energy density on the sample was approximately 1 mJ cm^{-2}, although the exact value is not known because of the cylindrical shape of the cuvette. The fluorescence kinetics were measured through a monochromator (Bausch and Lomb, focal length 500 mm, blaze 500 nm) by a silicon avalanche photodiode (RCA C30902E, sensitive area 0.2 mm^2). The photodiode, fitted in a coaxial mount, was connected to the input amplifier (7A19) of a Tektronix transient digitizer (R7912) by a 20 m long coaxial cable to prevent reflections. The system response was approximately 400 ps. The relative energy of the laser flashes was measured with a photodiode (Siemens BPX65) connected to an integrating amplifier. Test measurements showed that the observed fluorescence curves were within 15 % accuracy proportional to the flash intensity, when this intensity was varied over a factor of three. Therefore, the measurements could be normalized by dividing the signal by the laser energy.

For the low temperature measurements a perspex sample cuvette was used (2 mm path length, volume 1 ml), fitted with a temperature probe connected to a temperature controller. The cuvette was placed inside a narrow dewar through which flowed liquid nitrogen or cooled nitrogen gas, maintaining the cuvette at the desired temperature. At room temperature each measurement required a fresh sample, and therefore small cylindrical glass cuvettes were used with a capacity of 100 μl, which were fitted directly into the entry slit of the monochromator.

Absorption spectra of the samples were recorded on a split beam spectrophotometer as previously described (7).

3. RESULTS

The photoconversion of protochlorophyllide into chlorophyll is accompanied by large spectral changes. The photoactive complex absorbs at about 650 nm, while the formed chlorophyll has an absorption maximum around 675 nm. A total photoconversion is demonstrated in fig. 1, where the absorption spectra are shown of total membranes prepared from etioplasts of dark grown plants, taken before (fig. 1a) and after (fig. 1b) 0.5 minute of continuous illumination. Such membranes retain a large proportion of their protochlorophyllide reductase as the photoactive enzyme-substrates complex (9). Residual non-photoactive protochlorophyllide in the illuminated membranes causes an absorption maximum at 630 nm (fig. 1b).

Excitation of the dark membranes at room temperature by a single 35 ps laser flash (λ = 532 nm) causes fluorescence which has biphasic decay kinetics at 657 nm (fig. 2, curve a). The fluorescence can be fitted by two exponentials with a decay time of 1 ns and 5 ns, respectively. The initial amplitude of the fast component was 3 - 4 times higher than the amplitude of the slow component. The same exponentials could be used to fit the fluorescence at 635 nm, but at this wavelength the 5 ns component had the highest amplitude. After the sample had been illuminated, either by 1 min of continuous light, or by approximately 10 intense focused laser flashes, the fast phase in the fluorescence disappeared completely, resulting in a single

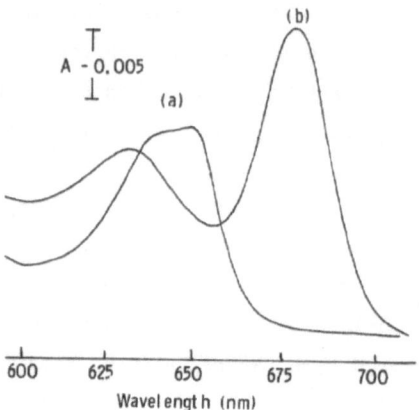

FIGURE 1. Absorption spectra of protochlorophyllide-enriched, purified membranes from oat etioplasts. Spectra of the membranes, resuspended in buffer at a protein concentration of 2.5 mg ml^{-1} were recorded before (a) and after (b) illumination for 30 s from a 60 W tungsten lamp.

exponential decay with a decay time of 5 ns (fig. 2, curve b).

Since the fluorescence emission at 657 nm is mainly caused by excited photoactive protochlorophyllide, these results indicate that this excited state disappears in about 1 ns, while the fluorescence lifetime of the

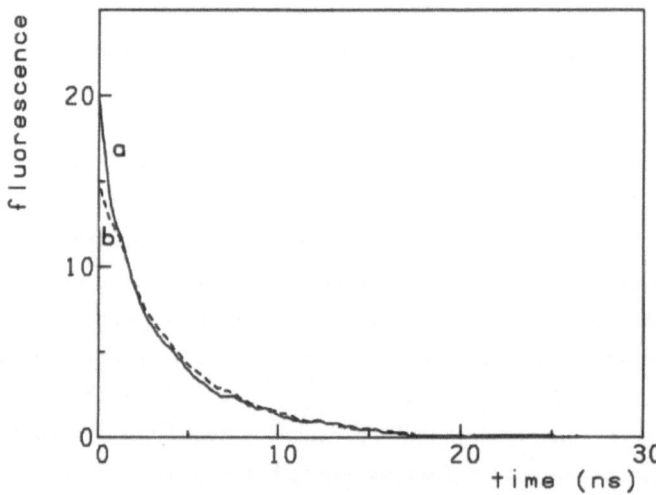

FIGURE 2. Fluorescence kinetics induced by a laser flash (arbitrary units) in membranes from etiolated oats, measured at 657 nm, without (curve a) and with (curve b) preillumination for 1 min by continuous light.

non-photoactive protochlorophyllide is about 5 ns. A similar decay time for protochlorophyllide in cholate solution has been observed (unpublished work).

The emission observed at 657 nm after complete photoconversion (fig. 2, curve b) is due partly to the chlorophyll a formed (emitting maximally at 685 nm) and the endogenous non-photoactive protochlorophyllide (emitting maximally at 635 nm).

To provide further evidence for the assignment of the fast fluorescence decay phase to the photoactive complex, a sample was illuminated for 1 min by continuous light. Fig. 3, trace a, shows the single exponential decay (at 657 nm) of this illuminated sample. 1 mM NADPH was then added followed by incubation in darkness for 2 min before re-recording the fluorescence characteristics (fig. 3, trace b). The biphasic nature of the resulting trace indicates reconstitution of the fast decay phase by the NADPH, probably as a result of regeneration of the photoactive complex from the added reduced coenzyme and the endogenous protochlorophyllide, in agreement with earlier observations (9). After several subsequent flashes (or continuous illumination) the fast decay phase disappeared again, indicating exhaustion, due to complete photoconversion of the endogenous protochlorophyllide.

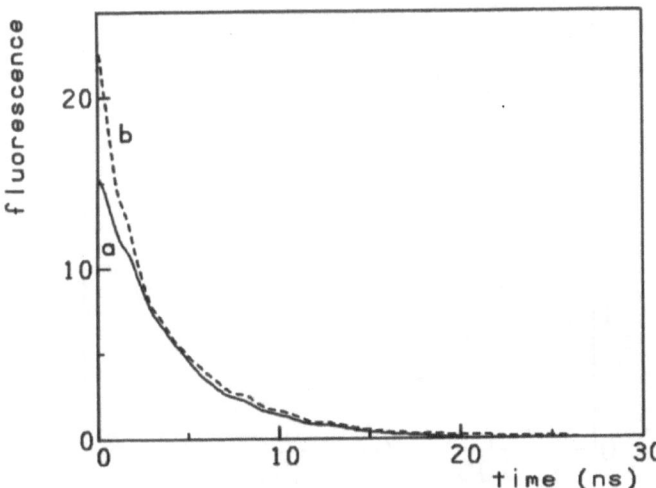

FIGURE 3. Fluorescence kinetics induced by a laser flash (arbitrary units) in membranes from etiolated oats, measured at 657 nm, after preillumination for 1 min by continuous light. Curve a, no additions; curve b, 1 mM NADPH added after illumination and sample incubated for 2 min before flash excitation of fluorescence.

118

To further correlate photoconversion with the fast fluorescence decay
phase the effect of detergent treatment on the fluorescence behaviour was
investigated. Non-illuminated membranes were incubated for 5 min with
100 mM Triton X-100, a treatment previously shown to completely abolish
photoconversion (9). Comparing the emission decay kinetics from the untreated
and Triton-treated membranes, it was apparent that the detergent treatment
had resulted in complete abolition of the fast component in the decay of the
657 nm emission. Another procedure which is known to inhibit the photoconversion is preincubation of the membranes in a water bath at 50 $^{\circ}$C for 1 min
(10). This treatment also caused the disappearance of the fast phase in the
fluorescence emission.

The fluorescence kinetics at 657 nm were also studied at lower temperatures. The membrane preparation was first incubated at room temperature
with 1 mM NADPH for several minutes to maximize the amount of photoactive
complex present. The membranes were then frozen to 77 K and gradually
warmed up, and the fluorescence was measured at various temperatures. A
kinetic analysis of the resulting data showed the fluorescence at 657 nm

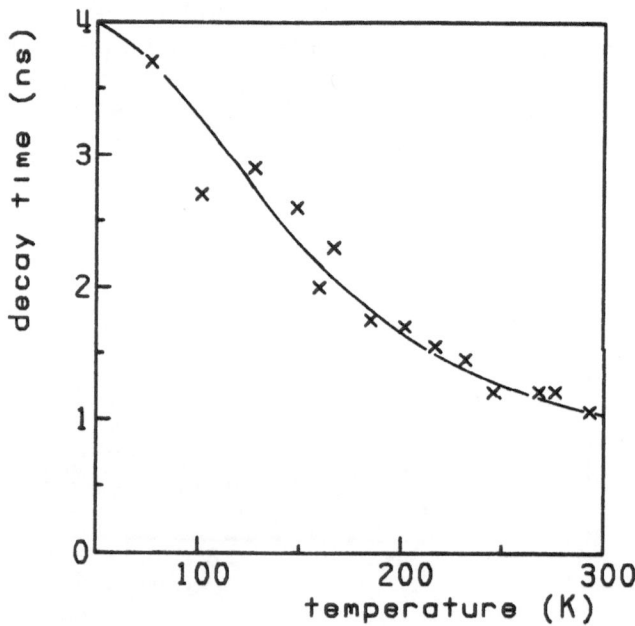

FIGURE 4. Decay time of the fast component of the biphasic emission at
657 nm in membranes from etiolated oats, as a function of temperature
(crosses). The curve shows the best fit, using a model which describes a
transfer process, coupled with nuclear vibrations (11).

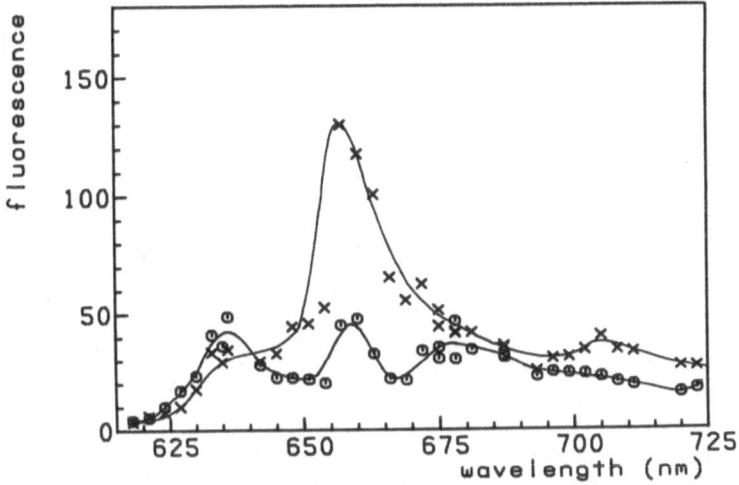

FIGURE 5. Spectra at 293 K of the fluorescence induced by a laser flash in membranes from etiolated oats, preincubated with 1 mM NADPH for 15 min prior to assay, fitted with two exponentials. x, component with 1 ns decay time; o, component with 5 ns decay time. The initial amplitudes of both components (arbitrary units) divided by the laser energy are plotted. For each wavelength a fresh sample was flashed with one single laser pulse, but for 657 nm seven samples were averaged, with a single laser pulse each.

to be biphasic at all temperatures studied, even at 77 K, where no photo-production of chlorophyll is observed (10). The decay time of the slow phase is only weakly dependent on the temperature, increasing from 5 ns at 293 K to 6 ns at 77 K. The decay time of the fast phase, however, increases from 1 ns at 293 K to about 3.5 ns at 77 K. The temperature dependence of the fast decay phase is given in fig. 4.

The spectra of the two components in the fluorescence emission were measured at three different temperatures, 293 K, 160 K and 77 K. Total membranes from oats were preincubated with 1 mM NADPH for 15 min prior to assay. At room temperature it was necessary to use a fresh sample for each measuring flash, to be sure that only dark adapted membranes were measured. The room temperature spectrum is given in fig. 5. The fluorescence was fitted with two exponentials with decay times of 1 ns and 5 ns. The initial amplitudes of both components were divided by the laser energy and plotted against the wavelength. The fast phase has a large peak at 657 nm and a small peak at about 705 nm. The spectrum of the slow phase is more complicated with peaks at about 635 nm, 657 nm and 677 nm.

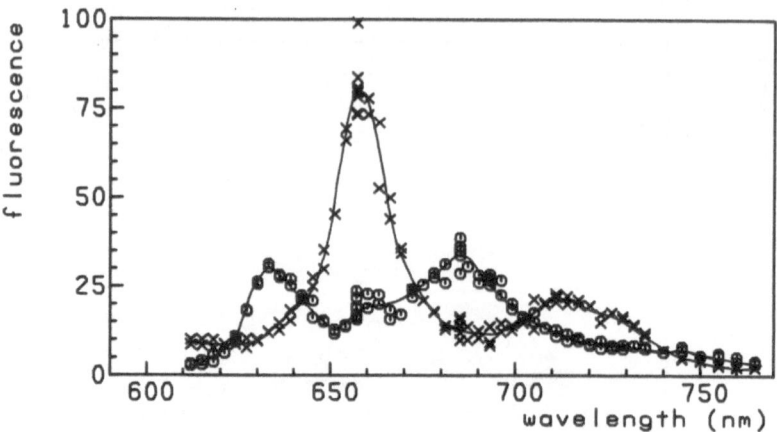

FIGURE 6. Spectra at 160 K of the fluorescence induced by a laser flash in membranes from etiolated oats, preincubated with 1 mM NADPH for 15 min at room temperature before cooling, fitted with two exponentials. x, component with 2 ns decay time; o, component with 6 ns decay time. The initial amplitudes of both components (arbitrary units) divided by the laser energy are plotted. The whole spectrum was recorded with one sample. Each measuring point represents a single laser flash.

For the spectra at 160 K and 77 K one single sample was used to measure fluorescence over the complete spectral region. This was possible because at these temperatures no chlorophyll (or only a small amount) was produced by the laser pulses. This was checked by monitoring the fluorescence emission at 685 and 657 nm before and after recording the spectrum, which showed no increase due to chlorophyll emission, and no depletion of the emission of photoactive protochlorophyllide. At 160 K (fig. 6) the fluorescence is fitted with two exponentials with decay times of 2 ns and 6 ns, in agreement with the data shown in fig. 4. The fast component has a main peak at 657 nm and a smaller peak at about 710 nm. The slow component has peaks at about 633 and 685 nm. The fluorescence at 77 K is also biphasic, but the decay times of the fast and the slow components (3.5 and 6 ns, respectively) are too close together to allow a reliable two-exponential fitting procedure. Some information on the fast and slow component can be obtained by plotting the spectrum of the average decay time, using a single exponential fitting procedure (fig. 7a), in combination with a spectrum of the total integrated fluorescence divided by the laser energy (fig. 7b).

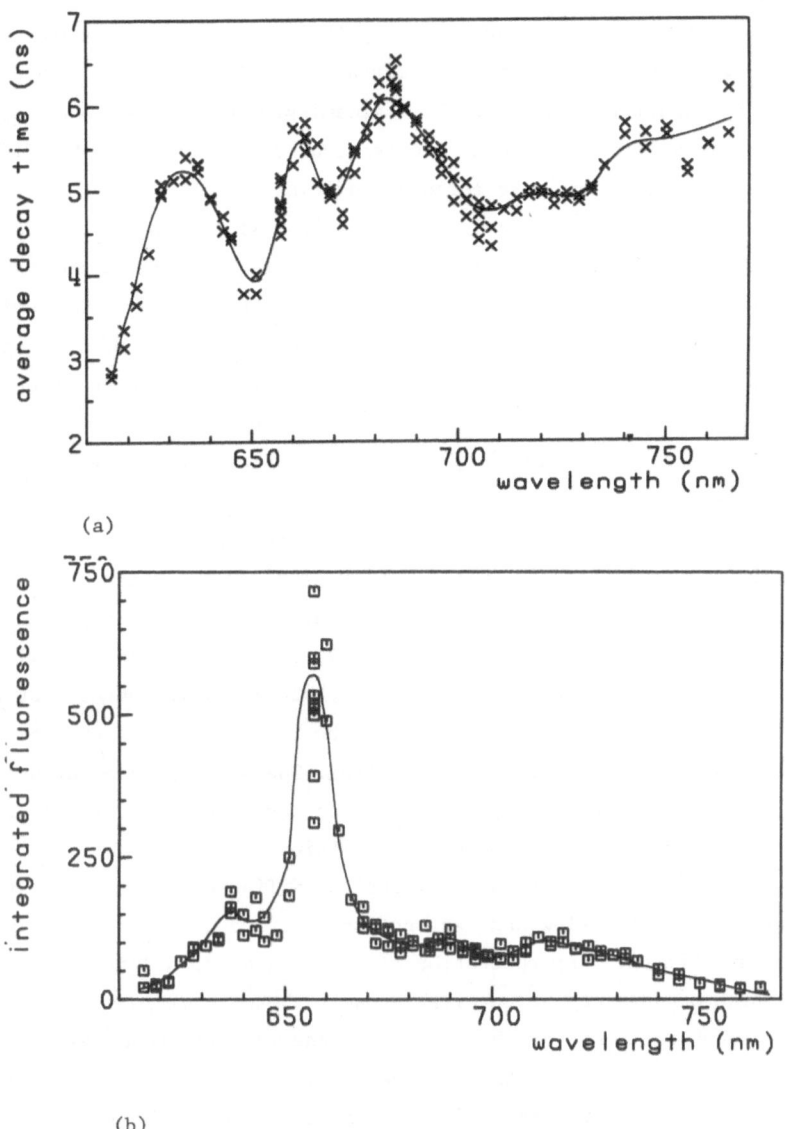

(a)

(b)

FIGURE 7. Spectra at 77 K of the fluorescence induced by a laser flash in membranes from etiolated oats, preincubated with 1 mM NADPH for 15 min at room temperature before cooling in liquid nitrogen. The spectra were recorded with one sample. Each measuring point represents a single laser flash. (a), Spectra of the average lifetime, calculated by fitting the fluorescence decay with a single exponential. (b), Spectrum of the total integrated fluorescence (arbitrary units) divided by the laser energy.

4. DISCUSSION

The presented results show clearly that the fast decay phase in the fluorescence emission of etioplast membranes is associated with the photoactive protochlorophyllide complex. The fast emission decay disappears after photoconverting all the endogenous photoactive protochlorophyllide (fig. 3a), but can subsequently be restored by addition of NADPH (fig. 3b), which, due to the activity of the enzyme protochlorophyllide reductase converts non-photoactive protochlorophyllide into its photoactive form. Furthermore, exposure of the membranes to conditions which had been shown previously to inhibit photoconversion, such as a mild heat treatment and the addition of Triton X-100, also led to the disappearance of the rapid emission decay.

The spectrum of the fast phase has a main peak at 657 nm and a vibrational peak at 705 - 710 nm. This indicates that the source of this fast fluorescence is probably the photoactive form of protochlorophyllide absorbing at about 650 nm (fig. 5 (x)). The spectrum of the slow phase in the fluorescence decay is rather complicated and the shape is dependent upon the temperature. The emission peak at about 635 nm is present at both 293 K (fig. 5 (o)) and 160 K (fig. 6 (o)) and probably due to emission from the non-photoactive form of protochlorophyllide, which absorbs at about 630 nm. The amplitude of this band may be reduced by self absorption, which takes place in this part of the spectrum. The slow phase emission band at 675 - 685 nm, which also appears at these same two temperatures, could well be a vibrational band of the 635 nm emission.

At 77 K, due to the relatively smaller difference between the fluorescence decay times of the slow and fast components, 3.5 and 6 ns, respectively, exact quantitative interpretation of the data becomes difficult as the curve fitting procedure may be unreliable under these conditions. Some general conclusions can, however, be drawn from the 77 K fluorescence data shown in fig. 7. The maxima in the average decay time spectrum (fig. 7a) at 635 and 685 nm, together with maxima at the same wavelengths in the integrated fluorescence (fig. 7b) indicate peaks in the spectrum of the slow phase, similar to those observed at 293 K and 160 K. The minimum in the average decay time around 710 nm corresponds to the peak in the integrated emission at the same wavelength in fig. 7b, thereby identifying the latter as being due to the fast phase. This is in agreement with the vibrational bands found at 293 K and 160 K. The situation around 657 nm is more complicated. The decay time has a minimum at 650 nm and a maximum at 663 nm, which

possibly suggests that the emission peak around 660 nm at 77 K is hetero-
genous, containing fast and slow emission components, the latter shifted
to the red. A preliminary fitting procedure using two exponentials leads
to the same conclusion. The physical meaning of this feature is unknown,
but the contribution to the slow component at 657 nm at room temperature
in fig. 5 (o) may be related to the same phenomenon. A maximum of 5.5 ns
at 662 nm in the decay time has also been found by Van der Cammen and
Goedheer (4), using a phase fluorimeter to measure the average fluorescence
decay time in etiolated bean leaves at 77 K.

Recent studies on the mechanism of chlorophyllide formation from protochlorophyllide concluded that several intermediates are involved in this
process (3). Since the reduction of protochlorophyllide is a light-driven
reaction, it may be assumed that the first intermediate is formed from
the excited state of the photoactive pigment. The data presented in this
paper show clearly that this excited state lives at room temperature only
about one nanosecond. This indicates that the first intermediate in the
photoconversion is formed from the excited protochlorophyllide in about
one nanosecond at room temperature. The fluorescence lifetime of non-
photoactive protochlorophyllide is approximately 5 ns, which is of the
same order as the emission decay time of protochlorophyllide in solution.
The most probable explanation of the short lifetime of the excited photoactive complex is the assumption that the fluorescence decay has to compete
with an efficient primary reaction in the photoconversion process. This
would mean that this reaction has an efficiency of about 80 % at room
temperature.

The observation that at lower temperatures the fast phase in the 657 nm
emission slows down, but does not disappear, suggests that at this temperature range the first intermediate is still formed but less efficiently as
at room temperature. As no significant chlorophyll accumulation or depletion
of the fast phase of the 657 nm emission is observed upon repetitive laser
flashes at 160 and 77 K, this suggests that at these temperatures most of
the intermediate formed reacts back on a time scale shorter than several
minutes. The accumulation of small amounts of intermediate cannot, however,
be excluded.

Thus, we conclude that the primary process of protochlorophyllide reduction involves the excitation of photoactive protochlorophyllide and the
subsequent formation of an intermediate with a rate constant k_r. If we

assume that the rate constants for internal conversion, fluorescence and triplet formation do not change when the pigment is bound to the enzyme, forming a photoactive complex, then the rate constant for the primary photochemical reaction is given by $k_r = \frac{1}{\tau_a} - \frac{1}{\tau_n}$, where τ_a and τ_n are the fluorescence lifetimes of the photoactive and non-photoactive protochlorophyllide. If we use the fact that τ_n only increases from 5 to 6 ns upon lowering the temperature from 293 to 77 K, it can be seen from fig. 6 that k_r decreases from 0.8×10^9 s^{-1} to 0.12×10^9 s^{-1}.

In the high temperature region, down to approximately 130 K, k_r shows an Arrhenius type of temperature dependence with an activation energy of about 0.03 eV. It is not possible to fit the whole temperature region with an Arrhenius plot, because at lower temperatures the activation decreases. Such a temperature dependence suggests that a transfer process coupled with nuclear vibrations may be involved in the primary photoreaction (12). The experimental data could be fitted using such a model with nuclear vibrational frequency of 200 ± 50 cm^{-1} (11). The calculated curve is plotted in fig. 4. No evidence is available on the nature of the species transported in this reaction. Future experiments with a deuterated proton donor will hopefully provide more information about the physical mechanism of the primary photoreaction in this system.

REFERENCES

1. Dujardin E, Correia M. 1979. Photobiochem. Photobiophys. 1, 25–32.
2. Franck F, Mathis P. 1980. Photochem. Photobiol. 32, 799–803.
3. Dobek A, Dujardin E, Franck F, Sironval C, Breton J, Roux E. 1981. Photobiochem. Photobiophys. 2, 35–44.
4. Van der Cammen JCJM, Goedheer JC. 1981. Photobiochem. Photobiophys. 3, 159–165.
5. Van der Cammen JCJM, Goedheer JC. 1982. Photobiochem. Photobiophys. 4, 145–152.
6. Van Bochove AC, Griffiths WT, van Grondelle R. 1983. Photochem. Photobiol., in the press.
7. Griffiths WT. 1975. Biochem. J. 152, 623–635.
8. Beer NS, Griffiths WT. 1981. Biochem. J. 195, 83–92.
9. Griffiths WT. 1978. Biochem. J. 174, 681–692.
10. Boardman NK. 1966. In Vernon LP, Seely GR, eds. The Chlorophylls, pp. 437–476. New York: Academic Press.
11. Jortner J. 1980. Biochim. Biophys. Acta 594, 193–230.
12. DeVault D. 1980. Quart. Rev. Biophys. 13, 387–564.

ACKNOWLEDGEMENTS

This investigation was supported in part by the Netherlands Foundation for Biophysics with financial aid from the Netherlands Organization for the Advancement of Pure Research (ZWO). One of us (WTG) acknowledges the financial support of the Science and Engineering Research Council (UK) (grant no. GR/A/72490) and the European Molecular Biology Organization for a short term travelling fellowship.

PROTOCHLOROPHYLLIDE AND
CHLOROPHYLLIDE FORMS.

PROTOCHLOROPHYLLIDE AND CHLOROPHYLLIDE-PROTEIN COMPLEXES
DURING GREENING

B. EL HAMOURI

Department of Biochemistry and Plant Physiology. Institut
Agronomique et Vétérinaire Hassan II, BP 704. Rabat. Maroc.

1. PROTOCHLOROPHYLLIDE-PROTEIN COMPLEXES IN THE ETIOLATED LEAF.

Higher plants do not synthesize chlorophyll in darkness.
They contain the precursor of this pigment protochlorophyllide
(PChlide). PChlide occurs in various complexes with proteins
characterized by their fluorescence emission and absorption
spectra. In these complexes they are, either photoreducible,
or non photoreducible. The major photoreducible PChlide is
P_{650}^{*}; this pigment is accompanied by a minor photoreducible
pigment, P_{637}; the overall total pool being called $P_{650,637}$.
(Shibata, 1957; Litvin and Krasnovsky, 1957; Sironval and
Michel, 1967). The main non photoreducible PChlide is P_{630}
(Litvin and Krasnovsky, 1957; Goedheer, 1967).

Over the past few years, many advances have been made in
the study of the PChlide-complexes. It has been demonstrated
that the photoreducible PChlide occurs as a ternary complex
which comprises NADPH-protein and PChlide (Griffiths, 1980).
The protein moeity exhibits enzymic properties. It consists
of two polypeptides of 35 and 37 Kd in oat (Oliver and
Griffiths, 1981: Beer and Griffiths, 1981) and in bean (Oliver
and Griffiths, 1981 and Gerday et al, 1982) or one polypeptide
of 36 Kd in barley (Apel et al, 1980) squash (Ikeuchi and
Murakami, 1982) and cucumber (El Hamouri and Sironval, 1983).
It is called NADPH protochlorophyllide oxidoreductase (EC

---

[*] The subscript refers to the wavelength of the red absorption
maximum.

C. Sironval and M. Brouers (eds.); Protochlorophyllide Reduction and Greening. ISBN 90 247 2954 8
© 1984, Martinus Nijhoff/Dr W. Junk Publishers, The Hague/Boston/Lancaster.

130

1.6.99,-). Active thiol groups are involved in the binding
of substrate in the enzymic complex. (Griffiths, 1980; Oliver
and Griffiths, 1981).

Recently, a non photoreducible PChlide complex with absorption and fluorescence emission maxima at 642 nm and 649 nm
respectively has been discovered in cucumber etioplasts in
addition to P_{630} (El Hamouri and Sironval, 1979). This additional complex P_{642} is formed when $NADP^+$ is added to the non
photoreducible P_{630}. P_{642} is transformed into the photoreducible $P_{650,637}$ by reducing $NADP^+$ in the dark (fig. 1) (El
Hamouri et al, 1981). The following sequence :

$$P_{630} \xrightarrow{NADP^+} P_{642} \xrightarrow[\substack{\text{reduction} \\ \text{(dark)}}]{NADP^+} P_{650,637} \xrightarrow{h\nu} \text{Chlide}$$

has been demonstrated in Cucumis sativus etioplasts (fig. 1),
in purified oat etioplast membranes (El Hamouri et al, 1983)
and in etioplasts suspensions isolated from ALA treated bean
leaves (Brouers and Wolwertz, 1981).

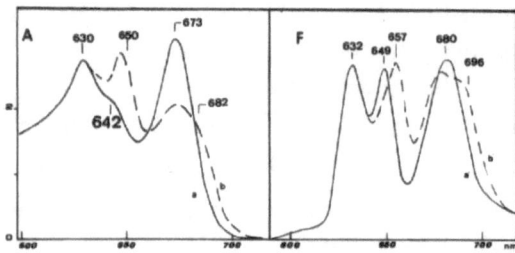

FIGURE 1. Cucumber
(*Cucumis sativus*) etio-
plasts were isolated
from etiolated cotyledons
(El Hamouri and Sironval,
1979). The etioplasts
were incubated in the
dark for 30 min at +10°C
(a treatment which trans-
forms part of photoredu-
cible P650,637 into non-
photoreducible P630).
After that time, the etioplasts were illuminated and incubated
in the presence of $NADP^+$ (0.5 mM) for 30 min, in the dark at
+10°C (spectra a and a'). The etioplast suspension was then
supplemented with glucose-6-phosphate (1 mM) and glucose-6-
phosphate dehydrogenase (0,4 µ/mg etioplast protein) and the
dark incubation was prolonged for 15 min at +10°C (spectra b
and b'). All spectra (absorption : A, fluorescence : F) were
recorded at liquid nitrogen temperature.

The formation of P_{642} from P_{630} in the presence of exogenous $NADP^+$ is inhibited by thiol group reagents such as
N-phenylmaleimide, implying that thiol groups are involved
in the process (Fig. 2). The inhibition by thiol group reagents
has been also reported in the case of the regeneration of the

photoreducible $P_{650,637}$ from P_{630} and NADPH (Griffiths, 1978).

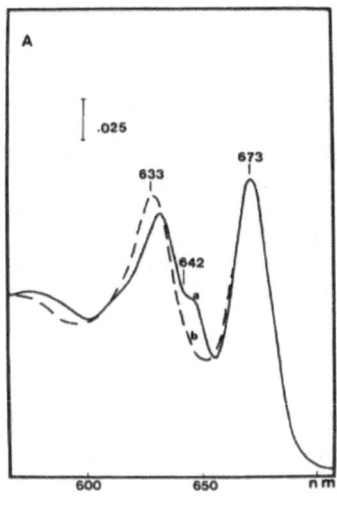

FIGURE 2. Purified oat (*Avena sativa*) etioplast membranes were illuminated and then were incubated in the presence of NADP$^+$ (2.5 mM) (curve a), or in the presence of N-phenylmaleimide (0.35 mM) and NADP$^+$ (curve b). Absorption spectra were recorded at liquid nitrogen temperature (taken from El Hamouri and Sironval, 1983).

The combination of spectrophotometric analysis, protein labelling techniques and inhibition studies led to the conclusion (I) that the 642 nm absorbing PChlide, P_{642}, is a ternary complex (NADP$^+$-protein-PChlide), and that (II) the same thiol groups are required for the formation of both the photoreducible NADPH-protochlorophyllide reductase, $P_{650,637}$ and the non photoreducible NADP$^+$-complex, P_{642}. Further investigations have established that the nucleotides, NADP$^+$ and NADPH, are in a dynamic equilibrium between the medium and the site of the reductase to which they bind (El Hamouri et al, 1983). An indirect measure of the relative stabilities of the NADP$^+$ and NADPH forms of the reductase is given in the data of fig. 3. This figure indicates a greater stability and a tighter binding of the nucleotide in the NADPH form (El Hamouri et al, 1981). It can be concluded from these results that the PChlide reductase is able to bind reversibly either NADP$^+$ or NADPH, producing complexes with different spectroscopic properties.

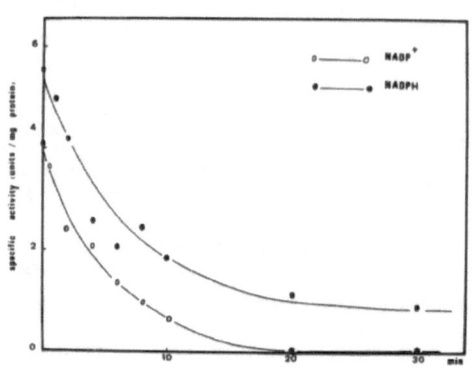

FIGURE 3. Purified oat (*Avena sativa*) etioplast membranes were incubated with NADP$^+$ (2.5 mM) or NADPH (1.5 mM). After 25 min dark incubation, excess cofactors were removed by gel filtration, N-phenylmaleimide (0.35 mM) was added and aliquots were taken for assaying the regeneration activity of the protochlorophyllide reductase after various time intervals. (Taken from El Hamouri et al, 1983).

However, until now, there is no explanation for the occurence of two different photoreductase complexes absorbing at 637 nm and 650 nm as these complexes are both photoreducible NADPH-PChlide-enzyme complexes. We must assume that other factors than nucleotide binding must play a role in the determination of the spectroscopic properties.

Furthemore, we are still unable to assign a molecular composition to the non photoreducible PChlide P_{630}. The transformation of P_{630} into $P_{650,637}$ *in vitro* after NADPH addition might be explained in two different ways.

a.- The pigment is already bound to the enzyme, and P_{630} is a binary complex enzyme-PChlide. This complex becomes ternary when it binds NADPH. This would mean that the enzyme-PChlide complex has spectroscopic characteristics near those of the free pigment.

b.- The enzyme does bind the pigment only when NADPH has been added. In this case, P_{630} is the free pigment.

2. THE CHLIDE COMPLEXES AFTER THE ACTION OF LIGHT

The first Chlide appearing in the first second which follows the action of light is P_{678}. In the leaf, the life-time of this Chlide is about 30 sec at room temperature and in darkness (Sironval and Michel, 1967; Gassman et al, 1968); it is transformed into another Chlide, P_{684}. This pigment has a life-time of about 30 min. It is progressively transformed into a chlorophylle-complex, P_{673} (the Shibata shift) according to the scheme :

$$P_{650,637} \xrightarrow{h\nu} P_{678} \xrightarrow{\frac{30 \text{ sec}}{\text{dark}}} P_{684} \xrightarrow{\frac{30 \text{ min}}{\text{dark}}} P_{673}$$

which applies to the etiolated leaf.

The molecular structure of P_{678} is derived from that of the PChlide oxidoreductase. According to Griffiths (1980), P_{678} is the ternary complex (NADP$^+$-enzyme-Chlide).

In vitro, in isolated etioplasts or membrane fractions, the scheme is different. P_{678} is still the first Chlide observed, but the long wavelength absorbing Chlide, P_{684}, is not formed.

Instead of this, P_{673} is formed directly from P_{678} within 30 min at room temperature in darkness. However, when the photoreduction takes place in the presence of exogenous NADPH, the Chlide P_{684} is seen immediatly after illumination and this pigment has then a much longer life-time than *in vivo* (five to six times longer). On the other hand, when the amount of NADPH is increased in the medium after the formation of P_{678}, a part of this Chlide is transformed into P_{684} in the dark (fig. 1).

It can be assumed from these results that the long wavelength absorbing Chlide, P_{684} is an (NADPH-enzyme-Chlide) ternary complex (El Hamouri et al, 1981; Oliver and Griffiths, 1982).

Some similarities between the photoreducible PChlide, P_{650} and the Chlide, P_{684} must be emphasized (Goedheer and Verhülsdonk, 1970). Both pigments are complexes made of :

(I) the same protein, the PChlide-reductase;

(II) NADPH;

(III) a pigment acting as a photoreceptor (PChlide or Chlide).

These similarities suggest that the long wavelength absorbing Chlide, P_{684} should possess some photoactivity properties as P_{650} does. This was already proposed by Jouy and Sironval (1979); see Franck and Inoue, Sironval et al, this volume). The new conclusion, here, is the statement that NADPH should play a role in the photoactivity of the pigment-protein complexes (compare P_{642} with $P_{650,637}$, and P_{678} with P_{684}).

134

3. THE POLYPEPTIDE PATTERNS DURING GREENING

Fig. 4 shows two polypeptide patterns obtained after SDS polyacrylamide gel electrophoresis (PAGE) of extracts from intact etioplasts of cucumber cotyledons. Four groups of peptides can be distinguished :

a.- peptides of high molecular weight, between 65 and 56 Kd;

b.- a major peptide of 36 Kd;

c.- a peptide of 25 Kd with a doublet of 20-21 Kd;

d.- peptides of lower molecular weight, in particular a 15 Kd peptide.

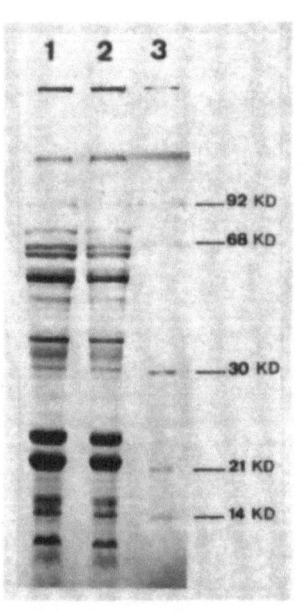

FIGURE 4. Polypeptide composition of cucumber etioplasts (SDS PAGE). Track 1 : etioplasts; track 2 : illuminated etioplasts (2 saturating flashes) and track 3 : standard proteins. The SDS-polyacrylamide gel electrophoresis was carried out as described in El Hamouri and Sironval (1983).

Among these peptides, the 36 Kd PChlide reductase, the 65 Kd and the 15 Kd were identified as NADPH-binding peptides (El Hamouri and Sironval, 1983).

Surprising is the presence of a 25 Kd peptide. Peptides of this molecular weight are reported to belong to the apoprotein of the light-harvesting-chlorophyll-complex in green leaves (Hoyer-Hansen et al, 1973; Hoyer-Hansen and Simpson, 1977; Ikeuchi and Murakami, 1982).

The comparison of track 2 with track 1, fig. 4, shows that the polypeptide pattern was not changed when etioplasts were flash illuminated. The continuous illumination of cucumber cotyledons for 1 and 6 hours did not provoke any obvious change of the polypeptide patte⌐ either (fig.5). After 18 hours, however, a new peptide of 30 Kd was seen.

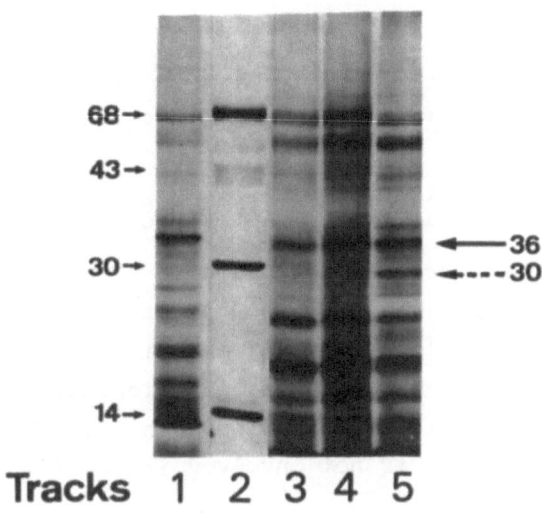

FIGURE 5. Peptide content of purified membranes from cucumber etioplasts and of cucumber etiochloroplasts. Track 1 purified etioplast membranes; track 2 standard proteins (Bovine Serum Albumin, 68000 Ovalbumin, 43000, Carbonic Anhydrase, 30000 and Cytochrome c, 14000). Tracks 3,4,5 etiochloroplasts from cotyledons illuminated continuously for 1, 6 and 18 h respectively. The arrows indicate the 36 Kd (PChlide-reductase) and the stippled arrows the 30 Kd peptide which is seen in the 18 h light treated sample.

The regeneration of the PChlide reductase (reconstitution of photoreducible P_{650} from added PChlide and NADPH) has been assayed on the samples used for the electrophoresis. This regeneration could not be detected after 1 h illumination; it was very low after 6 h. However after 18 h the regeneration of the reductase was as effective as in the etiolated sample (see Table 1). It must be noticed that in etiolated cucumber cotyledons, the regeneration was always low as compared with the values reported for other plant material (Oliver and Griffiths, 1981; Ikeuchi and Murakami, 1982).

Table 1.

	Illumination time (hour)			
	0	1	6	18
Regeneration activity of PChlide-reductase (μmole Chlide min^{-1} mg^{-1} protein)	0.76	0	0.12	0.80
PS II activity (Hill reaction, O_2 evolution)	-	-	+	++

Regeneration activity was measured following assays described by Griffiths (1978).
PS II activity was measured at an O2 electrode as a light dependent O_2 evolution in the presence of $K_3Fe(CN)_6$.
(- : undetectable ; + : low activity ; ++ : high activity).

PS II activity has been detected after 6 and 18 h continuous light treatment. This activity was not detected in the etiolated sample or after 1 h illumination (Table 1). Detection of a PS II activity in the 6 h illuminated etioplasts in the absence of the 30 Kd peptide seen after 18 h illumination indicates that this peptide is not involved in PS II activity.

As seen in fig. 5, the 36 Kd peptide was maintained during the illumination. It was observed after 18 h of illumination as well as at time 0 just before illumination. This constrasts with the results reported for cucumber cotyledons by Ohya et al (1981), and for other plant materials by Santel and Apel (1981) and Ikeuchi and Murakami (1982). The polypeptide pattern did not show any changes in the 25 Kd region.

The explanation for the lack of regeneration of PChlide reductase (P_{650}) after 1 hour illumination although the apoprotein was present might be the concentration decrease of the substrate and/or cofactors. This decrease could be correlated with the lag phase observed in chlorophyll synthesis during illumination of the etiolated leaves. The regeneration of the photoreducible enzyme after 18 h of illumination would then be due to an increase of the substrate and/or cofactor concentration.

We would like to stress that the maintenance of the 36 Kd polypeptide of PChlide oxidoreductase during the illumination of cucumber etioplasts support the conclusion that the photo-reduction step of chlorophyll biosynthesis is not restricted to the etiolated material. It is seen also that many proteins which take part in the photosynthetic activity seem to be present in the etioplast, as Hill activity becomes measurable without obvious changes of the polypeptide pattern.
This conclusion has been also drawn by Remy (1973) and by Lagoutte et al (1980) (see also Franck et al, this volume).

REFERENCES
1. Apel K, Santel HJ, Redlinger TE and Falk H. 1980. Eur. J. Biochem. 111, 251-258.
2. Beer NS and Griffiths WT. 1981. Biochem. J. 195, 83-92.
3. Brouers M and Wolwertz M-R. 1981. Proceedings of the fifth international photosynthesis Congress (G. Akoyunoglou ed.) vol. V pp. 185-196 Balaban Int. Science services. Phila-delphia.
4. El Hamouri B and Sironval C. 1979. FEBS Lett. 103, 345-347.
5. El Hamouri B, Brouers M and Sironval C. 1981. Plant Science Lett. 21, 375-379.
6. El Hamouri B and Sironval C. 1983. Photobiochem. Photobio-phys. 5, 263-272.
7. El Hamouri B, Oliver RP and Griffiths WT. 1983. Photobiol. Photobiochem. in press.
8. Franck F, Sironval C, Schmid GH. 1984. this volume.
9. Franck F and Inoue Y. 1984. this volume.
10. Gerday C, Michel-Wolwertz M-R and Brouers M. 1982. in Cell function and differentiation (G. Akoyunoglou et al Eds) part B pp 25-32 Alan R. Liss Inc. New York.
11. Gassman LM, Granick S and Mauzerall D. 1968. Biochem. Biophys Res. Commun. 32, 295-300.
12. Goedheer JC. 1967. in "Le chloroplaste, croissance et vieillissement (C. Sironval ed.) pp 77-85, Masson Paris.
13. Goedheer JC and Verhülsdonck CAH. 1970. Biochem. Biophys. res. com. 39, 260-266.
14. Griffiths WT. 1978. Biochem. J. 174, 681-692.
15. Griffiths WT. 1980. Biochem. J. 186, 267-278.
16. Hoyer-Hansen G, Machold O and Kahn 1. 1976. Carlsberg Res. Com. 41, 349-357.
17. Hoyer-Hansen G and Simpson DJ. 1977. Carlsberg Res. Com. 42, 379-389.

138

18. Ikeuchi M and Murakami S. 1982. Plant and Cell Physiol. 23, 575-583.
19. Jouy M and Sironval C. 1979. Planta 147, 127-133.
20. Lagoutte B, Setif P and Duranton J. 1980. Photosynthesis Res. 1, 3-16.
21. Litvin FF and Krasnovsky AA. 1957. Dokladij Akademii Nauk SSSR, 117, 106-109.
22. Ohya T, Naito K and Suzuki H. 1981. Z. Pflanzenphysiol. 102, 167-172.
23. Oliver RP and Griffiths WT. 1981. Biochem. J. 195, 93-101.
24. Remy R. 1973. FEBS Lett. 31, 308-312.
25. Santel HJ and Apel K. 1981. Eur. J. Biochem. 120, 95-103.
26. Shibata K. 1957. J. Biochem. (Japan) 44, 147-173.
27. Sironval C and Michel J-M. 1967. Book of abstracts. European photobiology symposium, Hvar, Yougoslavia, p.105.
28. Sironval C, Franck F, Gysembergh R, Bereza B and Dujardin E. 1984. this volume.

FLUORESCENCE EMISSION SPECTRA OF ETIOLATED LEAVES MEASURED AT 296 AND 77 K DURING THE FIRST SECONDS OF CONTINUOUS ILLUMINATION

C.BUSCHMANN* and C. SIRONVAL**
*Bot. Inst. Universität Karlsruhe.Kaiserstr.12, D-7500. Karlsruhe (FRG)
** Lab. Photobiol. Univ. de Liège. B22. Sart Tilman. 4000 Liège.Belgique.

1. INTRODUCTION

In the literature it is already well established that the illumination of an etiolated leaf results in the photoreduction of protochlorophyll(ide) $P_{657 - 650}^{*}$ to two forms of chlorophyll(ide) $P_{676 - 668}$ (present in minor amounts) and $P_{688 - 676}$ (6, 7, 12). The latter is subsequently transformed into $P_{695 - 682}$ (4, 10, 11) which then undergoes the Shibata shift (8) to $P_{680 - 672}$. These maximum shifts have already been described also for etiolated Raphanus cotyledons (3) after illumination with one flash. During continuous illumination at room temperature the fluorescence intensity of an etiolated leaf shows a fast rise and a subsequent slow decline (1, 2, 6). This decline was explained by a continuously decreasing fluorescence yield of $P_{695 - 682}$ (6).

Fluorescence emission spectra as well as absorption spectra served to study extensively the early steps of the light-induced chlorophyll biosynthesis in vivo. As many of these processes are light-dependent and proceed very rapidly, the scanning of a spectrum at room temperature, which takes minutes with the conventional fluorescence or absorption spectrometer, is impossible. Therefore the spectra have either to be taken at very low temperature (e.g. 77 K) to prevent further light-induced conversion processes or at room temperature overall measurements in a limited spectral range (adjusted with a monochromator or a filter) had to be applied. In the first case the spectra do not represent the real in vivo situation (spectra not taken at room temperature). In the latter case the manyfold measurements needed - always with different samples - can only give a rough idea of the whole spectrum.

Optical multichannel analyzers (OMA) have become available recently. They give the advantage to repeat spectra measurements within a short time delay (milliseconds). Thus room temperature spectra during rapid spectral changes and with one sample only can be taken. These advantages

---

* P = pigment-protein complex with a fluorescence maximum (1st number) and an absorption maximum (2nd number) at the wavelengths indicated (measured at 77 K).

C. Sironval and M. Brouers (eds.); Protochlorophyllide Reduction and Greening. ISBN 90 247 2954 8
©1984, Martinus Nijhoff/Dr W. Junk Publishers, The Hague/Boston/Lancaster.

make optical multichannel analyzers a valuable tool for studying the processes of the chlorophyll biosynthesis in vivo.

First promising results of fluorescence emission spectra of etiolated leaves taken with an OMA before and after a laser flash induced photoconversion were already presented (5). Fluorescence emission spectra of etiolated radish cotyledons during the decline of the fluorescence emission (starting 1 s after the onset of illumination) have already been published (2). No maxima shifts, as would have been expected from the literature, could have been observed. These spectra taken with the OMA 1 and filmed with a video camera from an oscilloscope screen were now re-examined for the events in the fluorescence rise (1st second of illumination) which were also contained on the video tape. We tried to study these processes again with a more sofisticated system (OMA 2).

2. MATERIAL AND METHODS

Radish seedlings (Raphanus sativus L. cv. Saxa Treib) were grown for 6 days in complete darkness on vermiculite with tap water. The manipulations before and during the measurement were carried out with a green safety light (λ = 540 nm). The experimental set up for measuring fluorescence emission spectra was as follows:

A) The spectra were taken at room temperature every 40 ms during illumination with a 632.8 nm laser light (He/Ne laser). The fluorescence was detected through a red cut-off filter (RG 645) mounted on a polychromator (Jobin Yvon, 1/4 m) connected to a SIT-vidicon detector in combination with an optical multichannel analyzer (OMA 1, Model 1205 A, PAR). The spectra were displayed in real time (33 ms/spectrum) on an oscilloscope (Tektronix) and filmed with a video camera, see Fig. 1. Further details see (2).

B) 1) The spectra were taken at room temperature every second during illumination with a 632.8 nm laser light (He/Ne laser) with an intensity reduced by 25 % (neutral filter).

2) Leaves illuminated at room temperature by a 632.8 nm laser light with an intensity reduced by 25 % (neutral filter) were frozen in liquid nitrogen (77 K) after 1, 3 and 30 s of illumination. For the subsequent measurement at 77 K the full intensity laser light was used.

The fluorescence was detected through a 662.5 nm red cut-off filter (transmission spectrum shown in Fig. 5) mounted on a polychromator (Jawell-Ash, 1/4 m) connected to a ISIT-vidivon detector in combination with an optical multichannel analyzer with computer equipment (OMA 2, Model 1215/16, PAR). The digital data of 30 spectra could be stored and transfered to floppy disks. The original spectra, their second derivative spectra and difference spectra were recorded on a HP 7044 A plotter. For further details see (9).

3. RESULTS

Etiolated radish cotyledons illuminated for the first time showed a rise in the room temperature fluorescence emission with a maximum around 691 nm. Within the accurancy limits of the measurements with the OMA 1 ($\pm$ 2 nm, see Fig. 1) neither a maximum shift nor a change in the half bandwidth could be observed (Fig. 2).

640   680   720   760 nm

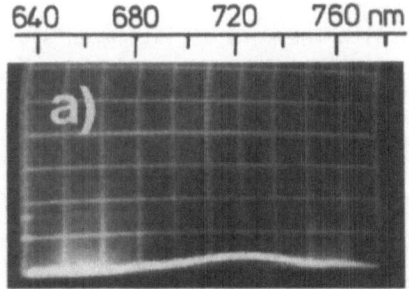

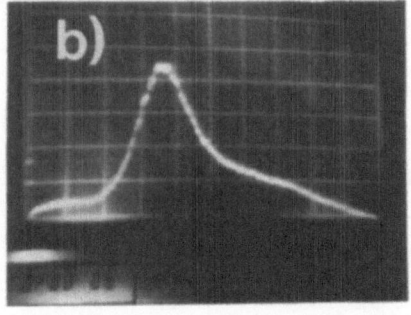

FIGURE 1. Photograph of the telescreen image showing the video-filmed oscilloscope screen during the measurement with an OMA 1 (experimental set up A) of chapter 2). a) baseline before measuring the fluorescence emission spectra. b) room temperature fluorescence emission spectrum of an etiolated radish cotyledon 6 s after onset of illumination (the illumination time in seconds is given by the stop watch on the lower left).

142

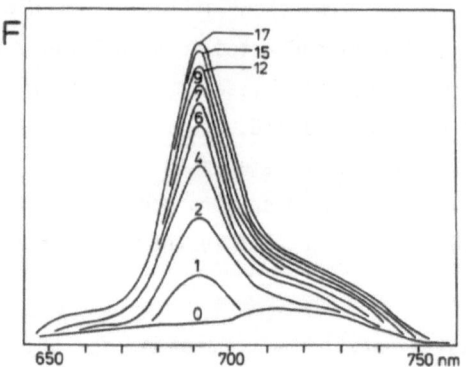

FIGURE 2. Room temperature fluorescence emission spectra of an etiolated radish cotyledon during the first second of illumination (fast rise of the fluorescence emission intensity) measured with an OMA•1 (experimental set up A) of chapter 2). The spectra were drawn from the telescreen image of a video tape (see Fig. 1). The numbers indicate the number of the video tape picture after the onset of illumination (0 = baseline, 1 picture/40 ms).

These first experiments were repeated with the more sofisticated system including the OMA 2. The illumination intensity was reduced by 25% (neutral filter) in order to make the 690 nm fluorescence kinetic maximum to be reached only after 2 to 3 seconds (Fig.3). During the first 30 seconds of illumination the room temperature fluorescence emission spectra were taken every second.

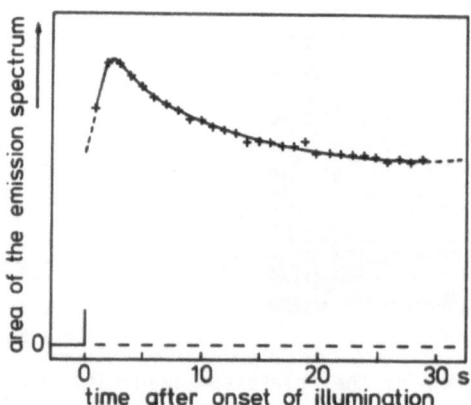

FIGURE 3. Area of the room temperature fluorescence emission spectra (see Fig. 4) of an etiolated radish cotyledon measured with an OMA 2 (experimental set up B)1 of chapter 2)every second during a 30 s light period after onset of illumination.

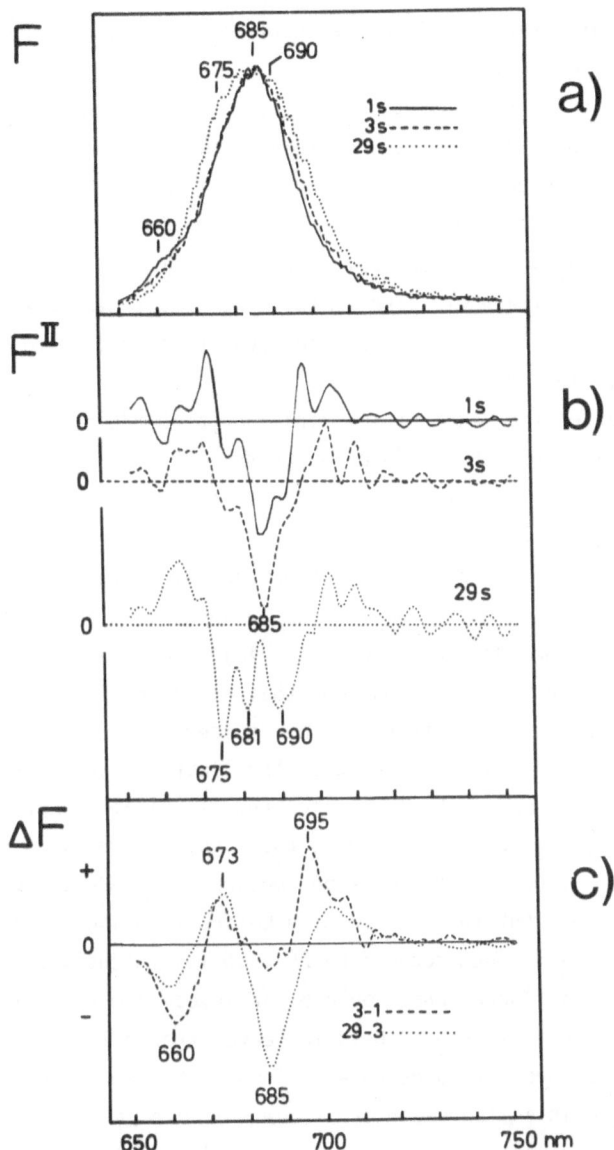

FIGURE 4. Room temperature fluorescence emission spectra of an etiolated radish cotyledon measured with the OMA 2 (experimental set up B)1 of chapter 2). a) Spectra taken 1 s (———), 3 s (– – – –) and 29 s (·············) after onset of illumination (normalized to the same maximum height). b) Second derivative of the spectra shown in a). c) Differences between the spectra shown in a) (normalized to the same area and smoothed once).

Fig. 4 shows the fluorescence emission spectra recorded after 1, 3 and 29 s of light. These were the spectra which differed significantly. The emission spectrum taken 1 s after the onset of illumination showed a small shoulder at 660 nm (Fig. 4 a and b).

2 seconds later (3 s light) this shoulder had decreased very much which is visible from the minimum of the difference spectrum 3 - 1 (Fig. 4c). Parallel to this the bandwidth became broader (Fig. 4a) due to the appearance of fluorescence emission on both sides of the main maximum (at 675 and 690 nm according to the derivative spectrum, Fig. 4b). The difference spectra calculated for the spectra normalized to the same area showed the highest relative increase at the long wavelength side (695 nm according to the difference spectrum 3 - 1, Fig. 4c).

29 seconds after the onset of illumination the bandwidth had again become broader (Fig. 4a). From the difference spectrum 29 - 3 (Fig. 4c) it is obvious that the relative fluorescence increase was now highest at the short wavelength side, at 673 nm. In parallel to this increase a strong decrease in the 685 nm maximum was observed. The broadening of the bandwidth during the first 29 s of illumination was not accompanied by a shift of the band maximum. But from the 2nd derivative (Fig. 4b) it can be seen that a new pigment form peaking at 681 nm replaced that at 685 nm.

In contrast to the observations at room temperature both band broadening and maximum shift could be detected when the fluorescence emission spectra were measured at 77 K after having frozen the illuminated leaf in liquid nitrogen (Fig. 5). In order to increase the signal/noise ratio the measurements of the 77 K fluorescence emission spectra were made using the full intensity laser light, but the illumination before freezing was done, as described above, with the laser light reduced by 25 % of its intensity.

Only little 77 K fluorescence could be observed from the etiolated leaves in the observed spectral range. The same leaves frozen 1 s after the onset of illumination showed a fluorescence maximum at 687 nm. Freezing after 3 s of light resulted in a fluorescence maximum shift to longer wavelengths (690 nm). The emission on the long wavelength side of the main maximum was found to be increased, but as the short wavelength side had not changed very much, the bandwidth had become broader. Leaves illuminated for 30 s before freezing showed a 77 K fluorescence emission spectrum with a clearly decreased short wavelength emission, but with no further maximum shift. The bandwidth had again become smaller.

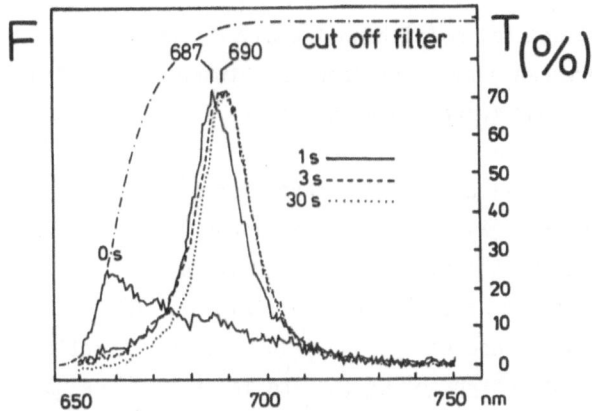

FIGURE 5. Fluorescence emission spectra of etiolated radish cotyledons measured with the OMA 2 at 77 K (experimental set up B)2 of chapter 2) and the transmission spectrum of the cut-off filter (—·—·—) placed in front of the detector system used for the experimental set up B. The leaves were frozen in liquid nitrogen before (0 s,———) and 1 s (———), 3 s (—————) or 30 s (···········) after the onset of illumination (spectra of the illuminated samples normalized to the same maximum height).

When comparing the spectra taken at room temperature with those taken at 77 K, two significant differences always showed up: a) the emission maximum was found at shorter wavelengths and b) the bandwidth was much broader (Fig. 6).

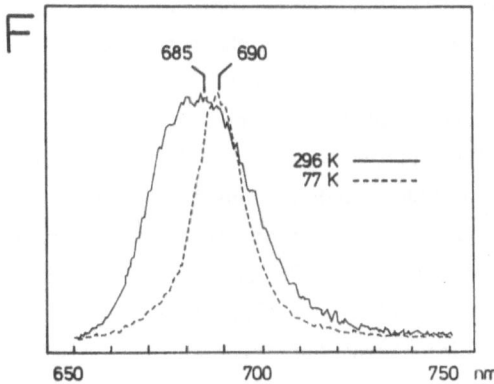

FIGURE 6. Fluorescence emission spectra of etiolated radish cotyledons after 30 s of illumination measured with the OMA 2 at room temperature (———) and at 77 K (—————), experimental set up B)1 and B)2 of chapter 2, respectively. The spectra were normalized to the same maximum height.

4. DISCUSSION

By means of an optical multichannel analyzer (OMA) we were able to measure the fluorescence emission spectra between 650 and 750 nm of an etiolated leaf at room temperature during the first 30 seconds of illumination.

After the onset of illumination a fast rise of one broad fluorescence band was observed (Fig. 2) followed by a slower decrease (Fig. 3). The rise time was determined by the quantum flux density of the incident light; it was 1 s with a full intensity laser light (Fig. 2) and 2 - 3 s with the laser light intensity reduced by 25 % (Fig. 3). This kinetic has already been described first for _Phaseolus_ (6) and later also for _Hordeum_ (1) and _Raphanus_ (2). The kinetic rise was attributed to the formation of a chlorophyll(ide) form ($P_{688 - 678}$); the subsequent kinetic decline was attributed to the gradual decrease of the fluorescence of this form as a consequence of its transformation into $P_{695 - 683}$, of energy transfer and of photobiochemical processes (6).

The light-induced appearance of chlorophyll(ide) in an etiolated leaf results from the photoreduction of protochlorophyll(ide). Protochlorophyll(ide) could be detected only as a decreasing shoulder around 660 nm in the fluorescence spectrum taken 1 s and 3 s after the onset of illumination (Fig. 4). The cut-off filter in front of the detector blocks the detection of most of the protochlorophyll(ide) fluorescence, which is expected to peak at 656 (phototransformable protochlorophyll(ide)) and 631 nm (non-phototransformable protochlorophyll(ide)) (3). The characteristics of the cut-off filter cause the sudden decline of the 77 K fluorescence spectrum taken from an etiolated leaf frozen before illumination (Fig. 5) at wavelengths smaller than 660 nm. Although the room temperature fluorescence emission spectra measured with the OMA appear to consist at first sight out of only one band (Fig. 4a), the 2nd derivative of the spectra showed three chlorophyll(ide) bands: a band peaking at 685 nm with two side bands at 675 and 690 nm (Fig. 4b). After 1 s of illumination the 685 nm maximum predominated. Later on, the main band decreased, and after 29 s of illumination it had shifted towards shorter wavelengths (maximum 681 nm). Simultaneously the side bands were rising first at 690 then at 675 nm, resulting in a broader bandwidth but not in a maximum shift (Fig. 4a). When the spectra were taken at 77 K after freezing the samples in liquid nitrogen the fluorescence band shifted from 687 (1 s light) to 690 nm (30 s light) (Fig. 5).

Similar observations of a chlorophyll(ide) form (absorption maximum at 680 nm) which is first gradually transformed in darkness at room temperature into a short wavelength form (absorption maximum at 670 nm) and then finally into a long wavelength form (absorption maximum at 685 nm) have already been described by Michel and Sironval (7) with leaves of Phaseolus having received a series of low intensity light flashes. After 30 s of light the long wavelength chlorophyll(ide) form is the predominant form in the spectra taken at 77 K but not in those taken at room temperature. This can be explained by a decreased fluorescence yield of this long wavelength form at room temperature which was also taken as an explanation for the fluorescence kinetic decline (see (6)).

In general, more chlorophyll(ide) forms are seen at room temperature than at 77 K (broader bandwidth). The room temperature emission spectra of etiolated leaves taken after different times of illumination during the first 30 seconds of light comprise emissions of three chlorophyll(ide) forms. Differences are also observed between spectra taken at 77 K and those taken at 296 K (room temperature) which have to be explained by changes in the fluorescence yield. At room temperature the long wavelength forms seem to be de-excited preferentially by other processes than fluorescence emission (including photobiochemical processes) as they emit much more light after freezing at 77 K, resulting in a shift of the fluorescence maximum towards longer wavelengths (Fig. 6).

ACKNOWLEDGEMENTS

The measurements with the OMA 1 were carried out at the Institute of Applied Physics of the Univ. of Karlsruhe in collaboration with Dr.H.Schrey.

We gratefully acknowledge the skillful technical assistance of R.Gysembergh during the measurements with the OMA 2.

This work was supported by the "Fonds National de la Recherche Scientifique" Belgium, and the"Ministère de la Politique Scientifique", "Actions Concertées" nr 80/85-1983, Belgium.

REFERENCES

1. Buschmann C. 1981. in: Photosynthesis V. Chloroplast development. Akoyunoglou G ed. Balaban Inetrnational Science Services, Philadelphia. pp. 417 - 426.
2. Buschmann C, Schrey H. 1981. Photosynthesis Research 1: 233 - 241.
3. Buschmann C, Sironval C. 1978. Planta 139: 127 - 132.

148

4. Gassman M, Granik S, Mauzerall D. 1968. Biochemical Biophysical Research Communications 32: 295 - 300.
5. Inoue Y, Kobayashi T, Ogawa T, Shibata K. 1981. in: Photosynthesis V. Chloroplast development. Akoyunoglou G ed. Balaban International Science Services, Philadelphia. pp. 55 - 64.
6. Jouy M, Sironval C. 1979. Planta 147: 127 - 133.
7. Michel J-M, Sironval C. 1977. Plant & Cell Physiology 18: 1223 - 1234.
8. Shibata K. 1957. Journal of Biochemistry 44: 147 - 173.
9. Sironval C, Franck F, Gysembergh R, Bereza B, Dujardin E. 1984, this volume.
10. Sironval C, Brouers M, Michel J-M, Kuiper Y. 1968. Photosynthetica 2: 268 - 287.
11. Sironval C, Kuiper Y, Michel J-M, Brouers M. 1967. Stud. Biophys. 5: 43 - 50.
12. Sironval C, Kuiper Y. 1972. Photosynthetica 6: 254 - 275.

THE REGULATORY EFFECT OF THE CYSTEINE/CYSTINE REDOX COUPLE ON
PROTOCHLOROPHYLLIDE PHOTOREDUCTION IN ETIOPLASTS.

B.BEREZA, M. LASKOWSKI and W. HENDRICH

Biophysics Laboratory, Institute of Biochemistry, Wroclaw
University, Tamka 2, 50 137 Wroclaw (Poland).

1. INTRODUCTION

Similarly to chlorophyll, its precursor, protochlorophyllide (Pide*) is in etioplasts complexed with proteins. Studies
of the low temperature absorption and emission spectra led to
the conclusion that there are few different types of these complexes. According to Dujardin and Sironval (1970) one can distinguish three main spectral forms of Pide: P_{628}^{632}, $P_{640}^{(645)}$ and
P_{650}^{657}. The two photoactive complexes consist at least of three
components, specific oxidireductase, Pide and NADPH (in the case of P_{650}^{657} form; Griffiths, 1978, El Hamouri et al., 1981).A
third spectral form P_{628}^{632} is either free Pide or a complex Pide-protein (El Hamouri, Sironval, 1980) and is photoinactive. El Hamouri et al.(1981) stated that etioplasts partially inactivated after incubation at 15°C may be reactivated by treatment
with NADP$^+$ and NADPH regenerating enzyme system. In such conditions the dark reaction P_{628}-->P_{650} takes place and is
followed by the photoinduced formation of the long wavelength
chlorophyllide form P_{682}^{696}. These results are consistent with
conception of Griffiths (1978): if NADPH is one of photoreduction substrates, it is not surprising that this compound activa-
tes the form P_{628}.

According to Bereza and Dujardin (1983) the increase of ratio NADPH/NADP$^+$ stimulates the Pide photoreduction in isolated
etioplasts. Simultaneously an increase of the red shift of the

* ABBREVIATIONS.

P^x - pigment-protein complex with absorption maximum at Y nm
and emission maximum at X nm;Pide -protochlorophyllide;
N - relative content of photoinactive Pide-protein form;
A - relative content of photoactive Pide-protein forms.

C. Sironval and M. Brouers (eds.); Protochlorophyllide Reduction and Greening. ISBN 90 247 2954 8
© 1984, Martinus Nijhoff/Dr W. Junk Publishers, The Hague/Boston/Lancaster.

chlorophyllide absorption band is observed. One can infer that
the Pide photoreduction is somehow regulated by the redox potential of microenvironment.

This paper deals with the confirmation of this assumption.
The influence of the redox couple cysteine/cystine was investigated on the state of Pide complexes, their photoreduction
and successive dark processes.

2. MATERIAL AND METHODS

2.1. <u>Plant material</u>: Bean seedlings (Phaseolus vulgaris var.
Bor) were grown in the dark at room temperature on the nutrient
solution (Ingenstad,1973), which was exchanged every third
day. 9 to 16 days old primary leaves were harvested and used
for etioplast isolation.

2.2. <u>Preparation of etioplasts</u>: Etiolated leaves and buffer
(1:2,w:v) were ground using an Unipan 309,Poland,homogeneizer
(full speed 2x5s, with break 60s). The composition of buffer
was as follows: 0.1 M Tris-HCl, 4 mM $MgCl_2$, 0.1% bovine serum
albumin , 10% glycerol and 0.6 M sucrose, pH 8.0. The homogenate was filtered through 2 layers of muslin and centrifugated
10 min at 500 g, followed by second centrifugation of supernatant at 2000 g, 10 min. The pellet,containing whole etioplasts was suspended in this same buffer (1 ml/4 g fresh leaves). All manipulations were performed under dim, green light
at 0 to 4°C.

2.3. <u>Preservation of samples</u>: Suspension of freshly prepared
etioplasts was divided into aliquots sufficient for single experiment, frozen and kept under liquid nitrogen. Samples were
thawed directly before experiment.

2.4. <u>Experimental scheme</u>: Aliquot of etioplasts was diluted
with buffer (without or with cofactors, as indicated in figures), divided into two parts placed in sample (a) and reference (b) cuvettes. After insertion of thermostated adapters with
cuvettes into spectrophotometer, samples were incubated 15 min
in dark at 5°C and the base line (dark sample/dark reference)
was registered. After next 15 min of incubation (period necessary for recording of base line and introducing of data into
computer memory), sample a was irradiated with two 1 ms poly-

chromatic flashes. Immediately after this the absorption difference spectrum was measured. To inspect successive dark processes, second difference spectrum was taken after additional
30 min dark incubation at 5°C.

Each 50 µl of irradiated sample and reference were taken
off just after sample irradiation and after next 30 min of
dark incubation. They were frozen immediately in liquid nitrogen and used for recording of low temperature emission spectra.
2.5. Illumination of the samples: Etioplast suspensions were
irradiated with two 1 ms polychromatic flashes from photographic lamp VEB Elgava, GDR. In preliminary experiments it was
found that in our conditions the energy of two flashes is sufficient to complete the reduction of photoactive Pide forms.
2.6. Measurements: Absorption difference spectra (sample irra-
diated/reference) were measured in the region 30.000 - 13.200
cm^{-1}, 75 - 125% transmittance, using a Specord UV-VIS, GDR
spectrophotometer, equipped with thermostated cuvette adapters.
Data were accumulated in the memory of KSR-4100, GDR, minicomputer. Programs, developed in Biophysics Laboratory, Institute
of Biochemistry, were used for further calculations to receive
the smoothed absorption difference spectra in desired spectral
region. Spectra were plotted using XY recorder.

Emission spectra were taken at the liquid nitrogen temperature using COBRABID, Poland, spectrofluorimeter equipped
with high pressure hydrogen lamp HBO-200 (exciting wavelength
436 nm), photomultiplier EMI 9558 QB, Great Britain, and K-
200, Zeiss, GDR, recorder. Samples were prepared for spectra
measurements, following the method described by Sironval et al;
(1968). Presented emission spectra were not corrected.
2.7. Redox buffers:Influence of redox potential on the Pide
photoreduction was investigated using etioplast suspension in
buffer for preparation completed with redox couple. The Nernst
equation and data after Clark (1960) were used for calculation
of initial redox potential of the reaction medium.

3. RESULTS

3.1. Properties of etioplasts with different initial proportions of photoactive to photoinactive Pide forms:

In etioplasts, content of previously listed Pide forms depends in some degree on the preparation and preservation conditions. Prolonged preparation and keeping in higher temperature results in conversion of active forms into inactive ones. However, the initial ratio of non-active forms to active forms (N/A ratio) is mainly determined by duration and conditions of etiolated plants cultivation. With "physiological ageing" of leaves the N/A ratio increases. The difference between the etioplasts, isolated from physiologically older ("o") and younger ("y") leaves are very distinct, when the low temperature emission spectra are considered. The spectra of "o" etioplasts (Fig.1,curve 1) and "y"

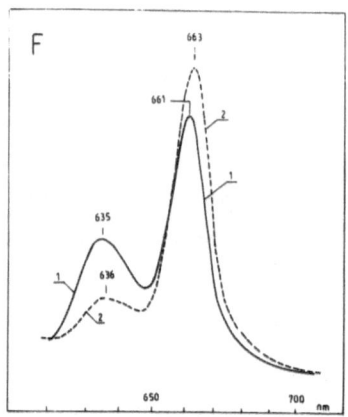

FIGURE 1. 77 K emission spectra of "physiologically older" (curve 1) and "physiologically younger" (curve 2) etioplast suspension at the beginning of dark incubation at 5°C.

etioplasts (Fig.1, curve 2) differ slightly in emission maxima, but remarkably in N/A ratios: these are 45% and 16% for "o" and "y" etioplasts, respectively.

These differences of initial state affect both the dark and photoinduced processes. Etioplasts, incubated in darkness, undergo thermal inactivation. In our experimental conditions after 60 min dark incubation at 5°C the increase of N/A ratio to 80% and 26% was observed for "o" and "y" etioplasts, respectively.

The irradiation causes the absorption spectra changes, characteristic for Pide photoreduction. In the red region of dif ference spectrum the negative band centred at 652 nm develops, connected to diminution of Pide photoactive forms. It is accompanied by appearance of chlorophyllide positive band.

FIGURE 2.Absorption difference
spectra of "o" etioplasts in:
red region (600-720 nm) - R
blue region (415-500 nm) - B

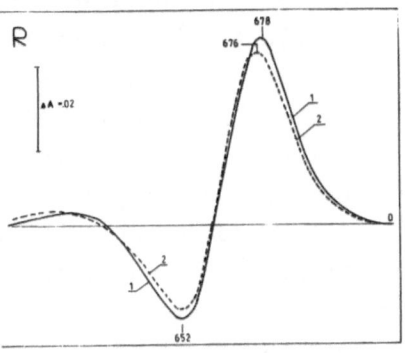

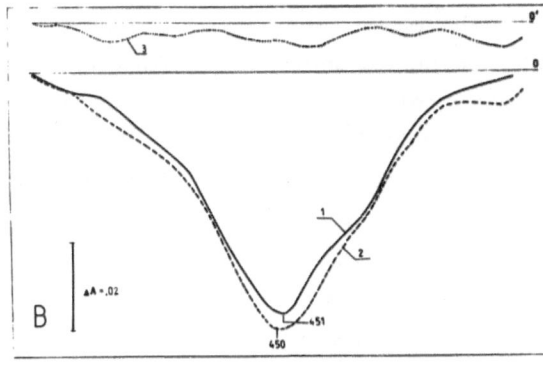

curve 1 - directly after
irradiations with two 1
ms flashes;
curve 2 - after 30 min
successive dark incubation at 5°C;
curve 3 - difference 2-1

Comparing the curves "1" in fig. 2R and 3R one can conclude
that the irradiation of "o" and "y" etioplasts in the same
experimental conditions leads to different chlorophyllide forms.

FIGURE 3. Absorption difference
spectra of "y" etioplasts.
All the indications - as in
figure 2.

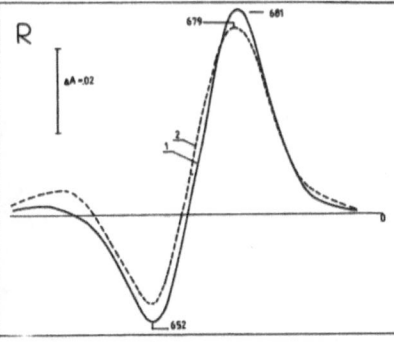

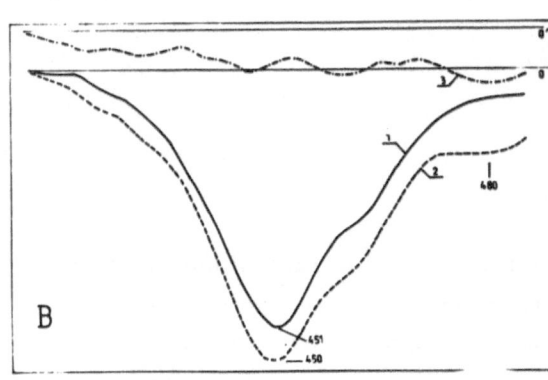

In the case of "y" sample the positive part of difference spectrum is shifted to the red.

Characteristic blue shift of positive absorption difference band (Shibata shift) in the spectra of both "o" and "y" etioplast suspensions (curves 2, Fig. 2R and 3R) is observed after subsequent dark incubation of illuminated samples.

Less remarkable and more difficult to explain are the differences, observed in the blue region of absorption difference spectra (Fig. 2B and 3B). The main negative band near 450 nm corresponds to disappearing of photoactive form P_{650}. Changes of band width and local deformations may be ascribed to the overlapping of this band with bands of other etioplast components (P_{640} with negative difference band at 440 nm, Soret bands of cytochromes, flavin derivatives, ferredoxin etc.).

Emission spectra of "o" and "y" etioplasts confirm the conclusion that their photoreduction results in different products. Fig. 4 presents the 77 K emission spectra of both the samples taken after irradiation and 30 min dark incubation at 5°C. One can distinguish the separated bands of photoinactive Pide (635 nm), regenerated photoactive Pide forms (655-660 nm) and chlorophyllide forms. The last ones differ both

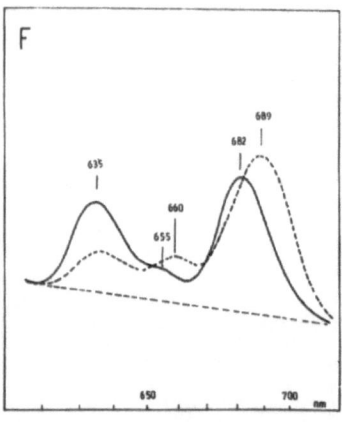

FIGURE 4. 77 K emission spectra of "o" etioplasts (curve 1) and "y" etioplasts (curve 2) after irradiation with two 1 ms flashes and 30 min dark incubation at 5°C.

in emission maxima (682 nm for "o" etioplasts, 689 nm for "y" etioplasts) and bandwidths. Is seems that "y" etioplasts accumulate more long-wavelength chlorophyllide forms than "o" etioplasts.

3.2. <u>Influence of cysteine/cystine redox couple on the state</u>
<u>of Pide-protein complexes</u>:Previously characterized "o" and
"y" etioplasts have been incubated in buffers containing cysteine and cystine in molar ratios giving initial redox potentials -350, -400 and -450 mV.

Fig.5 illustrates the influence of cysteine/cystine couple
on the properties of "o" etioplasts. As was mentioned previously at the start of experiment N/A = 45% (curve 1), which rises

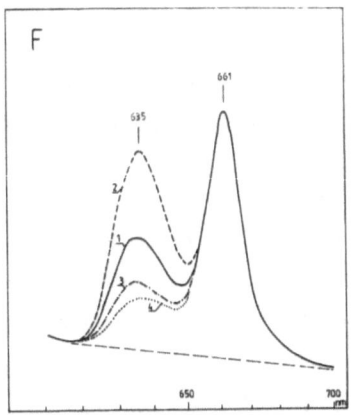

FIGURE 5. 77 K emission spectra
of "o" etioplasts.
1- in the buffer without cystei-
ne/cystine couple at the start
of experiment , 2- as 1, but after 60 min dark incubation at
5°C, 3- as 2, but incubated in
buffer E= -350 mV, 4- as 3, but
initial redox potential E= -400
mV.
All the curves were normalized
to 661 nm peak.

to N/A = 80% after 60 min dark incubation at 5°C (curve 2).
In the presence of investigated redox couple this inactiva-
tion process is inhibited and instead, activation of Pide com-
plexes is observed. Incubation of "o" etioplasts in the buffer
E= -350 mV causes lowering of N/A to 29%, while in the buffer
E= -400 mV N/A reaches 22% (curves 3 and 4, respectively).

Incubation of "y" etioplasts in a buffer with cysteine/cys-
tine couple results in less pronounced spectral changes. Because of low initial content of photoinactive Pide form (N/A=
8%) after 60 min dark incubation in the absence of cysteine/
cystine system as well as in its presence the value N/A=10-14%
was measured , the difference being of experimental error range.

These results indicate that N/A ratio depends on the microenvironment redox potential. For the sample with low N/A value ("y" etioplasts) there is no remarkable change of this
value corresponding to the change of redox potential in the

156

range -350 to -450 mV. On the other hand the initially high
N/A ratio (as for "o" etioplasts) is strongly influenced by
redox potential change. One can guess, that the initial endogeneous redox potential of "o" etioplasts was much higher than
-350 mV and rised permanently during incubation at 5°C in the
medium without cysteine/cystine system. Instead, the endogeneous redox potential of "y" etioplasts was of range -350 to
-400 mV (Fig. 6, insert).

3.3. The influence of cysteine/cystine couple on the Pide photoreduction process: In spite of lack of remarkable redox potential influence on fluorescence spectra changes of dark incubated "y" etioplast suspension, the photoreduction process
depends on this parameter.

Absorption difference spectra of "y" etioplasts exhibit
(in red region) minima near 652 nm and maxima near 678 nm, the
largest absorption changes being observed in the case of etioplasts incubated in buffer E= -400 mV (Fig.6R). Absorption

FIGURE 6. Absorption difference
spectra of "y" etioplasts, measured directly after irradiation
at 5°C.

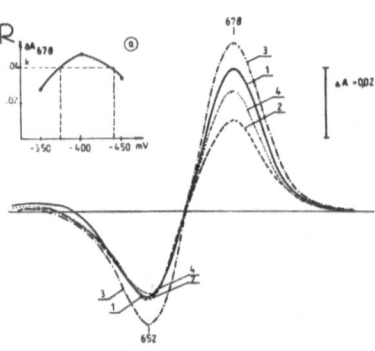

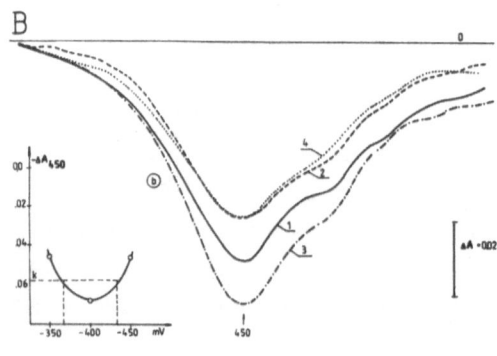

1- in buffer without
cysteine/cystine cou-
ple, 2- etioplasts in
a buffer E= -350 mV,
3- as in 2, E= -400 mV,
4- as in 2, E= -450 mV.
Insert: A (678) as a
function of E.Dashed horizontal line - A—(678) for control
Dashed vertical lines - possible endogeneous redox potentials.

spectra in blue region are more complicated, with main minimum
near 450 nm, a shoulder near 465 nm and local bands in the regions 415-430 nm and 470-500 nm. Also in this set of spectra
the largest absorption changes are seen for the sample in the
buffer E= -400 mV.

30 min dark incubation at 5°C leads to dimunution of chlorophyllide band (678 nm) and to stronger differentiation of
curves 1-4 in the region 652 nm (Fig. 7R) and 450 nm (Fig. 7B).
This effect can be ascribed to differences in dark regeneration of photoactive Pide complexes.

FIGURE 7. Absorption difference
spectra of "y" etioplasts, measured after irradiation and 30
min dark incubation at 5°C.

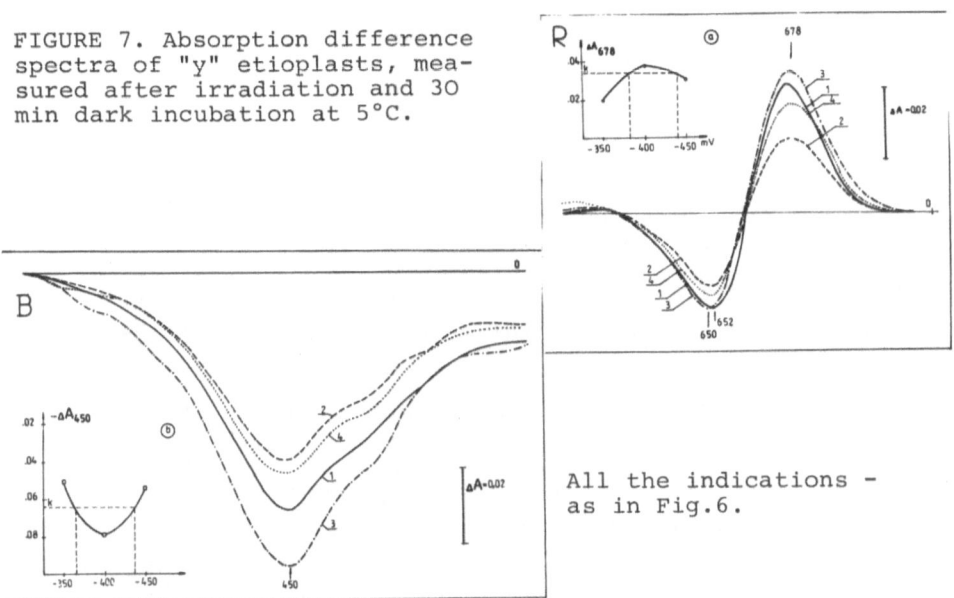

All the indications –
as in Fig.6.

These conclusions are supported by emission spectra measured after irradiation and 30 min dark incubation (Fig.8).
Inspection of these 77 K emission spectra indicates also,
that in the presence of cysteine/cystine couple the chlorophyllide form P_{678}^{688} is stabilized: in the absence of redox couple the emission maximum is shifted to 684 nm (Fig.8, curve 1).

158

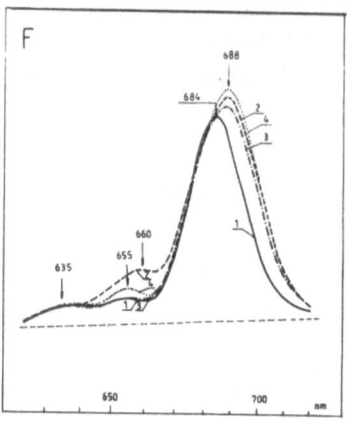

FIGURE 8. 77 K emission spectra of "y" etioplast suspension in buffer:
1- without cysteine/cystine couple, 2- with cysteine/cystine couple, E= -350 mV, 3- as in 2, but E= -400 mV, 4- as in 2, but E= -450 mV.
Spectra were measured after irradiation with two 1 ms flashes and 30 min dark incubation at 5°C.
All the curves were normalized to the peak 635 nm.

4. DISCUSSION

The results presented in this paper demonstrate the influence of cysteine/cystine redox couple on the state of Pide complexes in the dark, on the photoreduction process and on the properties of products.

It seems that the redox couple acts mainly by activating the enzyme, specific Pide oxidoreductase. Assuming after Griffiths (1978) that the main photoactive form is a ternary complex of the enzyme, Pide and NADPH, one can postulate an increase of the enzyme activity and of its affinity to the substrate with lowering of environment redox potential to some optimal value. This is consistent with the emission spectra changes of "o" etioplasts, incubated in the presence of cysteine/cystine couple (Fig.5). In this case lowering of redox potential favours the formation of the photoactive form at the expense of the photoinactive one.

In "physiologically younger" etioplasts the low ratio of non-active forms to active forms (N/A) depends rather insignificantly on the redox potential. However, the enzyme activity is still influenced by the cysteine/cystine couple and reaches the optimal value at redox potential corresponding to initial value -400 mV.

One can infer that the cysteine/cystine couple acts on the

following steps of chlorophyllide-protein complexes formation:

$$P_{628}^{632} \xrightarrow{\text{cysteine/cystine}} P_{650}^{657} \qquad \text{(I)}$$

$$\left.\begin{array}{l} P_{650}^{657} \\[4pt] P_{640}^{(645)} \end{array}\right\} \xrightarrow[\text{cysteine/cystine}]{h\nu} P_{678}^{688} \qquad \text{(II)}$$

In vivo this II step is followed by a 30 s dark transformation of P_{678}^{688} to the form P_{682}^{695} and much slower (30-45 min) Shibata shift leading to P_{672}^{685}. These dark steps are inhibited in the presence of cysteine/cystine couple.

ACKNOWLEDGMENTS

This work has been supported by fonds of the Technical University (nr R.1.9.02.03) Wroclaw, Poland.

REFERENCES

1. Bereza B, Dujardin E. 1984, this volume.
2. Clark W M. 1960, Oxidation-reduction potentials of organic systems. The Williams and Wilkins Co, Baltimore: 471-497
3. Dujardin E, Sironval C. 1970, Photosynthetica 4: 129-138
4. El Hamouri B, Sironval C. 1980, Photobiochem. Photobiophys. 1, 219-223
5. El Hamouri B, Brouers M, Sironval C. 1981), Plant Sc. Lett. 21, 375-379
6. Griffiths W T. 1978, Biochem. J. 174, 681-692
7. Ingenstad S. 1973, Plant Physiol. 52 , 332-338
8. Sironval C, Brouers M, Michel J M, Kuiper Y. 1968, Photosynthetica 2 , 268-287.

THE INFLUENCE OF THE NADPH/NADP$^+$ REDOX COUPLE ON THE PHOTOREDUCTION OF PROTOCHLOROPHYLLIDE AND ON THE SPECTRAL CHARACTERISTICS OF THE CHLOROPHYL-LIDE-PROTEINS IN ETIOPLASTS.

B. BEREZA and E. DUJARDIN.
Biophysics Laboratory; Institute of Biochemistry-Wroclaw University,
Tamka 2; 50137 Wroclaw-Poland.
Photobiology Laboratory; B22 Liège University; 4000 Sart Tilman;Belgium.

1. INTRODUCTION

When investigating the role of a reductant or of an oxidizer in a reaction, the concentration of the reactant is not the sole important factor to be considered. The redox potential of the medium is to be also considered as it determines primarily the equilibrium of the redox reactions.

It is known that NADPH is the specific reductant in the photoenzymatic reduction of protochlorophyllide into chlorophyllide in etiolated angiosperm leaves. The formation of a ternary complex has been postulated:

$$
\text{Apoprotein + protochlo-rophyllide} \quad \xrightarrow{\quad \text{NADPH} \quad} \quad \text{Apoprotein-protochlorophyllide-NADPH complex} \quad (1)
$$

The ternary complex containing the reduced nucleotide is photoactive, it is capable of synthesizing chlorophyllide when illuminated (Griffiths, 1978).

Previous works (El Hamouri, Sironval, 1980)have shown that the addition of NADPH, of NADP$^+$or of a mixture of them, to etioplasts, or to etioplast lamellae suspensions, influence the position of the absorption and fluorescence bands of the chlorophyllide-protein produced on illumination: a high concentration of the reductant favors the formation of a chlorophyllide-protein absorbing in the red around 682-684 nm, while a high concentration of NADP$^+$ induces the formation of a 677-678 nm absorbing chlorophyllide (see also Oliver, Griffiths, 1982).

In fact, when they add the reduced or/and oxidized cofactor(s)to the reaction medium, these authors modify at the same time both the redox-potential and the nucleotide concentrations. Consequently, the equilibrium of many redox reactions is changed in the etioplasts or in the etioplast lamellae.

Therefore we have tried to investigate the role of the redox potential created by adding various redox couples to etioplast suspensions(Bereza,

C. Sironval and M. Brouers (eds.); Protochlorophyllide Reduction and Greening. ISBN 90 247 2954 8
© 1984, Martinus Nijhoff/Dr W. Junk Publishers, The Hague/Boston/Lancaster.

Dujardin ,1984). This paper presents preliminary experiments in which we study the influence of the exogenous NADPH/NADP$^+$ redox couple on the process of photoreduction on the one hand, and on the other hand on the subsequent spectral dark shift which the chlorophyllides undergo as they have been synthesized. In other papers, we investigate the role of other exogenous couples, such as the cysteine/cystine couple (Bereza et al.,1984a) and of dithiothreitol (Bereza and Dujardin, in preparation).

2. MATERIAL AND METHODS

2.1. Plant material

Beans (*Phaseolus vulgaris* var. Bor) were grown at room temperature in the dark on a vermiculite/perlite 1:1 mixture moistened with tap water. All the operations were performed with a green low-intensity safelight transmitting only at 520 nm. The primary leaves were harvested after 14 days.

2.2. Preparation of etioplasts

40 g etiolated leaves were ground in a mixer during two times 5 s in 200 ml Tris buffer 0.1 M, pH 8.0, containing 4 mM MgCl$_2$, 0.1% bovine serumalbumine, 10% glycerol and 0.6 M sucrose. The mixture was filtered through 2 layers of muslin and centrifuged at 500 g during 10 min. The pellet was discarded and the supernatant was centrifuged during 10 min at 2000 g. The pellet was resuspended in 5 to 10 ml of Tris buffer. The etioplast suspension was divided into several parts which were kept in liquid nitrogen until utilization (= the stock).

2.3. The low temperature absorption and fluorescence spectra were recorded according to the method described earlier (Dujardin, 1976; Sironval et al., 1968).

2.4. Difference absorption spectra

The difference absorption spectra of the etioplast suspensions have been recorded from 325 to 740 nm in a Varian spectrophotometer, model Cary 17. The temperature was kept at 5°C by circulating a cryogenic liquid in the cooling jackets surrounding the two spectroscopic cells (optical path : 5 mm); (Bereza, Dujardin, 1980).

2.5. Actinic illumination

The sample cell was illuminated by a single 1 ms polychromatic flash (Multiblitz Report Porba; electric energy 125 J, colour temperature 5800 K); its intensity was enough for saturating the photoreduction of proto-chlorophyllide into chlorophyllide.

EXPERIMENTAL SCHEME

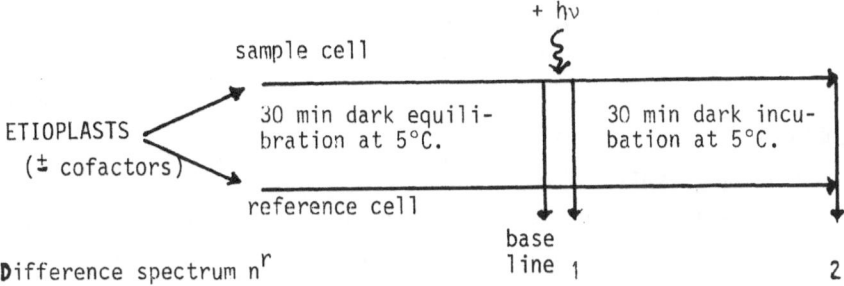

The etioplasts from the stock were thawed and diluted with Tris buffer, either without addition of the cofactors (control experiment),or with addition of the NADPH/NADP$^+$ redox couple. In each case, the etioplast suspension was divided into two parts, each of which was introduced in a 5 mm spectrophotometric cell. After equilibration of the temperature at 5°C, in the dark, the base line was registered using the Cary 17, and two 50 ul aliquots from each part were frozen in liquid nitrogen for recording the low temperature absorption and fluorescence spectra. One suspension was then illuminated with a 1 ms polychromatic flash, and the difference spectrum between this and the other part (light minus dark) was recorded; this yielded the difference spectrum nr 1; 50 ul aliquots from both suspensions were then frozen in liquid nitrogen. Both etioplast suspensions were subsequently incubated in the dark at 5°C for 30 min. The difference between them (spectrum nr 2) was also recorded, and 50 ul aliquots from both cells were again collected.

We have verified that a prolonged storage (several months) of the etioplats at 77 K, and their subsequent thawing did not modify their behaviour in our experiments.

2.6. Preparation of suspension medium for etioplasts with definite initial redox potentials.

The initial redox potential of the etioplast suspension medium was adjusted by adding various ratios of a NADPH/NADP$^+$ mixture at a total exogenous nucleotide concentration of 1.5, or 15 mM. The values for the redox potentials corresponding to 1/10, 1/1, and 10/1 ratios of NADPH/NADP$^+$ were calculated according to the Nernst equation (Clark,1960; E'° for the NADPH/NADP$^+$ couple = - 316 mV). They were respectively: - 286, - 316 and - 346 mV). (see Table I).

3. RESULTS

3.1. Control experiment.

Figure I reproduces the red part of the low temperature (77K) absorption spectrum (fig. Ia, spectrum 1, full line) and the 77K fluorescence spectrum (fig. Ib, spectrum 1, full line) of a control etioplast suspension. The dark absorption band has a maximum at 652 nm with a shoulder around 638 nm. The absorption at 652 nm is due to the main photoactive protochlorophyllide, while the shoulder corresponds to a mixture of photoinactive protochlorophyllide (absorbing near 628 nm) and photoactive protochlorophyllide-protein(absorbing at 640 nm).

The 77K fluorescence spectrum of the same sample has a main emission band at 658 nm due to the main photoactive protochlorophyllide. A minor 630 nm band is also seen due to the photoinactive protochlorophyllide.

The etioplast suspension remained stable for hours at 5°C in the dark. After a one hour dark incubation no noticable changes could be observed in the spectra.

After a 1 ms polychromatic actinic flash given at 5°C, the low temperature absorption spectrum had its 77K maximum shifted to 679 nm, as the photoactive protochlorophyllide had been reduced completely into chlorophyllide; consequently, the 77K fluorescence emission shifted to 691 nm (spectra 2 of fig. Ia and Ib). After a further 30 min incubation in the dark at 5°C, the chlorophyllide spectra shifted slightly towards shorter wavelengths (spectra 3 of figure Ia and Ib).

Difference absorption spectra (light minus dark) recorded at 5°C just after the flash are presented in figure II. The negative bands of spectrum 1 (full line) are due to the disappearance of the photoactive protochlorophyllide-proteins as a result of the actinic flash. They peak

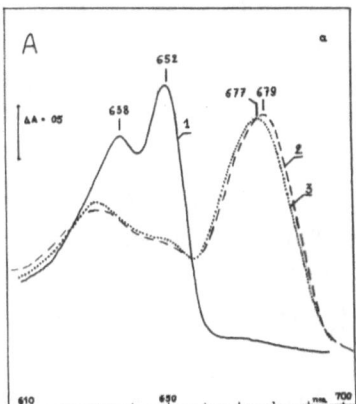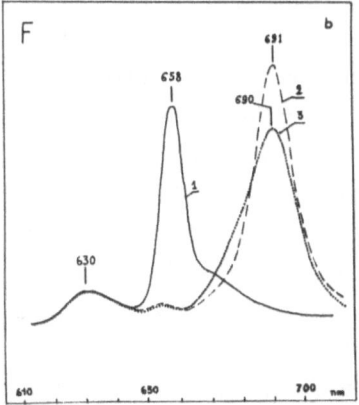

FIGURE I. 77K absorption (A) and fluorescence (F) spectra of an etioplast suspension incubated at 5°C (=control).0 (full line): before illumination. 1 (dashed line): after a 1 ms polychromatic actinic flash,2 (dotted line): as 2 but after a 30 min dark incubation at 5°C.

at 651 nm and at 640 nm (shoulder) in the red, and at 448 nm and at 465 nm (shoulder) in the blue . The positive band in the red at 679 nm is due to the formation of the chlorophyllide-protein(s). After a 30 min dark incubation at 5°C this chlorophyllide band had shifted to 677 nm (spectra 2 of fig. II R).

This experiment will be referred to as the control experiment of all experiments with an exogenous supply of NADPH/NADP$^+$. All the experiments described below were performed with etioplasts from the same stock as this control.

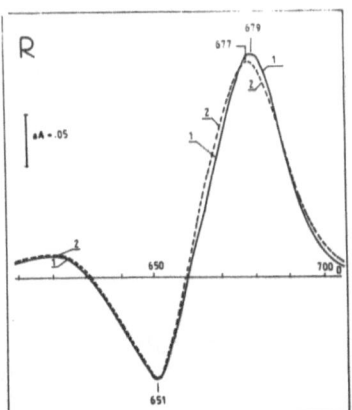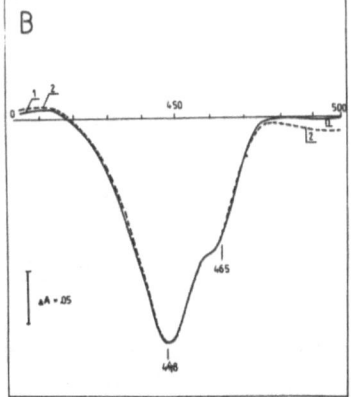

FIGURE II. Difference absorption spectra in the red (R) and in the blue (B) region of the control etioplast suspension at 5°C.1(full line):after a 1 ms polychromatic actinic flash. 2 (dashed line): as 1,but after a 30 min dark incubation.

3.2. Experiments with suspensions containing 1.5 mM exogenous nucleotides.

Figure III reproduces besides the control (curve 1) the difference absorption spectra which were found immediately after the actinic illumination of etioplast suspensions to which various NADPH/NADP$^+$ redox couples had been added at 5°C (curves 2 and 3). Compared to the control, the illumination of a suspension with a 1/10 ratio of NADPH/NADP$^+$ (spectrum 2) provoked a larger decrease of the protochlorophyllide absorption at 651 nm and a larger increase of the chlorophyllide absorption. The latter peaked at 681 nm (spectrum 2) instead of 679 nm in the case of the control (spectrum 1).

These changes were intensified when the ratio of NADPH/NADP$^+$ was 10/1 (spectrum 3 of figure III). The amount of chlorophyllide formed in this case was still higher and the absorption maximum was found at 683 nm. The incubation of the suspension with a 1/1 ratio of NADPH/NADP$^+$ gave a similar result (not shown in the figures; see Table I).

In the blue region of the spectra (figure III B) the absorption changes at 447 nm were also found greater in the case of a nucleotide supply than in the control(compare spectra 2 and 3 with spectrum 1).

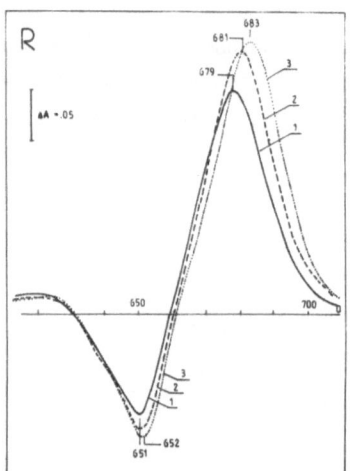
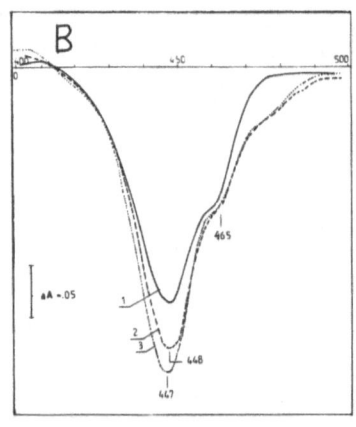

FIGURE III. Difference absorption spectra of etioplast suspensions incubated with nucleotides, recorded at 5°C immediately after a polychromatic actinic flash:
1 (full line); control . 2 (dashed line): etioplast suspension + exogenous NADPH/NADP$^+$ 1/10.
3 (dotted line): etioplast suspension + 10/1 NADPH/NADP$^+$.

When the proportion of NADPH was higher the decrease of the absorption of the main band at 447-448 nm was found more important (compare curve 3 to curve 2). However, the shoulder at 465 nm was practically not modified by the addition of the nucleotides.

During a dark 30 min incubation at 5°C (figure IV,R) a shift towards longer wavelengths of the chlorophyllide red band was observed (spectra 2 and 3 of figure IV). This shift was found to be increased as the proportion of NADPH increased. Comparing with the control (spectrum 1 of figure IV) the bathochromic shifts were : 4 nm, 6 nm and 7 nm for NADPH/NADP$^+$ ratios of 1/10, 1/1 and 10/1 respectively (i.e. for increasing negative initial redox potentials). There were no significant changes seen in the blue region of the spectra during the 30 min dark incubation.

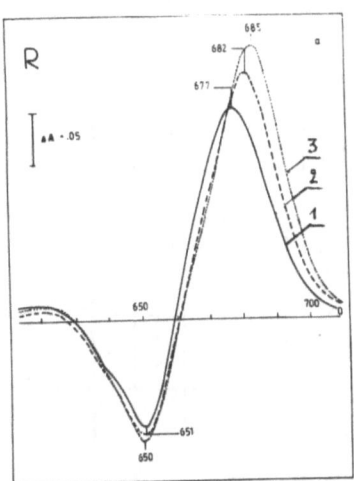

FIGURE IV. Difference absorption spectra of etioplast suspensions incubated with nucleotides recorded at 5°C after a 1 ms polychromatic actinic flash and a 30 min dark incubation. 1 (full line): control, 2(dashed line): etioplast suspension + NADPH/NADP$^+$ 1/10. 3 (dotted line): etioplast suspension + NADPH/NADP$^+$ 10/1.

The 77K fluorescence spectra of figure V confirm the changes observed in the difference absorption spectra. As the redox potential became more electronegative (i.e. as the NADPH/NADP$^+$ ratio increased), the position of the maximum of the emission band of chlorophyllide was found at longer wavelengths after the actinic flash (at 691, 692, 694 nm for the control, for {NADPH}/{NADP$^+$} = 1/10, and for { NADPH} / {NADP$^+$} =10/1 respectively).

After a 30 min dark incubation at 5°C, the fluorescence maxima had shifted towards shorter wavelengths for the control (2 nm, spectrum 1 of figure V), and towards longer wavelengths for the suspensions with a supply of nucleotides. The shift was 2 nm in the case of a ratio of {NADPH} / {NADP$^+$ } = 1/10 and to 3 nm for a ratio of 10/1. The shift was interme-

TABLE I.

NADPH/NADP+ ratio (1.5 mM)	Nucleotide concentration (mM) (NADPH)	(NADP+)	Calculated initial redox potential (mV)	Position of the maximum of the chlorophyllide absorption band at 5°C (nm)		Absorption difference (optical density; path = 0.5 cm) Protochlorophyllide	Chlorophyllide	Position of the 77 K fluorescence emission of chlorophyllide (nm)
Control	not measured		-	After the flash	679	0.10	0.22	691
				+ 30 min darkness	677			690
1/10	0.14	1.36	- 286	After the flash	681	0.11	0.25	692
				+ 30 min darkness	682			694
1/1	0.75	0.75	- 316	After the flash	683	0.12	0.27	693
				+ 30 min darkness	684			696
10/1	1.36	0.14	- 346	After the flash	683	0.12	0.27	694
				+ 30 min darkness	685			697
(15 mM) 1/10	1.4	13.6	- 286	After the flash	682	0.09	0.25	693
				+ 30 min darkness	683			697*
10/1	13.6	1.4	- 346	After the flash	683.5	0.10	0.23	695.5
				+ 30 min darkness	685			697
Fresh leaf	not measured		unknown	After the flash	684	0.018	0.039	not measured
				+ 30 min darkness	685			not measured

* This important shift of the maximum towards longer wavelengths is due to efficient energy transfer at 77 K from the short wavelength-absorbing chlorophyllides to the long-wavelength-absorbing ones.

diate in the etioplast suspension containing exogenous nucleotides in a
1/1 ratio. Table I summarizes all the data.

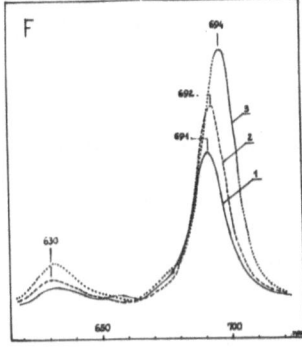

FIGURE V. 77 K fluorescence spectra of
etioplast suspensions incubated with
nucleotides after a 1 ms polychromatic
actinic flash.1(full line): control, 2
(dashed line): same suspension + 1/10
NADPH/NADP+ couple. 3(dotted line):same
suspension + 10/1 NADPH/NADP+ couple.

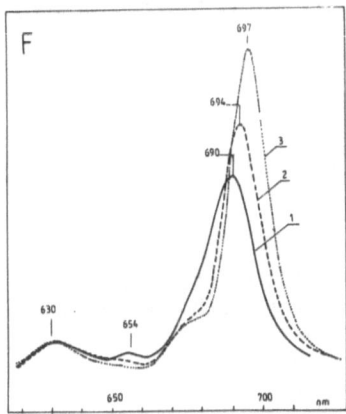

FIGURE VI. 77K fluorescence spectra of
etioplast suspensions incubated with
nucleotides, after a 1 ms polychromatic
actinic flash and a 30 min dark incubation at 5°C. 1 (full line):control,
2 (dashed line): same + 1/10 NADPH/NADP+
couple, 3 (dotted line): same + 10/1
NADPH/NADP+ couple.

3.3. Experiments with a supply of 15 mM exogenous nucleotides.

In order to evaluate the effect of the total nucleotide concentration on the phenomena, a 10 fold higher nucleotide concentration (15 mM)
was used at the same 1/10 and 10/1 ratios of the NADPH/NADP+ redox couple.
In this case, the calculated initial redox potentials of the medium were
the same as in the case of the supply of 1,5 mM nucleotides.

The results are reported in Table I. The increase of the absorption
of the chlorophyllide band after the actinic flash amounted to about
10%, by comparison with the control,(against 20% in the case of the
1.5 mM supply of nucleotides). After the flash, the maxima of the chlorophyllide red band were at 682 and 683.5 nm for the NADPH/NADP+ ratios of
1/10 and 10/1 respectively (against 681 nm and 683 nm respectively in the
case of the 1.5 mM supply in nucleotides).

During the subsequent dark 30 min incubation at 5°C, the maximum shifted to longer wavelengths : from 682 to 683 nm when the reduced to oxidized nucleotide ratio was 1/10, and from 683.5 to 685 nm when the ratio was 10/1; it was a little more marked than in the case of the 1.5 mM concentration. The maximum of the chlorophyllide 77K emission shifted similarly.

These results show that the 10 fold increase of the exogenous nucleotide concentration affected only slightly the initial positions and the dark shifts of the red chlorophyllide band. Both positions and shifts depended much more on the reduced to oxidized nucleotide ratio.

3.4. "In vivo" experiments with fresh etiolated leaves.

The two primary leaves of an etiolated plant were introduced into the spectrophotometer cells. They were incubated in the dark at 5°C until the temperature remained constant. After the recording of the base line, one of these leaves was illuminated with a 1 ms polychromatic flash and the difference absorption spectrum (1 of figure VII) was recorded.

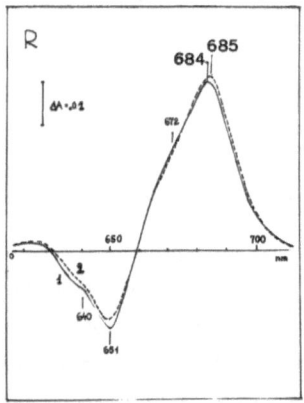

FIGURE VII. Difference absorption spectra of an etiolated leaf incubated at 5°C. 1 (full line) : after a 1 ms polychromatic actinic flash. 2(dashed line) : as 1 but after a 30 min dark incubation at 5°C.

The maximum of the chlorophyllide band was located at 684 nm. It shifted to 685 nm after a 30 min dark incubation at 5°C (spectrum 2 of figure VII).

In addition, a shoulder at 672 nm appeared after the flash. This shoulder did not change during the 30 min dark incubation. It is worth to note that this band at 672 nm was also observed in the etioplast suspensions, for example in the control (curve 2 of figure II).

4. DISCUSSION

4.1. Stability of the isolated etioplasts.

We have prepared isolated etioplasts which had the same spectral characteristics as etiolated leaves,i.e. which contained the same proportion of photoactive/photoinactive protochlorophyllides as leaves (see fig. IA and F). They remained stable during hours in the dark at 5°C. However, the fluorescence and absorption maxima of the chlorophyllide formed by a flash shifted very slowly to shorter wavelengths at 5°C . This contrasts with etiolated leaves in which the absorption and fluorescence maxima of chlorophyllide are found at longer wavelengths after the flash and undergo a red shift during the subsequent 30 min dark incubation at 5°C (see fig.VII).

To observe a similar shift of chlorophyllide in isolated etioplasts as in etiolated leaves a supply of the NADPH/NADP$^+$ redox couple was required. In this case, a long-wavelength-shift was seen whatever the total concentration of exogenous nucleotides, and whatever the ratio of the reduced to oxidized form of the nucleotide could be. This result confirms that the isolation of etioplasts in an aqueous buffer is accompanied by a loss of pyridine nucleotides as in green chloroplasts (Kraus, Heber,1976).

When supplied with the NADPH/NADP$^+$ redox couple, our etioplast preparations reproduced very satisfactorily the changes occuring in fresh leaves as can be seen in fig VII. Therefore, the samples used in our experiments have been further tested for the occurence of the Franck and Inoue cycle (Franck and Inoue, in preparation). It has appeared that this cycle was dependent on the NADPH/NADP$^+$ ratio, as well as on the total nucleotide concentration (Sironval et al., 1984).

4.2. Influence of the NADPH/NADP$^+$redox couple.

When the NADPH/NADP$^+$ redox couple was added to the suspension in a variable ratio and in a variable concentration, both the redox potential of the suspension and the concentration of the specific cosubstrate(NADPH) and coproduct (NADP$^+$) of the photoenzyme (protochlorophyllide photoreductase) were modified.

These additions had several effects:

A.- Chlorophyllide synthesis.

In our experiments, the amount of the chlorophyllide formed was higher in all the etioplast suspensions containing exogenous nucleotides than in the controls. (Table I). The optimal chlorophyllide synthesis was obtained by adding 1.5 mM total nucleotides in the 1/1 and in the 10/1 ratio. Increasing the concentration of exogenous nucleotides 10 fold decreased slightly this synthesis. We suggest that the activation of the photoenzyme,-i.e. the formation of the photoactive ternary complex "protochlorophyllide-NADPH-apoprotein" (reaction (1))- is better favoured by a 1.5 mM supply of nucleotides than by a 15 mM supply, and that this activation increases as the $NADPH/NADP^+$ ratio increases.

The same interpretation has been proposed by Bereza et al.(1984)who have incubated etioplasts in the dark, in the presence of the exogenous cysteine/cystine redox couple. They have found that the synthesis of chlorophyllide was at the optimum when the initial redox potential of the medium was near - 400 mV, i.e. when the concentration of cysteine was 10 fold higher than that of cystine. They have also found that the redox potential of the medium, as determined by the ratio of cysteine to cystine, influenced the activation of the photoreductase (see reaction 1).
Our data suggest that the optimal synthesis of chlorophyllide is reached at a potential between - 320 and - 350 mV.

B.- <u>Absorption and fluorescence spectra of chlorophyllide after the actinic flash.</u>

For technical reasons, the maxima reported in Table I were obtained only 90 seconds after the actinic illumination of the etioplast suspension. Table I shows,however, that 90 s after the actinic flash, the positions of the maxima were much more affected by the redox potential of the medium, than by the concentration of the exogenous NADPH. For example: increasing the NADPH concentration ten fold (from 0.14 mM up to 1.4 mM): changed the shift of the absorption and fluorescence maxima of chlorophyllide by 1 nm only. For a given total nucleotide concentration, the ratio of reduced to oxidized nucleotide had a much greater influence.

C.- <u>Long-term dark shifts of chlorophyllide at 5°C.</u>

After a 30 min dark incubation at 5°C, the displacement of the red absorption and fluorescence maxima of chlorophyllide towards longer wavelengths was found the largest when the potential was the most negative. For the lowest potential (- 346 mV), the absorption and fluorescence maxima reached the extreme values of 685 nm and 697 nm respectively.

This effect seems, however, to be specific of the presence of the $NADPH/NADP^+$couple. It does not occur in etioplast suspensions incubated with other redox couples like the cysteine/cystine couple (Bereza et al., 1984), or incubated with dithiothreitol (Bereza and Dujardin, in preparation). In these cases, the red absorption maximum of the chlorophyllide band remained fixed at 678 nm.

REFERENCES

1.- Bereza B, Dujardin E. 1980, in"Photosynthesis V. Chloroplast Development". Akoyunoglou G, ed. Balaban Intern. Science Services Philadelphia. Pa, 16-20.
2.- Bereza B, Dujardin E. 1984, in Proceedings of the VIth Internat. Photosynthesis Congress 1-6 Aug. 1983 . Sybesma C. ed., Martinus Nijhoff Dr Junk W, Publishers, The Hague, The Netherlands, in press.
3.- Bereza B, Laskowski M, Hendrich W. 1984a, in Proc.VIth Intern.Photosynthesis Congress. 1-6 Aug. 1983. Sybesma C. ed. Martinus Nijhoff, Dr Junk W,.Publishers, The Hague, The Netherlands , in press.
4.- Bereza B, Laskowski M, Hendrich W. 1984b, in "Protochlorophyllide reduction and greening". Sironval C, and Brouers M, eds. Martinus Nijhoff , Dr Junk W, Publ. The Hague, The Netherlands , this volume.
5.- Clark W M. 1960. Oxidation-reduction potentials of organic systems. The William and Wilkins Co, Baltimore: 471-497.
6.- Dujardin E. 1976; Plant Sci. Letters 7, 91-94.
7.- El Hamouri B, Sironval C, 1980; Photobiochem. and Photobiophys. 1, 219-223.
8.- Griffiths W T, 1978; Biochem. J. 174, 681-692.
9.- Oliver R P, Griffiths W T. 1982 ; 70, 1019-1025.
10. Kraus G H, Heber U. 1976, in "The Intact Chloroplast", Barber J, ed. Elsevier Scientif. Publ. Company. Amsterdam, New-York, Oxford, 2, 268-287.
11. Sironval C, Brouers M, Michel J M and Kuiper Y. 1968; Photosynthetica 2, 268-287.
12. Sironval C, Franck F, ıGysembergh R, Bereza B, Dujardin E.1984, in "Protochlorophyllide reduction and greening", Sironval C, Brouers M, eds. Martinus Nijhoff, Dr Junk W, Publishers, The Hague, The Netherlands, this volume.

ACKNOWLEDGMENTS

This work has been supported by Funds of the Technical University (nr R.1.9.02.03) Wroclaw, Poland; by the FNRS (Fonds National de la Recherche Scientifique Belge); by the "Actions Concertées" nr 80/85-1983, Belgium, and by a grant of the "Patrimoine de l'Université de Liège" to B.Bereza.

THE PHOTOREDUCTION OF ZINC-PROTOCHLOROPHYLLIDE BY ISOLATED
ETIOPLAST LAMELLAE.

P. BOMBART, E. DUJARDIN.
Photobiology Laboratory; Liège University, B22, B4000 Sart-
Tilman, Belgium.

INTRODUCTION

The experiments of Griffiths have shown that the presence
of a central Mg^{2+} ion in the tetrapyrrole ring is required to
observe the photoenzymatic reduction of protochlorophyllide
(Griffiths, 1980). As the replacement of Mg^{2+} by Zn^{2+} does
not modify greatly the absorption spectrum of Zn-protochlorophyllide (figure 1) and the fluorescence properties of the
pigment (Bombart, 1979; Bombart et al., 1983), it may be
thought that the physico-chemical properties of Zn-protochlorophyllide (Zn-Pide) are sufficiently close for this pigment
to behave as an analogue of Mg-protochlorophyllide (Mg-Pide)
towards the photoreductase.

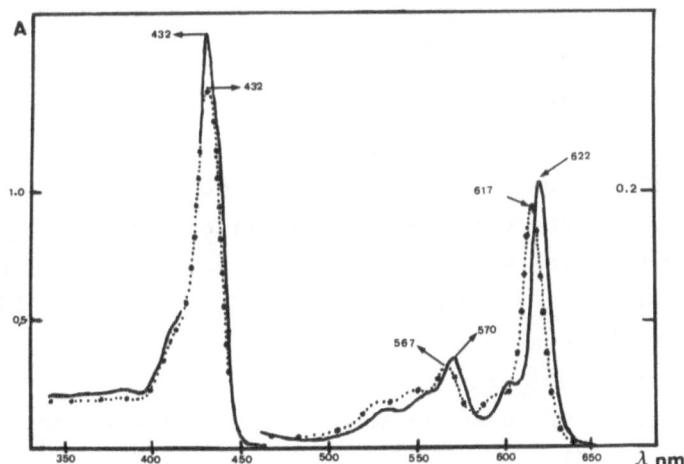

FIGURE 1. Absorption spectra of purified Mg-protochlorophyllide (full line) and Zn-protochlorophyllide (dotted line) in
ether (Bombart, 1979).

C. Sironval and M. Brouers (eds.); Protochlorophyllide Reduction and Greening. ISBN 90 247 2954 8
©1984, Martinus Nijhoff/Dr W. Junk Publishers, The Hague/Boston/Lancaster.

For this reason, it is tempting to cheat the photoenzyme by replacing its natural substrate by Zn-protochlorophyllide and to test the formation of the corresponding Zn-chlorophyllide (Zn-Cide) in the light. Griffiths (1980) has incubated etioplast lamellae in a medium to which NADPH and Zn^{2+}, Ni^{2+}, Cu^{2+} or Co^{2+}-substituted protochlorophyllides, or protopheophorbide had been added. Only the Zn-Pide did yield a compound having the spectral characteristics of a chlorophyllide.

Griffiths has drawn the conclusion that the pigment was Zn-chlorophyllide on the basis of the position of the red absorption maximum. This might not seem sufficient, however. For proving the validity of this conclusion, it is worth providing additional data. In order to photosynthesize Zn-chlorophyllide, etioplast lamellae should form first a photoactive complex comprising the photoreductase, Zn-Pide and NADPH. This complex should absorb light and emit a fluorescence at shorter wavelengths than the corresponding Mg-Pide-protein, and the same shift should be shown by the Zn-chlorophyllide appearing on illumination.

We show in this paper that two photoactive Zn-protochlorophyllide-protein complexes are formed in etioplast lamellae. We provide a chemical evidence that they produce Zn-chlorophyllide when illuminated.

MATERIAL AND METHODS

Plant material

Bean (Phaseolus vulgaris Var. Commodore)were grown at room temperature in the dark on a vermiculite/perlite (1:1)mixture moistened with tap water. All the manipulations were made with a green safelight (transmission around 520 nm). The primary leaves were harvested after 12 days.

Preparation of etioplasts

The etioplasts were isolated by the method described by Griffiths (1974).

Preparation of etioplast lamellae

The etioplasts were suspended in 3.3×10^{-2} M Tris-HCl buffer, pH:7.65, containing 0.6 M sucrose, 4.10^{-3} $MgCl_2$, 10^{-3} M EDTA-Na_2; BSA (fraction V) 0.2% (w/v); glycerol 10% (v/v). The suspension was diluted at 4°C with 9 volumes of distilled water, and allowed to rest during 5 minutes for completion of the lysis of the etioplasts. After centrifugation at 18000 g during 30 min, the lamellae were resuspended in 2-3 ml of Tris-HCl buffer and centrifuged again at 18000 g during 30 min. The pellet was stored in the dark at -20°C until utilization.

Preparation and purification of Mg-Pide and Zn-Pide.

a.- Mg-protochlorophyllide.

The pigment was extracted by soaking the leaves from 14-day-old, dark-grown seedlings during 24 h in diethylether. After evaporation of the solvent, the pigment was washed with n-hexane following the procedure described by Bombart et al (1981). The pigment was used without any further purification.

b.- Zn-protochlorophyllide.

Zn-Pide was prepared following the method of Führhop and Smith (1975).

Solubilization of Mg-Pide and Zn-Pide.

The pigment were dried and dissolved in a cholate/methanol solution (2 mg cholate per ml methanol). The methanol was vacuum evaporated during several hours (10^{-2} mm Hg).
The residue was dissolved in Tris-HCl buffer, pH:7.65; the pigment concentration was about 1 to 2 mM.

Preparation of Pide-depleted lamellae.

The lamellae stored at -20°C in the dark were thawed up to 5°C and suspended in Tris-HCl buffer, pH:7.65. After addition of Triton X-100 (either 60.10^{-3}% or 1.10^{-3}%) and 1 mM NADPH to the suspension, the protochlorophyllides were converted into chlorophyllides by illumination with repeated 1 ms polychromatic flashes. The lamellae usually appeared to be depleted from endogenous protochlorophyllide after 6 to 12 flashes, as controlled by recording the 77 K fluorescence spectra (= "depleted lamellae").

Experimental scheme.

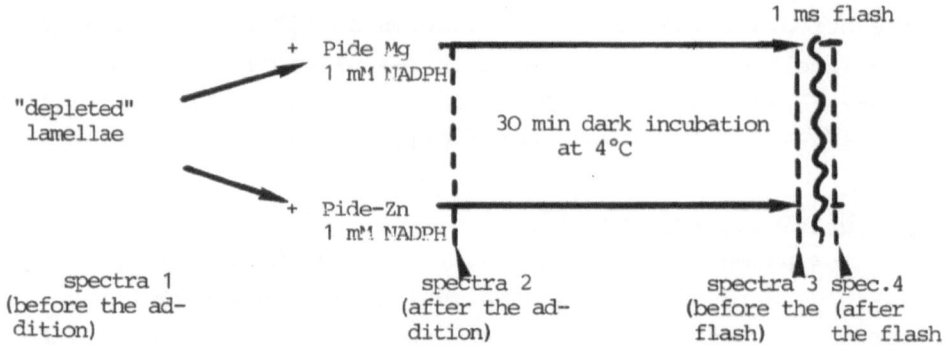

Illumination of the lamellae.

A polychromatic flash of 1 ms duration (Multiblitz Report Porba; electric energy 125 J, colour temperature 5800 K) was used. It was fixed at 40 cm from the suspension of lamellae.

Acetone extract of the lamellae.

1 ml of Zn-Pide or Mg-Pide-treated suspension of etioplast lamellae was diluted with 4 ml acetone. The mixture was clarified by a 60 min centrifugation at 48000 g.

Selective pheophytinization of the acetone extracts.

A KCl-HCl buffer 0.6 M, pH:2.2, was added to the suspension of lamellae until a pH value of 3.0 was reached. The suspension was incubated at 4°C in the dark during 30 min.

Registration of the 77 K fluorescence spectra.

The sample were fixed in liquid nitrogen and the fluorescence spectrum was recorded as previously described (Sironval et al., 1968; Brouers and Sironval, 1974). The spectra were not corrected for the response of the photomultiplier (EMI 9558B, S-20 response).

Registration of the low-temperature absorption spectra.

The method for recording the low-temperature spectra has been published elsewhere (Brouers and Sironval, 1974).

RESULTS

Formation of photoactive Zn-Pide-protein complexes.

Our first aim was to demonstrate that protochlorophyllide-protein complexes were formed when etioplast lamellae are incubated with Zn-Pide in the presence of NADPH.

The "depleted lamellae" (i.e. in which all the endogenous Mg-Pide had been exhausted by repeated flash illumination) were initially characterized by the low-temperature fluorescence spectra 1, fig.2. The stock of these lamellae was divided into two parts supplied with 1 mM NADPH, plus either exogenous Mg-Pide, or Zn-Pide. Drops of each part were frozen in liquid nitrogen just after the addition of nucleotide and pigment for recording absorption and fluorescence spectra 2, fig.2. The suspensions were incubated in complete darkness at 4°C during 30 min. A new series of drops were then frozen again in liquid nitrogen for recording spectra 3 (see experimental scheme in Material and Methods).

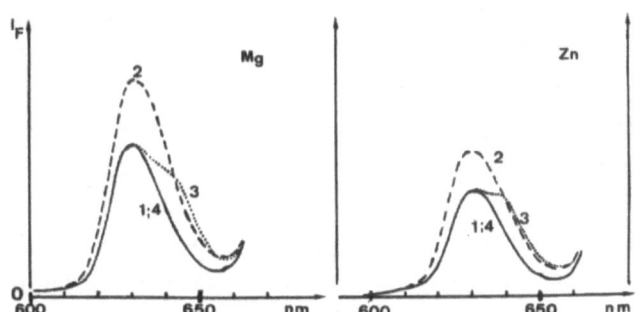

FIGURE 2. 77 K fluorescence spectra of suspensions of etioplast lamellae incubated at 4°C in the dark.
1. After depletion of the endogenous Mg-Pides.
2. Immediately after addition of exogenous Mg-Pide (left) or Zn-Pide (right) in presence of 1 mM NADPH.
3. After a 30 min dark incubation at 4°C.
4. After a 1 ms polychromatic flash.

Compared with spectra 2, spectra 3 show a decrease of the fluorescence emitted at 630 nm, and an increase of the fluorescence around 640-650 nm for the suspension supplied with Mg-Pide, or around 635-645 nm for the suspension supplied with

Zn-Pide. The decrease at 630 nm may be explained by some reorganization favouring energy transfers, by a destruction of
free protochlorophyllide, or by the complexation of the free
pigment to the enzyme.As spectral characteristics of protochlorophyllide-proteins are clearly seen in spectrum 3, the
third alternative applies certainly to a part of the pigment
pool.

This conclusion was confirmed by testing the effect of light.
A 1 ms polychromatic flash supressed completely the shoulders
around 645 nm (Mg-Pide) or around 640 nm (Zn-Pide) (spectra 4,
fig.2). The difference spectrum (light-dark) (not shown) also
indicates that spectral species emitting around 638 nm had
disappeared in the Zn-Pide supplemented lamellae.

Two Mg-Pide and two Zn-Pide-photoactive complexes.

It was found that the complexation of the exogenous pigment was better observed when the concentration of detergent
was decreased in the medium. The level used originally by
Griffiths (0.06%) favoured a spectral shift towards shorter
wavelengths. Using Triton X-100 1.10^{-3}%, we obtained the difference spectra of figures 3a (absorption) and 3b (fluorescence). These difference spectra have been calculated by substrac-
ting spectrum (2), recorded immediately after a flash given
at the end of the 30 min dark incubation , from the spectrum 3,
recorded just before this flash (see the experimental scheme
in Material and Methods).   The differences (figure 3,
full line) prove that two spectral species disappeared when
illuminating the suspensions, both in the case of Mg-Pide and
of Zn-Pide. New emission and absorption bands appeared due to
the formation of chlorophyllides.

The two Mg-Pide-protein complexes, designated as $P^{Mg}_{655-650}$
and $P^{Mg}_{645-640}$, had the same spectral characteristics as in the
etiolated leaf. The Mg-chlorophyllide seen after illumination
emitted at 692 nm and absorbed at 678 nm   ($C^{Mg}_{692-678}$).The emission difference located at 670 nm in fig. 3b (full line) originates from some of the Mg-chlorophyllide remaining after
exhaustion of the endogenous protochlorophyllide from the lamellae (see Material and Methods). The disappearance of this

fluorescence after flashing is presumably due to an increased
energy transfer towards the long-wavelength chlorophyllide formed by the flash.

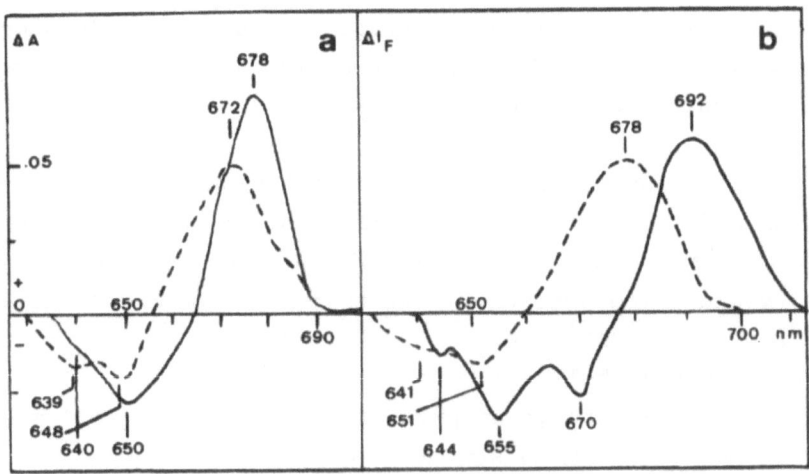

FIGURE 3. Difference (a) absorption and (b) fluorescence spectra of a suspension of etioplast lamellae incubated during 30
min at 4°C in the dark, either with exogenous Mg-Pide,(full
line), or with Zn-Pide (dashed line). The spectra represent
the difference between the spectra recorded at the end of the
incubation period immediately after a 1 ms flash, and the spectra recorded just before the flash.(see experimental scheme in
Material and Methods).

Two photoactive Pide-complexes were also distinguished in
the suspension supplied with Zn-Pide. We label them $P_{650-645}^{Zn}$
and $P_{640-635}^{Zn}$ respectively. These positions have been determined
in a series of repetitions of the experiment. They represent
approximate mean values. These complexes are transformed by
light into a chlorophyllide emitting at 678 nm and absorbing
at 672 nm (dotted line, figs 3a and 3b).

Repeated synthesis of Zn-Cide.

The method used in the previous section yielded a too low
amount of a chlorophyllide presumed to be Zn-chlorophyllide to
allow any senseful chemical analysis. Therefore, we repeated
the illuminations (by repeating the flashes) in order to accumulate more chlorophyllide. In this way, a repeated synthesis
of chlorophyllide was observed in the lamellae to which Zn-Pide
had been added. These lamellae were able to renew the synthesis

several times accumulating chlorophyllide in the suspension.

Chemical proof of the synthesis of Zn-Cide.

The extraction of the pigments in organic solvents from aqueous suspensions containing detergents (Triton X-100 and cholate) did not yield any extract suitable for chromatography. The pigments did not separate properly on the chromatographic plates. This cicumstance obliged us to work out another type of proof.

We took advantage of the different behaviour of Mg-Cide and Zn-Cide against mild acidification. It is well known that the central ion is removed from Mg-Cide by lowering the pH to 3.0, yielding pheophorbide, but that the same treatment does not remove the Zn^{2+} ion.

We proceeded as follows:

1.- A four fold volume of pure acetone was added to 1 ml of the suspensions previously flashed in the presence of Mg-Pide or Zn-Pide in excess and NADPH.

2.- The suspension was clarified by centrifugation.

3.- The acetone supernatants containing either Mg-Cide, or Zn-Cide plus some Mg-Cide, were acidified at pH:3.0, at 4°C during 30 min.

The acetone extract from the Mg-Pide supplemented suspension had the absorption spectrum 1, fig.4b, with red maximum at 665 nm. Spectrum 2 is the spectrum of the same extract after acidification. The acidified pigment is pheophorbide with an absorption band extending sligthly to the longer wavelengths. The ratio between the absorbancies at 665 nm before (spectrum 1) and after acidification (spectrum 2) is 1.7; this value is near the ratio of the extinction coefficients of chlorophyll a and pheophytin in ether (1.6). Some amount of Mg-Pide was also converted into protopheophorbide. This pigment originates from a rest of the substrate added in excess (spectrum 3, fig. 4b).

The non acidified acetone extract from the Zn-Pide supplemented suspension gave absorption spectrum 1, fig.4a. Spectrum 2a was recorded after acidification. In this case the decrease of the absorbance at 665 nm was less important than in the case

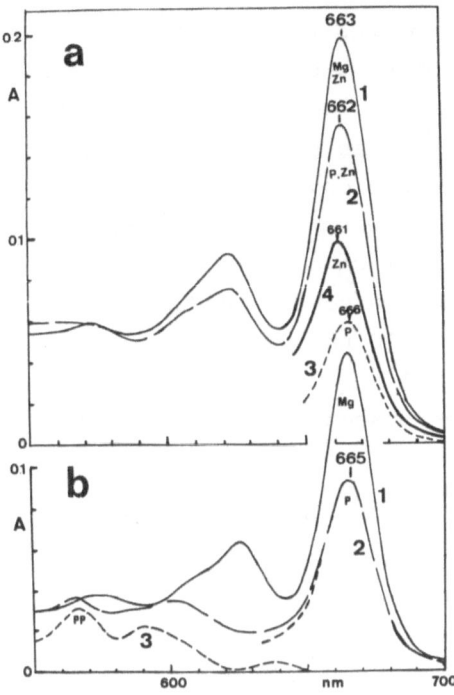

of the extract containing Mg-Cide only. The ratio between the red absorbance before acidification and the red absorbance after acidification amounted to 1.3, indicating that some of the chlorophyllides were not pheophytinized.

FIGURE 4.Absorption spectra of acetone extract:

1: extract from etioplast lamellae incubated either with Mg-Pide (b) or Zn-Pide (a) and 1 mM NADPH, and illuminated.

2a and 2b: the same extracts after incubation at pF:3.0 at 4°C during 30 min. •

3a: spectrum of the pheophorbide present in the Mg-treated extract.

3b: spectrum of protopheophorbide.

4a: spectrum of Zn-chlorophyllide (= spectrum 2a - spectrum 3a).

The red absorption band of Zn-Cide with maximum at 661 nm in acetone was obtained by subtracting the spectrum of pheophorbide from the spectrum of the mixture (pheophorbide + Zn-Cide), i.e. by subtracting spectrum 3a from spectrum 2a. This proves that the acid-stable Zn-Cide had been synthesized in the light by the lamellae supplemented with Zn-Pide. Zn-Cide amounted to about 60% of the total chlorophyllide in the acetone extract.

The behaviour of the newly synthesized Mg and Zn-Cides.

It is known that the addition of NADPH favours the formation of long-wavelength-absorbing Mg-chlorophyllides in isola-

ted etioplasts or etioplast membranes. (El Hamouri, Sironval,
1980; Oliver, Griffiths, 1982). Indeed, in our experiments,
most of the Mg-chlorophyllides did first absorb and emit at
long wavelengths.

However, in the suspensions containing a high concentration of Triton X-100 (60×10^{-3}%) the chlorophyllides underwent
later a blue shift. Figure 5 compares the 77 K fluorescence
spectra of etioplast lamellae after the synthesis of Mg-Cide
(left) and Zn-Cide (right) with the spectra after a subsequent
dark incubation for 30 min at 4°C, in presence of 1 mM NADPH.

At the beginning of the incubation, spectra 1 show important emissions at 680 nm and at 690 nm in the suspension containing Mg-Cide (left). The longer wavelength emission is reduced to a shoulder in the suspension containing Zn-Cide (right).
Most of the pigments emitted at 680 nm in the Zn-Cide containing suspension. The half bandwidth was smaller in this case
(23 nm) than in the case of the suspension containing Mg-Cide
(27 nm).

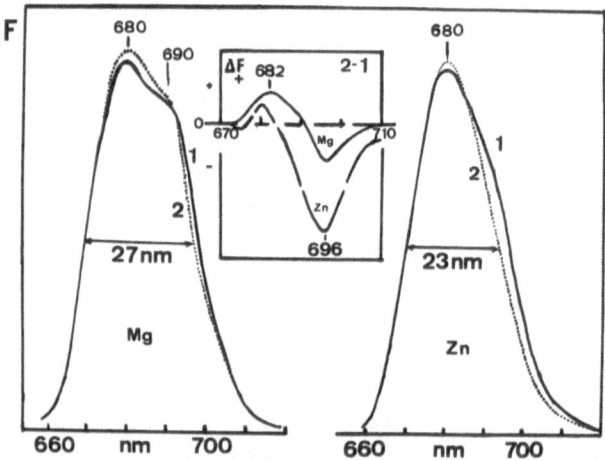

FIGURE 5. Fluorescence spectra of the chlorophyllides in etioplast lamellae incubated in the dark at 4°C during 30 min after
illuminating in presence of either Mg-Pide (left) or Zn-Pide
(right), and NADPH 1 mM. 1: spectra recorded immediately after
the flash. 2: spectra recorded 30 min after the flash.
Insert: difference fluorescence spectra (= spectra 1 minus spectra 2). Detergent concentration 60×10^{-3}%.

After a 30 min dark incubation at 4°C part of the long-wavelength-emitting Mg-chlorophyllides were transformed in the dark into short-wavelength-emitting Cides, as demonstrated by the difference spectrum "Mg" in the insert, fig.5. In the suspension containing Zn-Cide plus Mg-Cide the disappearance of the long-wavelength chlorophyllides was particularly obvious. However, in this case only a minute emission appeared at shorter wavelengths (at 680 nm; see the difference spectrum:"Zn" in the insert, fig.5).

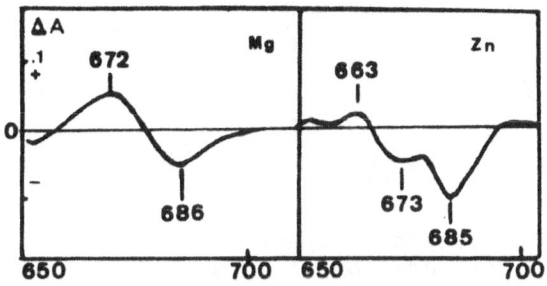

FIGURE 6. Difference absorption spectra of the suspensions of etioplast lamellae incubated either with Mg-(left) or with Zn-Pide and 1 mM NADPH during 30 min at 4°C after flashing both suspensions. The difference is : spectrum recorded immediately after the flash minus spectrum recorded at the end of the dark incubation. Detergent concentration $60x10^{-3}$%.

Similar facts were seen in the absorption spectra (fig.6): long-wavelength-absorbing chlorophyllides disappeared in both suspensions during the dark incubation (fig.6 a and b);in the Mg-Pide supplemented lamellae, a shorter wavelength chlorophyllide absorption band was formed (at 672 nm); but in the Zn-Pide lamellae, the long wavelength chlorophyllide disappeared without formation of a shorter wavelength chlorophyllide.

Thus, in the presence of a high detergent concentration ($60x10^{-3}$%),the behaviour of the chlorophyllides was very different when incubating lamellae in which either Mg-Cide or Zn-Cide had been formed;in the latter case, the pigments were apparently destroyed.

It must be emphasized that, when the detergent concentration was sufficiently low (1.10^{-3}% , in the medium) the results differed from what was seen with a high detergent concentration.

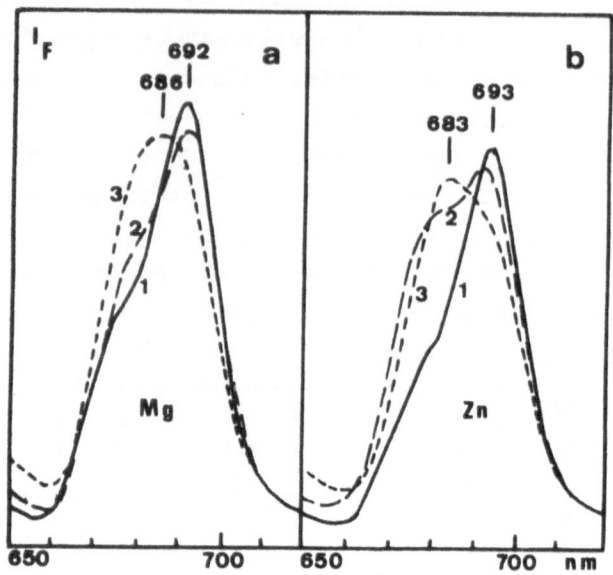

FIGURE 7. 77 K fluorescence spectra of depleted lamellae .
1(full line): after addition of 1 mM NADPH plus either Mg-Pide
(a) or Zn-Pide (b). 2 (interrupted line   ): after repeated flash illuminations. 3(dashed line): after a subsequent
3 h dark incubation at 5°C. Detergent concentration 1×10^{-3}%.

Spectra 1, fig.7, reproduce the emissions of the chlorophyllide seen in this case, using dark depleted lamellae supplemented with NADPH and either Mg-Pide (1a), or Zn-Pide (1b);
the main emission peaks at 692-93 nm in both suspensions,with
a shoulder at shorter wavelengths. Spectra 2 were recorded
after repeated illuminations: in the Mg-supplied suspension,
the short-wavelength shoulder at 686 nm increased at the expense of the 692 nm emission; in the Zn-Pide-supplied suspension, the main emission stayed at 691-692 nm, while the short-
wavelength shoulder increased around 683 nm. Spectra 3 were
registered after a 3 h dark incubation at 5°C: they show that
both Mg-Cides and Zn-Cides behaved in the same way; the main
emission was found at shorter wavelength (686 nm in the suspension containing Mg-Cide, at 683 nm in the suspension containing Zn-Cide), although the contributions of the long-wavelength chlorophyllides remained important.There was no destruction of the pigments.

The emissions at 686 nm or 683 nm did not shift towards shorter wavelengths in this experiment with low detergent concentration. When the detergent concentration was high, the emission peaked at 680 nm in both Mg-Cide or Zn-Cide suspensions. This indicates clearly a deleterious effect of a too high Triton X-100 concentration on the organization of the chlorophyllides inside the lamellae.

DISCUSSION.

A.- We have found two complexes between exogenous Mg-protochlorophyllide and the photoreductase in etioplast lamellae incubated in darkness with Mg-Pide and NADPH. These complexes have the following properties:

1.- the main complex absorbs at 650 nm and emits a low-temperature fluorescence at 655 nm ($= P^{Mg}_{655-650}$).

2.- the second complex absorbs at 640 nm and its 77 K fluorescence is maximal at 645 nm (= $P^{Mg}_{645-640}$).

3.- when illuminated, both $P^{Mg}_{655-650}$ and $P^{Mg}_{645-640}$ yield a chlorophyllide absorbing around 678 nm and emitting at 692 nm ($C^{Mg}_{692-678}$).

The behaviour of these complexes is essentially the same as the behaviour of $P_{657-650}$ and $P_{645-640}$ in the etiolated leaf.

B.- The formation of two complexes between exogenous Zn-protochlorophyllide and the photoreductase has been demonstrated in etioplast lamellae incubated with Zn-Pide and NADPH. The low temperature absorption and fluorescence maxima of these complexes are shifted towards shorter wavelengths, as expected, in comparison with the Mg-complexes:

1.- the main complex absorbs at 645 nm and emits a low-temperature fluorescence at 652 nm (= $P^{Zn}_{650-645}$).

2.- the second complex absorbs at 635 nm and its 77 K fluorescence peaks at 640 nm (= $P^{Zn}_{640-635}$).

3.- both $P^{Zn}_{650-645}$ and $P^{Zn}_{640-635}$ are phototransformed into Zn-chlorophyllide absorbing at 672 nm and emitting at 678 nm ($C^{Zn}_{678-672}$).

We have observed these complexes only when lowering drastically the detergent concentration . The concentration used by Griffiths (1980) provoked a shift of the Mg-and Zn-proto-chlorophyllides towards shorter wavelength. Then, the Mg-or Zn-Pide complexes absorbing and emitting at longer-wavelengths were not preponderant, or even were absent in our preparations.

C. We give a chemical proof that Zn-chlorophyllide is synthesized in etioplast lamellae from exogenous Zn-Pide, as the shorter wavelength-absorbing Zn-chlorophyllide resisted to acid incubation at pH 3.0. The lamellae could accumulate repeatedly up to 60% Zn-Cide among the total chlorophyllide. Griffiths reached a similar result, but his conclusion was based on the characteristics of the visible spectrum of the suspension (i.e. position of the red absorption maximum).

D.- The chlorophyllides of the lamellae suspensions containing a high amount of detergent underwent a blue shift and did never show any important contribution of long-wavelength-absorbing or emitting components (figs. 5a, 5b; 6a,6b). The blue shift is apparently linked to the damaging of the structure of the lamellae. This blue shift was also observed by Griffiths (Oliver, Griffiths, 1982). When the lamellae produced Zn-Cides, it seemed that a release of free Zn-Cide and a destruction of the chlorophyllides occured (see figs 5 and 6b).

E.- We have seen that the blue shift and the other deleterious effects did not occur, at least not to a detectable extent, if the detergent concentration was sufficiently low. In this case, the Pide-photoreductase did not distinguish between the Zn^{2+} and Mg^{2+} ion for the binding of the substrate protochlorophyllide, for the reduction process and for the binding of the newly formed chlorophyllide. This means that the enzyme is capable to handle the Zn-analogue in the same way as the natural substrate.

Further experiments are needed to study the fate of Zn-Cide in the lamellae. Is this pigment capable to play the same role as Mg-Cide once it has been synthesized in the lamellae? The answer could help to solve the fundamental question: why did nature select Mg^{2+} so exclusively and eliminate Zn^{2+} so completely from the chlorophylls which function in photosynthesis ?

ACKNOWLEDGMENTS

The authors thank the F.N.R.S. (Fonds National de la Recherche Scientifique), the Ministère de la Politique Scientifique Belge, and the "Actions Concertées nr 80/85-18.1983, Belgique", for financial support.
P. Bombart is a Research Associate at the F.N.R.S.

REFERENCES

1. Bombart P. 1979, Mémoire de Licence Sc. Bot. Université de Liège, 1979 - p.20-84.
2. Bombart P, Dujardin E and Brouers M. 1981, in "Photosynthesis V Chloroplast Development" , Akoyunoglou G ed. Balaban Intern. Science Services, Philadelphia, Pa, pp.87-92.
3. Bombart P, Dujardin E, Grandjean J., Laszlo P and Sironval C. 1983, Proceedings of the 6th Photosynthesis Congress, in press.
4. Brouers M, Sironval C. 1974, Plant Sc. Lett. 2, pp.67-72.
5. El Hamouri B, Sironval C. 1980, Photobiochem. Photobiophys. 70, pp.1019-1025.
6. Führhop J H and Smith K M. 1975, in "Porphyrins and Metalloporphyrins " K M Smith ed. Elsevier Scientific Publishing Company, Amsterdam,The Netherlands, pp.757-869.
7. Griffiths W T. 1974 , FEBS Lett. 46 (1), pp. 301-304.
8. Griffiths W T. 1980, Biochem. J. 186, pp 267-278.
9. Oliver R P, Griffiths W T. 1982, Plant Physiol., 70, pp 1019-1025.
10.Sironval C, Brouers M, Michel J M, Kuiper Y. 1968, Photosynthetica, 2, pp.268-287.

A MODEL FOR THE REDUCTION OF PROTOCHLOROPHYLL(IDE) CONTAINING
THREE DIFFERENT FORMS OF PHOTO-ACTIVE PROTOCHLOROPHYLL(IDE)

J.C.J.M. VAN DER CAMMEN AND J.C. GOEDHEER
Phys. Lab., Princetonplein 5, 3508 TA Utrecht,The Netherlands.

INTRODUCTION

In order to interpret the results of our fluorescence
lifetime studies, we adapted and extended the existing models
for the reduction of protochlorophyll and arrived at a
reduction scheme, which contains three different photo-active
forms of protochlorophyll(ide) [4]. In this paper we compare
this scheme with some recent findings of other investigators
and present an extended version of our model, with which these
latest results can also be explained.

DISCUSSION

In order to interpret the results of our initial
spectral lifetime measurements, we postulated the model of
hydrogen bridges for the connection between protochlorophyll-
(ide) and its carrier [1]. At that moment there were three
important arguments in favour of this model.

1. The hydrogen atoms, required for the reduction of proto-
   chlorophyll(ide), have to be situated in close proximity
   to the protochlorophyll(ide) molecule, since at room
   temperature and under sufficient light, the reduction
   proceeds within milliseconds [2].

2. Protochlorophyll(ide) has to be connected to a site of
   limited size, as the initial formation of chlorophyll-
   (ide) upon illumination of etiolated leaves and the
   subsequent resynthesis of new protochlorophyll(ide) take
   place in fixed quantities [3].

3. After hydrogenation of protochlorophyll(ide), the newly
   formed chlorophyll(ide) moves away from its original
   site [3], which implies that coincident with or shortly
   after the reduction, protochlorophyll(ide) has to be
   released from its site.

C. Sironval and M. Brouers (eds.); Protochlorophyllide Reduction and Greening. ISBN 90 247 2954 8
© 1984, Martinus Nijhoff/Dr W. Junk Publishers, The Hague/Boston/Lancaster.

The results of further fluorescence lifetime investigations as a function of wavelength made it necessary to extend the initially proposed model for the reduction of protochlorophyll(ide). We arrived at the model which is given in Scheme I of [4]. Support for this model is provided by the recent findings of Dobek et al [5]. They measured the room temperature fluorescence spectrum of protochlorophyllide *in vivo*, which was found to have emission bands at 633, 642·, 645, 649 and 657 nm. We ascribe the bands at 633, 645, 642 and 657 to respectively the monomeric photo-inactive P(628,630), photo-active P(635) and the dimeric species PP(639,642) and PP(650,656).

For the interpretation of the emission at 649 nm, we consider the *in vitro* results of El Hamouri and Sironval [6,7]. These authors report that a certain amount of the inactive protochlorophyllide-protein complex P(642,649) is formed, besides inactive P(628,633), simultaneously with the disappearance of the photo-active P(650,657) and P(637,645) complexes during the incubation of a suspension of cucumber etioplasts in the dark at 283 K. Addition of $NADP^+$ to the incubate favours the formation of P(642,649). The photo-inactive forms could be restored to the photo-active forms by adding NADPH to the incubate. These results are in close agreement with the findings of Griffiths [8-10], who demonstrated the direct role of NADPH in providing the reductant for chlorophyllide synthesis. Griffiths concludes that the etioplasts' photo-active protochlorophyllide should be a ternary complex of protochlorophyllide reductase-NADPH + protochlorophyllide.

The above mentioned results strongly suggest that P(642,649) represents the "free" dimeric form of protochlorophyllide.

Under natural conditions the formation of PP(650,655) is a result of ageing and seems to proceed via P(635,645) and PP(639,642). In 2 day old etiolated leaves one finds mainly P(635,645), in 3 day old etiolated leaves one finds both P(635,645) and PP(639,642) and in 5 day old etiolated

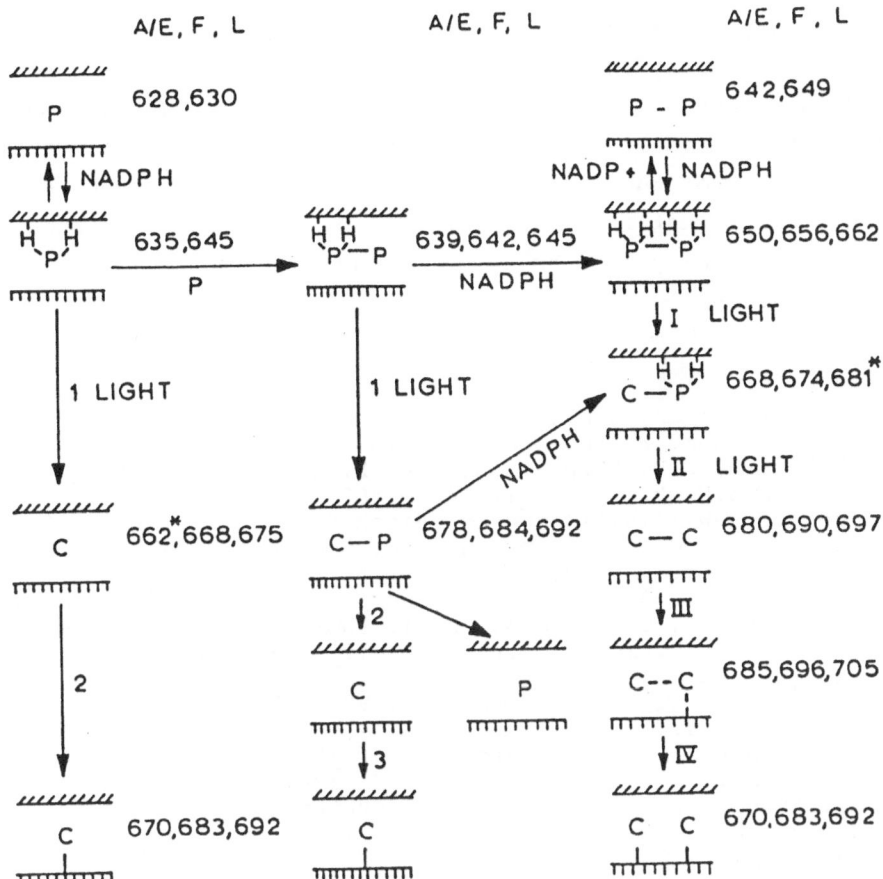

SCHEME I. Model for the reduction of protochlorophyll(ide).
P, Protochlorophyll(ide); H, Hydrogen; C, Chlorophyll(ide).
Oblique and vertically hatched: two different protein or
membrane sites. Drawn conclusions mean stronger bonds than
interrupted lines. Along each picture the wavelengths
(in nm) are given of absorption/excitation (A/E), fluorescence
intensity (F) and fluorescence lifetime (L) respectively.
*Wavelength estimated.

leaves the definite form PP(650,655) dominates [11,12].
PP(639,642) if probably formed by the association of
P(635,645) and P(628,630), and the formation of PP(650,655)
proceeds from PP(639,642) by the action of NADPH.

Further, as Litvin et al [13] report the species
PC(678,684) as a precursor of CC(680,690), it seems probable
that instead of dissociation into P(628,630) and C(662,668)

PC(678,684) can also be transformed to PC(668,674) by NADPH.

On including the species PP(642,649) and the action of NADP$^+$ and NADPH in the model which we postulated in [4], we arrive at Scheme I. Scheme I accounts for all of our findings and for many of the findings of other authors. Although we do not claim that it is complete, we believe that this scheme can be fruitfully used as a working hypothesis for further investigations.

REFERENCES

1. Van der Cammen, J.C.J.M. and J.C. Goedheer (1982) Photobiochem. Photobiophys. 3, 159-165
2. Akulovich, N.K., T.N. Godnev and K.I. Orlovskaya (1970) Dokl.Akad.Nauk. SSSR 191, 1406-1409
3. Thorne, S.W. (1971) Biochim. Biophys. Acta 226, 128-134
4. Van der Cammen, J.C.J.M. and J.C. Goedheer (1982) Photobiochem. Photobiophys. 4, 145-152
5. Dobek, A., E. Dujardin, F. Franck, C. Sironval, J. Breton and E. Roux (1981) Photobiochem. Photobiophys. 2, 35-44
6. El Hamouri, B. and C. Sironval (1979) FEBS Lett. 103, 345-347
7. El Hamouri, B. and C. Sironval (1980) Photobiochem. Photobiophys. 1, 219-223
8. Griffiths, W.I. (1975) Biochem. J. 152, 623-635
9. Griffiths, W.I. (1978) Biochem. J. 174, 681-692
10. Griffiths, W.I. (1981) In: Photosynthesis V Chloroplast Development pp. 65-71 (G. Akoyunoglou,ed.) Balaban Int. Science Services, Philadelphia
11. Akulovich, N.K., T.N. Godnev and K.I. Orlovskaya (1970) Dokl. Akad. Nauk. SSSR 191, 1406-1409
12. Klein, S. and J.A. Schiff (1972) Plant Physiol. 49, 619-626
13. Litvin, F.F., N.V. Ignatov, E.I. Efimtsev and O.B. Belyaeva (1978) Photosynthetica 12, 375-381

THE FIRST PHOTOACTIVITIES APART
FROM PROTOCHLOROPHYLLIDE
PHOTOREDUCTION

THE FRANCK-INOUE CHLOROPHYLLIDE MICROCYCLE II *IN VIVO* AND *IN VITRO*

C. SIRONVAL, F. FRANCK, R. GYSEMBERGH, B. BEREZA* AND E. DUJARDIN*
Laboratory of Photobiology, Department of Botany,
4000 Sart Tilman, Liège, BELGIUM.

1. INTRODUCTION

K. Shibata (1957) opened a new avenue to research when, having introduced the opal glass technique, he recorded absorption spectra of etiolated and greening leaves. He discovered how these spectra were changing, and he showed that the chlorophyll(ide) forms formed in the light were not everlasting, but were replaced one by another. Since this discovery, a number of chlorophyll(ide) forms have been characterized in the etiolated leaf. The life-time of some of these chlorophyllides is short. Some are transformed reversibly into another; some are transformed irreversibly. These transformations need light in some cases. Their temperature dependence varies greatly.

The course of the events is conveniently described in a first approximation by the protochlorophyllide-chlorophyllide cycle (P-C cycle, fig. 1). This cycle is empirical as far as the successive pigments are designated by observable characteristic features : their spectra, their life-time and the order of their appearance as the leaf has started to be illuminated.

The function of producing chlorophyll in the light does not seem sufficient for justifying the arrangement of the cycle. A shorter path by-passes the P-C cycle, which fits more strictly this function (see Griffiths, this volume). This by-pass is most often observed *in vitro*. In the etiolated leaf,

*These authors have contributed to the second part of this paper (from section 3.2 onwards). B. Bereza was on leave from the Biophysics Laboratory of the University of Wroclaw, Poland.

C. Sironval and M. Brouers (eds.); Protochlorophyllide Reduction and Greening. ISBN 90 247 2954 8
© 1984, Martinus Nijhoff/Dr W. Junk Publishers, The Hague/Boston/Lancaster.

however, it is followed by a minority of pigments. In this
case, the great majority of the chlorophyllides are transformed into $P_{694-682}$ within 20 to 30 seconds and follow a longer
path. This suggests an additional function of the P-C cycle
pigments.

When an etiolated leaf is illuminated at room temperature
for the first time by continuous light, the intensity of the
fluorescence emitted at 690 nm by the chlorophyllides rises up
to a maximum, and then falls down to a steady level at half
of the maximum (Jouy and Sironval, 1979). The rate of the rise
is dependent on the intensity of the light; it is still considerable at temperatures of the order of -20°C. This rise is
due to the formation of $P_{688-678}$ as a result of protochlorophyllide reduction. By contrast the fall of the fluorescence
is suppressed completely below 0°C and does not depend so much
on light intensity. It is due to the transformation of $P_{688-678}$
into $P_{694-682}$, which may take 30 sec to 1 min after saturation of $P_{688-678}$ formation at room temperature. Absorption
spectra prove that $P_{694-682}$ is the main pigment of the leaf
after 1 to 2 min.

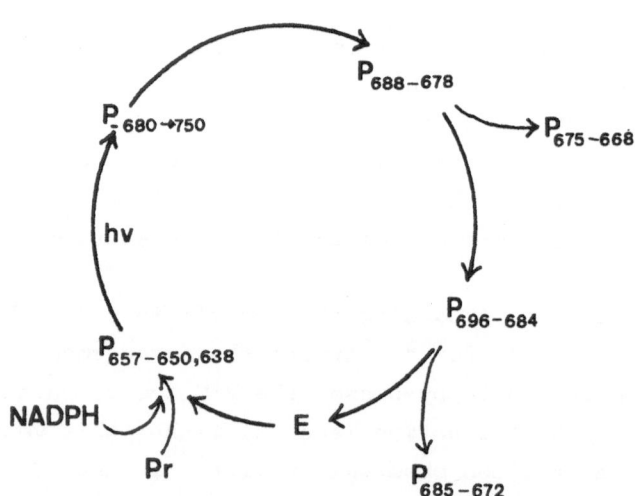

FIGURE 1. The P-C cycle in the leaf. Note that, in the present
paper, P696-684 has been called P694-682.

When a leaf with $P_{694-682}$ is left in darkness for a long
time, the Shibata shift transforms $P_{694-682}$ into $P_{685-672}$.
This process takes 30 to 40 min. In the light, however,
$P_{694-682}$ may have another fate. To observe this fate it suf-
fices to illuminate a leaf in which $P_{694-682}$ has been accumulated in the dark. Franck and Inoue have proved that in this
case $P_{694-682}$ is transformed back into $P_{688-678}$. There exists
a couple of reactions (Franck and Inoue, 1984) :

$$P_{688-678} \longrightarrow P_{694-682} \qquad \text{(a)}$$

$$P_{694-682} \xrightarrow{h\nu} P_{688-678} \qquad \text{(b)}$$

These reaction form a microcycle inside the P-C cycle. We
label it II (see the discussion).

In this paper we confirm the findings of Franck and Inoue.
We give further informations on the composition of the fluorescence emission of the leaf, and on the changes which it
undergoes at room temperature as the Franck-Inoue fluorescence
kinetics goes on. We show that this kinetics may be seen in
isolated etioplasts if an adequate supply of pyridine
nucleotides is provided to the medium.

2. MATERIAL AND METHODS

2.1. The leaves

Seedlings of *Phaseolus vulgaris* (var. Commodore in the first
part of this paper, var. Bor in the second part) were grown
on a 1:1 vermiculite/perlite mixture in complete darkness at
room temperature. When the plants were 12 to 15 days old, the
primary leaves were harvested in darkness and fixed to a holder
described in Sironval et al (1968).
They were illuminated using an intense photographic flash
(Multibliz Report Porba; electric energy 125 J) and were left at
room temperature for 2 min before the start of the experiments.

2.2. Simultaneous records of room temperature emission spectra and fluorescence kinetics

The holder was fixed in the device of fig.2 at the place
labelled H, where the leaf was illuminated by an He-Ne laser

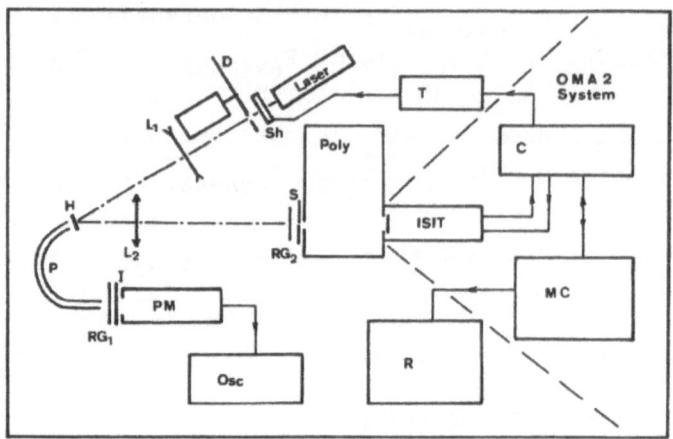

FIGURE 2. Block diagram of the device for recording simultaneously both the fluorescence spectra and fluorescence kinetics.
The fluorescence spectra emitted at H are displayed by the
polychromator (Poly) on the target of the ISIT detector of
the OMA 2 system. The fluorescence received by the photomultiplier (PM) is used to follow the kinetics on the memory
oscilloscope (Osc). The excitation light is furnished by a
He-Ne laser (8 mW). The ISIT, the controller (C) and the microcomputer (MC) are parts of the OMA 2 system, which dis-
plays the spectra on the xy recorder (R) - D : the rotating
disk - H : the leaf, or the cuvette containing the etioplast
suspension - P : a light pipe - The electronic shutter (Sh)
in front of the laser is triggered by T - RG1 and RG2 are
cut-off filters cutting the light below 664,5 nm and 662,5 nm
respectively. I is a 690 nm interference filter (large band
width). The lense L1 disperses the excitation light on the
leaf (or on the etioplast suspension). The lense L2 concen-
trates the fluorescence on the slit (S) of the polychromator.

(8 mW). The fluorescence was received by an ISIT detector
connected to an OMA 2 system (after dispersion by a polychromator, POLY), and by a photomultiplier (PM) connected to an
oscilloscope (OSC) (after filtration through an interference
filter, I). The laser light was cut off by filters RG_1 and
RG_2. The computer of the OMA 2 memorized the spectra and made
the calculations for normalized, difference and derivative
spectra.

The spectra were recorded at the scanning rate of 140 μs.
target channel^{-1}. The fluorescence of each single spectrum
was collected using 500 channels of the ISIT target during
350 ms (5 scans of 70 ms duration). Series of spectra were

also recorded at 6 sec intervals; in this case, 300 channels
were utilized, and the duration of the record of one spectrum
was 210 ms (5 scans of 42 ms duration).

2.3. Excitation of the fluorescence

Usually the fluorescence was recorded during the illumination of the leaf (or of the etioplast suspension) by the He-
Ne laser. In this case the excitation was due to the actinic
light which provoked the changes under investigation.
However, analytic chopped light was used in some cases. To
record an emission spectrum scans of 42 ms duration were then
repeated during the entire excitation period (10 to 20 sec).
The excitation light was chopped using the rotating disc (D)
which gave 30 μs flashes at a frequency of 330 Hertz.

2.4. Preparation of the etioplasts

The etioplasts were prepared from etiolated bean seedlings
of the Bor variety (12 to 15 day-old) following the method
described by Bereza and Dujardin (this volume).

The room temperature emission spectra were recorded as
indicated here above, the etioplast suspensions being introduced with a syringe into the spectroscopic cell (1 mm optical
path) which was illuminated at place H, fig. 2.

2.5. 77 K emission and absorption spectra

77 K emission spectra were sometimes recorded using the
OMA 2, in the conditions described in the previous sections
In this case, liquid nitrogen was poured into a Dewar around
the holder at place H, fig. 2.

Conventional 77 K emission spectra were registered from
600 to 780 nm using the apparatus described in Sironval et al(1968).
A mercury lamp equipped with a Wood 436 nm interference filter
provided the excitation light. 77 K absorption spectra were
recorded using a Cary 17 spectrophotometer (Dujardin, 1976).

3. RESULTS

3.1. Microcycle II in the etiolated leaf

3.1.1. The Franck-Inoue fluorescence kinetics. Figure 3
reproduces the kinetics of the fluorescence which was emitted
by a 14-day-old, etiolated leaf illuminated by a He-Ne laser

beam during 3 min at room temperature. The leaf had received
a lms, strong white flash, and then had stayed for 2 min in
darkness, in order to be enriched with $P_{694-682}$ before starting the experiment. This kinetics is composed of two parts :
an initial increase phase, from time zero to point M, and a
subsequent, slower decrease phase, to point S.

When such a leaf was plunged into liquid nitrogen at time
zero, or a short while before, the 77 K spectrum (I) of fig. 4
was observed. This spectrum is representative of the leaf
frozen at the end of the 2 min dark period. It shows that
$P_{694-682}$ was present at that time. A similar leaf plunged into
liquid nitrogen 36 s later emitted spectrum (M), fig. 4.

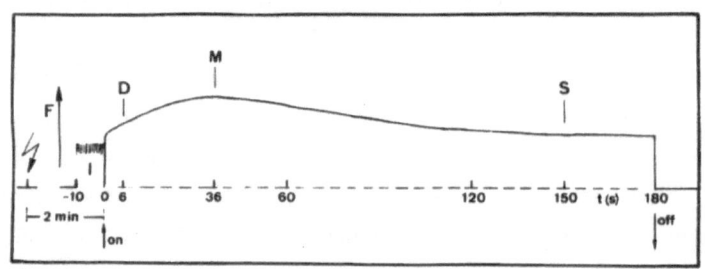

FIGURE 3. Kinetics of the intensity of the fluorescence emitted at 690nm by an etiolated leaf which had accumulated P694-
682 at time O. In ordinate, the fluorescence F; in abscissa,
the time t. Before time O the leaf received a strong lms flash
of white light (⚡) and then stayed for 2 min in darkness (=
preilluminated leaf). A pulsed laser light was given during
period I (10s) and this was followed by starting at time O
a continuous laser illumination. The fluorescence was seen by
a photomultiplier through an interference filter which transmitted the 690 nm light. Other explanations in the text.

The comparison of spectrum (M) with spectrum (I) shows that
the 77 K fluorescence shifted towards shorter wavelengths,
from 694 to 690 nm, during the initial increase phase of the
kinetics. Both this shift and the increase phase are related
to photoreaction (b) :

$$P_{694-682} \xrightarrow{h\nu} P_{688-678}$$

as Franck and Inoue (1984) have shown that the red absorption
band of a leaf with $P_{694-682}$ shifted from 682 to 678 nm during

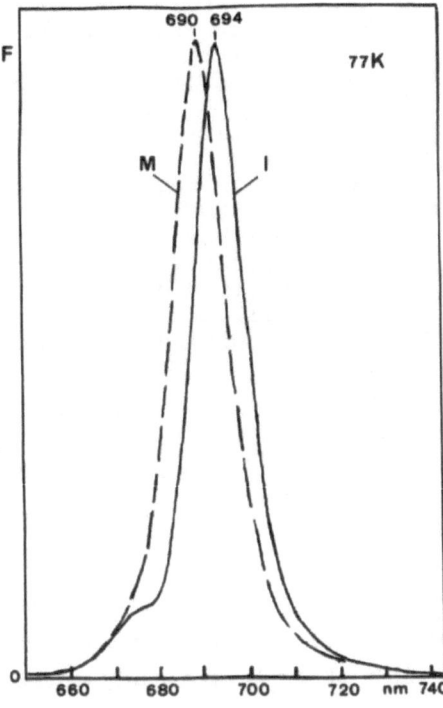

FIGURE 4. Emission spectra of preilluminated etiolated leaves plunged into liquid nitrogen .
Spectrum (I) : at the end of a 2 min dark period following a 1 ms strong flash of white light (= during period I, fig.3). Spectrum (M) : at point M of the kinetics, fig. 3.

this phase. Reaction (b) was found to be overcome by reaction (a) during the next phase, from (M) to (S).

3.1.2. <u>Changes in the composition of the room temperature emission</u>. Using an optical multichannel analyser, OMA 1, Franck and Inoue (1984) have shown further that the spectral shift seen in fig. 4 and its reversal, could be observed at room temperature as well. The analysis of the emission spectra which we have made using an OMA 2 has revealed, however, a greater complexity of the spectra at 293 K than at 77 K.

Figure 5 reproduces the emission spectra recorded at room temperature at the points labelled I, D, M and S in fig. 3. These spectra have been normalized to a same value of the emission area from 660 to 720 nm. It is obvious that the room temperature fluorescence shifted, first towards shorter wavelengths during the increase phase of the kinetics (from time zero to M), and then towards longer wavelengths during the decrease phase (from M to S). In order to get spectrum (I) the

204

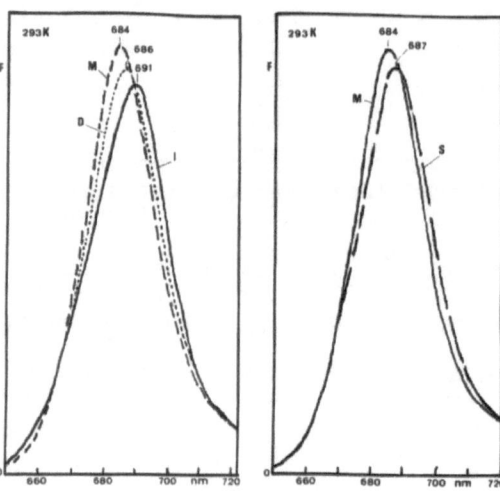

FIGURE 5. Room temperature fluorescence emissions recorded by the OMA 2 during period I and at the points D, M and S of the Franck-Inoue kinetics (fig. 3). The spectra have been normalized to a same total area from 660 to 720 nm. Spectrum I has been obtained by cumulating the emissions received during a chopped illumination extending over period I of the kinetics.

OMA 2 has been asked to cumulate the emissions due to a series of very weak flashes before the continuous laser illumination. This chopped excitation light did not allow to form any detectable amount of $P_{688-678}$ (by reaction b), and did not change the kinetics. Therefore, the fluorescence excited by this light reflected the state of the leaf enriched with $P_{694-682}$ before the onset of the illumination.

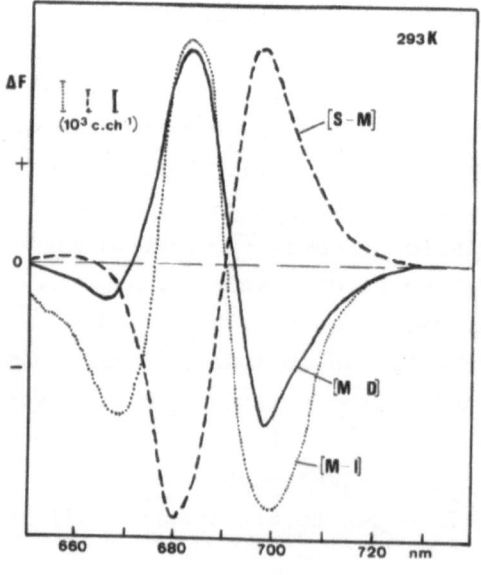

FIGURE 6. Differences between the room temperature spectra of fig. 5. These differences show how the composition of the emitted light changed from one of the spectra to another. The bars give the scale for a difference of 10^3 counts per 1 channel of the ISIT target after normalization of the spectra to 10^6 counts.

The comparison of any spectrum with any later spectrum
shows how the fluorescence changes as a function of time.
Comparing spectra normalized to a same area elicits composition changes of the emitted light. Such changes are given in
fig. 6 as difference spectra. From I to M, during the fluorescence increase phase, the emission became relatively more
intense in the wavelength region around 680nm, and relatively
less intense around 695-700nm. The reverse was seen from M
to S . The isosbestic region around 690nm shows that the
part which the 690nm light shared within the overall fluorescence remained nearly constant, implying that the kinetics of
the intensity of this light gave a reasonable picture of the
time dependence of the fluorescence. In fact, the intensities
of a series of spectra, as measured by their non-normalized
areas, matched the kinetics of fig.3.

3.1.3. <u>Components of the emission</u>. The 4th derivatives of
the spectra recorded at I, D and M, fig. 3, are reproduced
in fig. 7 and 8. These derivatives resolve the room temperature
spectra into three main components, in sharp contrast with the
corresponding 77K spectra which show one main component only.
The long-wavelength component of each room temperature spectrum
and the corresponding component of the 77K spectrum were found
at the same wavelengths : 694nm at the start of the fluorescence kinetics, and 690nm at the end of the increase phase.
As shown by Franck and Inoue (1984) this shift is due to the
transformation of $P_{694-682}$ into $P_{688-678}$. Two shorter wavelength components observed at room temperature are located
around 685 and 676nm respectively. They shifted also towards
shorter wavelengths during the increase phase of the kinetics,
but to a lesser extent than the longer wavelength components
(compare fig. 7 with fig. 8). As a result of the replacement
of the emission at 694nm by an emission at 690nm, the room
temperature spectrum became poorer in 695-700nm light during
the increase phase, as seen in fig. 6.

A concomitant event is demonstrated in fig. 9. This figure
shows the absolute difference between spectrum D , registered
6 s after the start of the continuous illumination, and spectrum

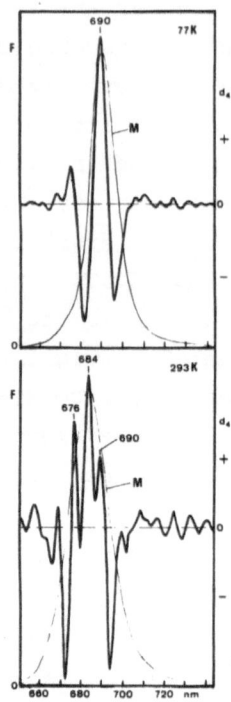

FIGURES 7 AND 8. Fourth derivative of fluorescence emission
spectra registered by the OMA 2 : below, at room temperature
at the points labelled D, M and during period I of the kinetics, fig. 1; above, after having plunged the leaf in liquid
nitrogen during period I and at point M.

M᾽ , registered 30 s later. The [M-D] difference is "absolute"
in the sense that it is concerned with non-normalized spectra.
The 4th derivative of this difference shows that the two short-
wavelength components were essentially involved in the increase
of the overall fluorescence besides the 690 component $(P_{688-678})$.
These components are more clearly defined in fig. 9 than in
figs. 7 and 8. The [M-D] difference yielded a very precise
spectrum, as it eliminated the background noise nearly completely. The emissions of the short wavelength components are
located at 673 and 681nm respectively; their contribution to
the [M-D] spectrum resulted in a peak at 681nm.

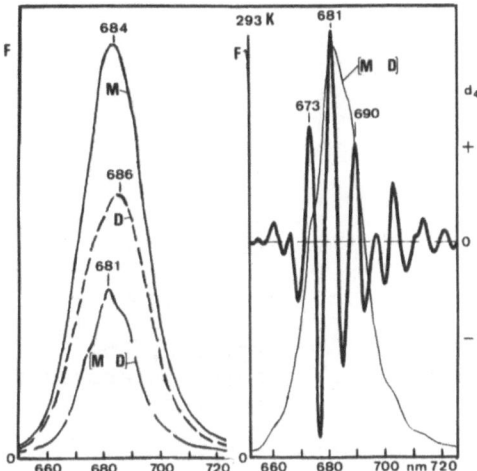

FIGURE 9. Emission spectra
registered at points M and
D of the kinetics, and the
absolute difference [M-D]
between these spectra.
The right part of the fig.
shows the 4th derivative
of the absolute difference

It would thus be erroneous to deduce from fig. 6 that, as
it could seem at first sight, the Franck-Inoue kinetics is
due to the transformation of a pigment emitting around 695-
700nm into a pigment emitting around 680nm, and vice versa.
In fact, the composition of the fluorescence changes as a
result of the conjunction of two distinct, though not necessarily independent processes. The long-wavelength components
of the spectra are involved in the first of these processes,
consisting in the transformation of $P_{694-682}$ into $P_{688-678}$,
or in the reverse transformation. The short-wavelength components are involved in the second process, consisting in a
decrease, or in an increase, of the intensity of the light
emitted at 675-680nm.

3.1.4. <u>Narrowing and broadening of the spectrum</u>. The width
of the overall room temperature emission spectrum was seen
to narrow, and then to broaden during the kinetics of fig.3.
This is shown in fig.10 which gives the kinetics of the half-
spectrum-width. A maximum value of 30nm was found by extrapo-
lating to zero time, or by using chopped light to excite fluorescence just before the start of the kinetics (during period
I, fig.3). A minimum of 23nm was reached 20 to 30 s later, as
the fluorescence intensity was near its maximum M, i.e. when
the short-wavelength components contributed maximally to the
emission. Later, there was a slow rise to a steady value of

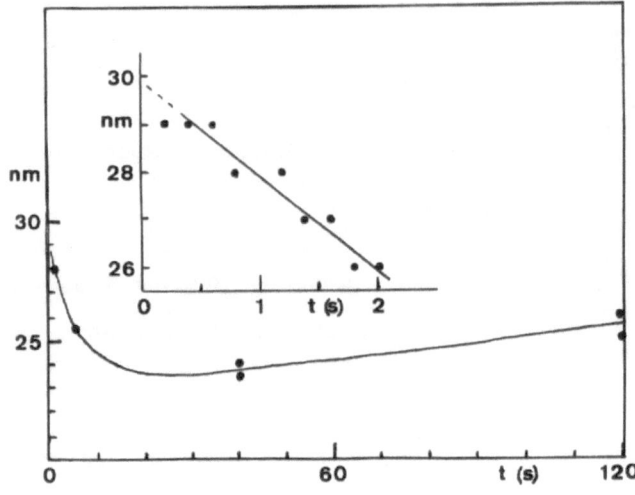

FIGURE 10. Evolution of the half width of the emission spectra registered at room temperature during the Franck-Inoue kinetics, fig.3. In ordinate, the half-width; in abscissa, the time

28nm. These features are well reproducible.

The role of the decay of the long-wavelength 694nm emission in narrowing the spectrum is well illustrated by considering the fluorescence added during the increase phase of the intensity kinetics. This is done in the [M-D] absolute difference spectrum of fig. 9. Any contribution of $P_{694-682}$ is excluded from this spectrum, as the emission of this pigment fades away during the increase phase. Accordingly, the half-width of the [M-D] spectrum is 21nm only.

3.1.5. <u>Effect of freezing at 77K.</u> Freezing enhances the intensity of the fluorescence. The effect is demonstrated in fig. 11. The room temperature emission spectrum I_{293K} of a leaf enriched with $P_{694-682}$ (as indicated in section 3.1.1.) was registered by the OMA 2 during period I, fig.3, using chopped excitation. Liquid nitrogen was then poured around the leaf, and spectrum I_{77K} was registered in the same excitation and reception conditions.

The 77K emission arose mainly from $P_{694-682}$, as already mentioned. This is due to the exceptional enhancement of the fluorescence yield of this pigment by freezing. The intensity of the 77K emission was multiplied by 7.5 at 694nm, against 4.5 at 690nm and 1.5 only at 670-680nm. In fig. 11, the room

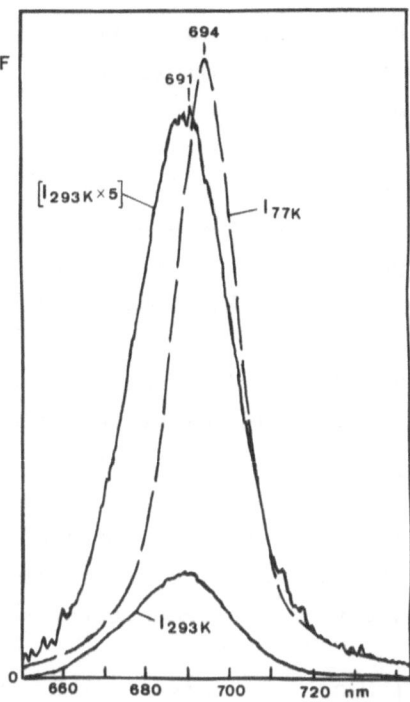

FIGURE 11. Enhancement of the leaf fluorescence emission by freezing at 77K. As soon as spectrum I_{293} had been registered under chopped excitation at room temperature, the leaf was frozen in liquid nitrogen, and the emission was again registered without changing the other conditions ($=I_{77K}$).

temperature spectrum has been reproduced at a 5 time expanded scale to underline the fact that the emission of the short wavelength components of the spectrum was much less enhanced by freezing than the emission of $P_{694-682}$.

This property of $P_{694-682}$ has already been demonstrated previously (Jouy and Sironval, 1979). It is shared to some extent by $P_{688-678}$, a feature which stresses the distinction to be made between short-wavelength and long-wavelength components of the overall emission.

3.1.6. <u>Microcycle II</u>. All the events discussed in the previous sections could be repeated several times. After turning the light off at the end of the kinetics of fig. 3, a 2 to 3 min dark period was sufficient to reproduce the state of the pigments found at the start of the kinetics, -i.e. during period I-, with $P_{694-682}$ as the main pigment. Turning the light on again triggered a second Franck-Inoue kinetics; and so on... The amplitude of the phases tended to decrease slowly,

210

however, from one repetition to the next (see Franck et al., this volume).

The first repetitions of the kinetics were concerned with an almost constant amount of pigments. For this reason the events appeared to fit a cycle in which reaction (a) alternates with reaction (b). We call this cycle a microcycle because it involves only two pigments of the P-C cycle ($P_{694-682}$ and $P_{688-678}$), and we label it II (see the Discussion).

3.1.7. <u>Sensitivity of microcycle II to heat</u>. The following experiment shows that microcycle II is distinguished by its sensitivity to heat :

A 14 day old etiolated bean leaf received a 1 ms flash of light, stayed for 2 min in darkness at room temperature in order to become enriched with $P_{694-682}$, was then heated at 53°C for 5 min and finally was returned to room temperature for 5 min, before recording the kinetics of the fluorescence emitted at 690nm under an He-Ne laser excitation. This record, curve b, fig. 12 B, does not show any Franck-Inoue kinetics.

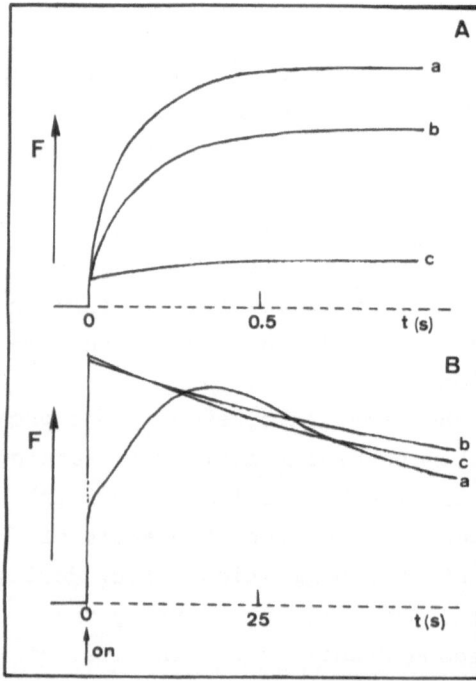

FIGURE 12. Effect of a pretreatment with heat on the photoreduction of protochlorophyllide (A) and on the Franck-Inoue kinetics (B). Control leaves (a) are compared with leaves heated for 5 min (b) and for 10 min (c) at 53°C. The treatment with heat was started 10 min before putting the laser light on. In (A) this treatment was applied to an etiolated leaf; in (B) it was applied to a preilluminated, etiolated leaf enriched with P694-682. Abscissa : the time t. Ordinate : the intensity of the 690nm fluorescence F. The fluorescence was seen through an interference filter which transmitted the 690nm light. Other explanations in the text.

The heated leaves emitted maximally at the onset of the excitation light; a steady decrease of the fluorescence followed.
The prolongation of the heat treatment did not change this
result (curve c, fig. 12 B). A typical Franck-Inoue kinetics
was obtained, however, when the experiment was done using a
non-heated control leaf of the same age (curve a, fig. 12B).
The suppression of microcycle II by a pretreatment with heat
was thus obvious.

The photoreduction of protochlorophyllide was also inhibited by heat, but to a lesser extent (Sjronval and Brouers,1970).
To compare this effect with the effect of heat on microcycle
II, a 14 day old etiolated leaf was heated at 53°C for 5 min
in darkness, and was then brought back to room temperature in
darkness for 5 min. The kinetics of the photoreduction of
protochlorophyllide, recorded using the same experimental set
as for the Franck-Inoue kinetics, is curve b, fig. 12 A. This
curve shows only a slight inhibition of the photoreduction
by comparison with a control (curve a, fig. 12 A). It was
necessary to prolong the heat treatment for over 10 min to
suppress the photoreduction (curve c, fig. 12 A).

However, although it hindered the photoreduction to a rather limited extent, a 5 min heat treatment inhibited totally
the subsequent formation of $P_{694-682}$. The decrease phase of
the fluorescence kinetics due to the formation of this pigment
was not seen, although it was seen as expected in non heated
controls (not shown in fig. 12 A).

3.2. Microcycle II *in vitro*

El Hamouri et al. (1981), and Oliver and Griffiths (1982)
have shown that etioplasts are able to perform reaction (a)
in a buffer to which NADPH has been added. The following experiment proves that in this case, reaction (a) alternates
with reaction (b) to close microcycle II. We have done this
experiment with etioplast suspensions prepared as in the
paper of Bereza and Dujardin (this volume).

The nucleotides were added to start with a total concentration of 15 mM   and a [NADPH]/[NADP$^+$] ratio of 10/1.

3.2.1. <u>The Franck-Inoue kinetics in isolated etioplasts.</u>
Two etioplast suspensions, -a control and a NADPH enriched
suspension [NADPH]/[NADP$^+$] = 10/1-, were let to cool down to
5°C. They received a strong, 1 ms light flash in order to
photoreduce totally protochlorophyllide, and they were then
incubated in darkness for 30 min. $P_{694-682}$ was formed in the
NADPH enriched suspension, while $P_{688-678}$ was formed in the
control, as proved by the spectra of the samples frozen at
77K after the end of the 30 min dark period (fig. 13).

When the pigments were in the state defined by fig. 13, the
suspensions were warmed up to room temperature and the kinetics of the intensity of the fluorescence was tested as described in section 3.1.1. In the case of the NADPH enriched suspensions of etioplasts which contained $P_{694-682}$, the kinetics
comprised 2 phases : (I) an increase phase during the first

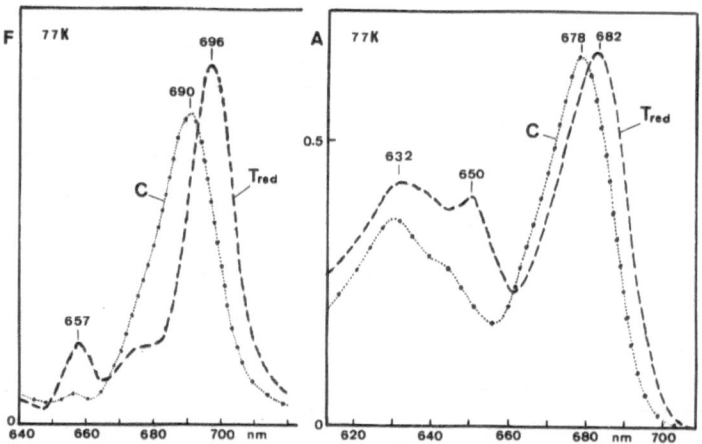

FIGURE 13. Effect of the addition of [NADPH]/[NADP$^+$] (10/1)
on the spectral properties of preilluminated,isolated etioplasts. (A) is the absorption and (F) the fluorescence spectrum of control etioplasts (curves C) and of etioplasts suspended in the NADPH enriched buffer [NADPH]/[NADP$^+$] = 10/1
(curves T red). The samples have been frozen at 77K after an
incubation period of 30 min at 5°C, which followed the illumination by a strong, 1 ms white flash (= preilluminated
etioplasts).

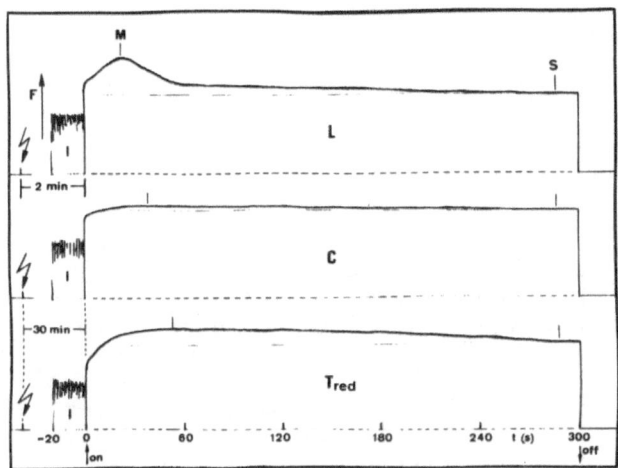

FIGURE 14. Kinetics of the intensity of the fluorescence emitted at 690nm by preilluminated etioplast suspensions characterized by the spectra of fig. 13 at time 0. C = the control etioplasts; Tred = the etioplasts of the NADPH enriched suspension ([NADPH]/[NADP$^+$] = 10/1), L = a leaf of the bean variety used for preparing the etioplasts (var. Bor); this leaf was treated as in fig.3. Other explanations in the text (see also fig. 3).

40 sec of illumination, up to a maximum M ; (II) a slower decrease phase, down to a steady level S (fig. 14, T$_{red}$). In the case of the control, without nucleotide addition, the increase phase was much depressed and the decrease phase was extremely slow (fig. 14, C).

Etiolated leaves which had been pre-treated to accumulate P$_{694-682}$ showed a kinetics of the 690 nm intensity similar to that shown by etioplasts in the NADPH enriched suspension (fig. 14, L). This similarity was the most obvious during the first 20 s. In leaves, however, the decrease phase started earlier, and it was more rapid than in the suspension.

3.2.2. Alternation of reaction (a) with reaction (b) *in vitro*. The alternation of reaction (a) with reaction (b) in NADPH enriched etioplasts was demonstrated by the changes of the composition of a series of emission spectra memorized by the OMA 2 during the Franck-Inoue kinetics. Differences between area normalized spectra are shown in fig. 15. The overall emissions of both the leaf and the NADPH enriched etioplast suspensions became richer in light emitted around 685nm, and poorer in light around 695-700nm, during the increase phase, -i.e. from I to M (fig. 15). The composi-

214

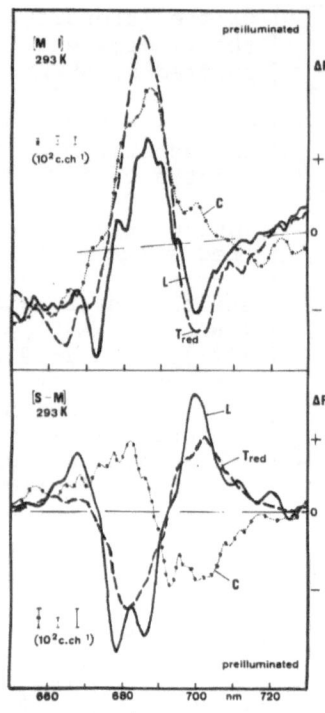

FIGURE 15. Composition changes of the spectra emitted by a preilluminated leaf (L) and by preilluminated etioplasts,- control etioplats (C) and etioplasts in the [NADPH]/[NADP$^+$] (10/1) enriched buffer (T_{red})-, during the kinetics shown in fig. 14. The figure gives differences between area normalized spectra, registered during period (I) (chopped excitation) and at the points (M) and (S). Above : increase phase of the kinetics; below : decrease phase.

tion of the emission changed inversely during the subsequent decrease phase (from M to S ; fig. 15).

These difference spectra are interpreted as showing that, in the NADPH enriched suspension of etioplasts as in the etiolated leaf, $P_{694-682}$ was changed into $P_{688-678}$ by reaction (b) during the increase phase of the Franck-Inoue kinetics, and that during the decrease phase, reaction (a) started to form $P_{694-682}$ again, initiating in this way a second turn of microcycle II.

Such an alternation of reaction (a) with reaction (b) was not seen in control etioplasts. In this case, the amplitudes of both the increase and the decrease phases of the fluorescence kinetics were feeble, and the composition of the emission behaved quite differently than in the NADPH enriched suspension, or in the leaf. During the increase phase, the overall emission became richer in 680nm light, but did not became poorer in long wavelength light (fig. 15, above, curve c). These changes cannot be ascribed to microcycle II.

3.2.3. <u>Repetitions of the alternation</u>. We have seen(in 3.1.6.)
that the alternation of reaction (a) with reaction (b) repeated itself several times in etiolated leaves submitted to
a series of light-dark cycles of a few minutes duration. When
they were enriched with NADPH, etioplast suspensions repeated
this alternation two to three times. In general, the composition of the spectrum changed in the same way at a second, as
at a third light-dark cycle. These changes remained significant even if the amplitudes of the phases of the kinetics
were depressed by repeating the illuminations, as it is the
case in the example given in Table 1.

Table 1. Influence of the $[NADPH]/[NADP^+]$ ratio on the relative
amplitude of the decrease phase of the Franck-Inoue kinetics
in a suspension of isolated preilluminated etioplasts.

$[NADPH]/[NADP^+]$ (15 mM total concentration).	first repetition of the light-dark cycle	second repetition of the light-dark cycle
1/10	6,7	0,13
10/1	23,8	4,3

(the relative amplitude is equal to $\frac{F_M - F_S}{F_M}$, where F_M and F_S
are the intensities of the 690nm fluorescence at points M
and S respectively; kinetics of fig. 12).

3.2.4. <u>The need for a high proportion of NADPH to total</u>
<u>nucleotides</u>. The nucleotide content of the medium was crucial
in the experiments with isolated etioplasts. The comparison
of NADPH enriched suspensions with controls evidenced this
effect, which was also met by considering suspensions containing a same amount of total nucleotides, but with different
$[NADPH]/[NADP^+]$ ratios.

For a same total nucleotide content, lowering the $[NADPH]/$
$[NADP^+]$ ratio from 10 to 0.1 reduced the relative amplitudes
of the phases of the Franck-Inoue kinetics by a factor of
three at the first repetition of the light-dark cycle, and
this figure increased at subsequent repetitions (Table 1).
Likewise, the composition of the fluorescence did no more
change when the $[NADPH]/[NADP^+]$ ratio was 0.1 (fig. 18B, curve
T_{ox}).

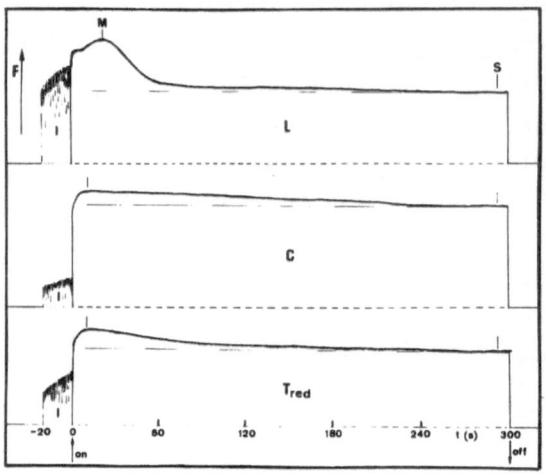

FIGURE 16. Kinetics of the intensity of the fluorescence emitted at 690 nm by etioplasts containing only protochlorophyllide at the start of the illumination with laser light (non preilluminated etioplasts). C = the control etioplasts; T_{red}= the etioplasts in the [NADPH]/[NADP$^+$] (10/1) enriched buffer. L = a non-preilluminated etiolated leaf of the bean variety used for preparing the etioplasts. Other explanations in the text (see also fig. 3).

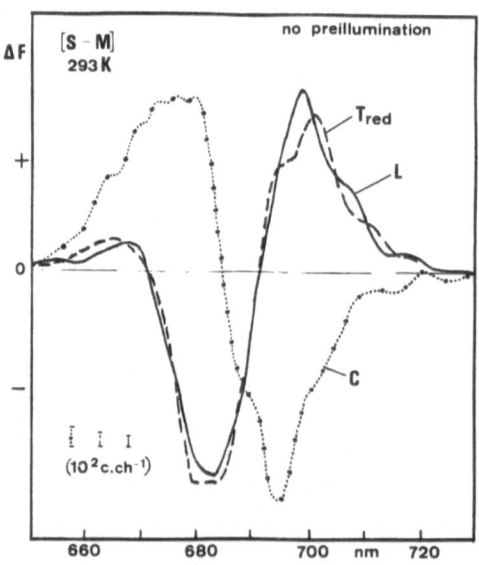

FIGURE 17. Composition changes of the spectra emitted by a non-preilluminated leaf (L) and by non-preilluminated etioplasts, -control etioplasts (C) or etioplasts in the [NADPH]/[NADP$^+$] (10/1) enriched buffer (T_{red})-,during the kinetics shown in fig.16. The figure gives differences between area normalized spectra from M to S, i.e. during the decrease phase of the kinetics.

3.2.5. Effects inherent in the addition of nucleotides.

The addition of the nucleotides at a [NADPH]/[NADP$^+$] ratio of 0.1 stopped the unilateral enrichment of the spectrum with 680 nm light characteristic of the control. This let us suspect that there was also an effect of the nucleotides as such (fig. 18B). The following experiment demonstrated another

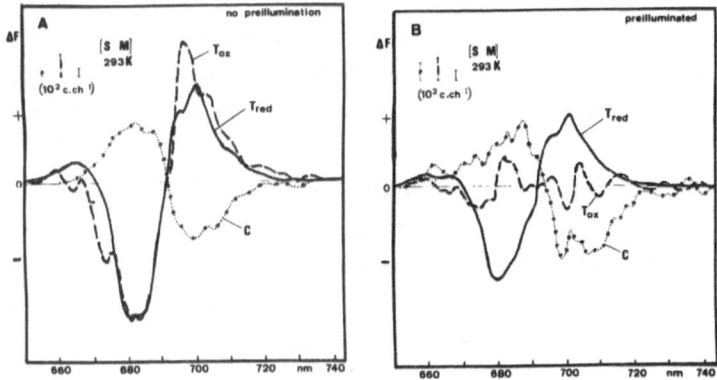

FIGURE 18. Composition changes of the spectra emitted by etioplasts suspended in a buffer during the decrease phase of the
fluorescence intensity kinetics, fig. 16 (non preilluminated
etioplasts) or fig. 14 (preilluminated etioplasts). (C) control without addition; (T_{red}) the buffer was enriched with
NADPH([NADPH]/[$NADP^+$] ratio = 10); (T_{ox}) the buffer was enriched with $NADP^+$ ([NADPH]/[$NADP^+$] ratio = 0.1). The total concentration in nucleotides in (T_{red}) + (T_{ox}) was 15mM. The
etioplasts contained $P_{688-678}$ at point M of the kinetics.
(see the text).

circumstance showing an effect of nucleotide addition, irrespective of the reduction state of the nucleotides :

Three etioplast suspensions, -a control and two nucleotide
enriched suspensions with [NADPH]/[$NADP^+$] ratios of 10 and
0.1 (15 mM),- thus NADPH and $NADP^+$ enriched respectively-,
were let to cool down to 5°C in darkness. They were incubated
in darkness for 30 min at 5°C. At this stage, samples were
frozen at 77K. Their absorption and fluorescence spectra were
similar; they showed the bands of the photoactive and inactive
protochlorophyllides, and of no other pigments (= non preilluminated suspensions of figures 16 to 18).

The suspensions were then warmed up to room temperature,
and the kinetics of the intensity of the fluorescence was
tested by a continuous laser illumination as usually. This
illumination did not result in any Franck-Inoue kinetics : it
triggered protochlorophyllide reduction. The formation of
chlorophyllide caused a rapid rise of the emission during the
first seconds of the illumination; this rise was followed by

a decrease to a steady level (fig. 16).

Fig. 16 shows that the kinetics were alike in the NADPH enriched and in the control suspension. This similarity is misleading, however. The differences between area normalized spectra (fig. 17) proved that during the decrease phase, $P_{694-682}$ was formed in the NADPH enriched suspension (like in a leaf), but not in the control. This result agrees with the spectra of fig. 13. Surprisingly, however, the $NADP^+$ enriched suspension did not follow the control; it followed the NADPH enriched suspension and produced $P_{694-682}$ (fig. 18 A).

Fig. 18 compares the effect of the degree of reduction of the nucleotides on the formation of $P_{694-682}$ after protochlorophyllide photoreduction in a non preilluminated suspension, with this effect on the reformation of $P_{694-682}$ initiating a second turn of the microcycle II in a preilluminated suspension. In both cases, the precursor of $P_{694-682}$ was $P_{688-678'}$ -the pigment present at point M of the fluorescence kinetics. But in part A of fig. 18, this pigment was a direct product of protochlorophyllide reduction while in part B, it had been regenerated by reaction (b). The formation of $P_{694-682}$ did not depend on the $[NADPH]/[NADP^+]$ ratio* in the first case; it was then dependent on the addition of the nucleotides as such, irrespective of their reduction state (fig. 18 A). By contrast, in the second case (fig. 18 B), the reformation of $P_{694-682}$ depended on the reduction state of the nucleotides.

4. DISCUSSION

4.1. The data of the first part of this paper suggest the next comments :

4.1.1. In the first place, we notice that the chlorophyllides of the etiolated, illuminated leaves share spectral properties with chlorophylls of the fully green leaves. This appears to be particularly true in the case of the pigments

---

* It should be notice that we refer to the initial ratio,- the ratio between reduced and oxidized nucleotides as supplied in the suspension medium.

which absorb light at long wavelengths These pigments are
responsible for bands of the fluorescence spectrum whose
emissions are strongly enhanced by freezing in both the etiolated and the green leaves. In green leaves, the 77K emission
bands of the chlorophyll are located at 688, 696 and 730-
740nm respectively, while the emissions of the chlorophyllides
of the etiolated leaves are at 688 and 694nm. These leaves
lack the broad 730-740nm band.

4.1.2. On the other hand, the data of Franck and Inoue
(1984) and the data presented in this paper, prove that
reaction (a) transforms one of the long wavelength components
of the spectrum into the other, and vice-versa for reaction
(b). The transformation of $P_{694-682}$ into $P_{688-678}$ (reaction
b) is light dependent, while the reverse transformation (reaction (a)) does not need light. For this reason, one of
these transformations cannot be conceived as the mirror image
of the other. Their interplay causes two stationary states of
the pigment system, depending on whether light is on (state
S), or off (state I). At state S, the level of $P_{694-682}$ is
lower than at state I; the reverse applies to $P_{688-678}$. This
situation is described by a light driven microcycle like the
following :

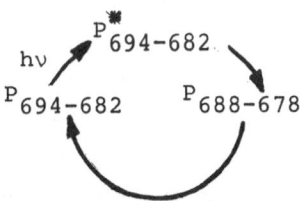

As the light dependent part of this microcycle is the second
photoreaction appearing in the illuminated etiolated leaves,
we call it microcycle II.

4.1.3. The qualitative argument by which one explains the
enhancement of the intensity of the fluorescence emitted by
long-wavelength pigments in frozen green leaves, applies to
frozen etiolated leaves as well. In both cases, freezing inhibits photoactivities, and this is thought to benefit the
fluorescence. Jouy and Sironval (1979) and Jouy (1980) have

produced evidences showing that $P_{694-682}$ does not emit much
light at room temperature because this pigment is then involved in photoactivities.

Moreover, as well at room temperature as at 77K, $P_{694-682}$
quenches the fluorescence of neighbour pigments and behaves
as a centre towards which the energy flows (Jouy, 1980). The
domains of energy transfer around the centres, -the so-called
"transfer units"-, are precursors of units appearing later in
green leaves. Both types of units share the organization prin-
ciples of an [antenna/centre] pigment system (Brouers et
Sironval, 1978; Sironval, 1982; Sironval and Brouers, 1983).

These comments bring us to conclude that (1) the nature,
(2) the function, and (3) the organization pattern of the
chlorophyllide-proteins in etioplasts, do not differ markedly
from the nature, function and organization pattern of at least
some of the chlorophyll-proteins in green plastids.

4.1.4. The sensitivity of microcycle II to heat (section 3.1.
7) points to the distinction to be made between the molecular
organization of the chlorophyllides and the molecular organization of the protochlorophyllides. The latter is less sensitive to heating than the former. One should conclude that
the physical structures which act in microcycle II differ in
some manner from the structures which act in protochlorophyllide photoreduction proper. The effects of a 5 min heat treatment reported in section 3.1.7 suggest that the difference may
lie in some temperature sensitive molecular arrangement, as
for example a certain protein intervening in microcycle II.
This idea is particularly supported by the fact that the pre-
treatment of an etiolated leaf with 5 min heating stopped
microcycle II to set in at room temperature, although protochlorophyllide photoreduction still proceeded nearly normally.
Something specifically needed to the function of microcycle II
appeared to have been destroyed by heat in this instance.

4.2. The data of the second part of this paper point to the
following conclusions :

4.2.1. The etioplasts appear to have the basic elements
needed to build up microcycle II. However, they do not seem
to have the right nucleotide content when isolated from the

leaf; they loose apparently nucleotides during isolation.

. However, at most 3 to 4 turns of the cycle could be repeated in NADPH enriched etioplast suspensions, the amplitude of the fluctuations of the Franck-Inoue kinetics diminishing at each repeat. By comparison, leaves repeated more turns. We surmise that, despite the enrichment with NADPH, all conditions necessary for the maintenance of microcycle II were not met in our experiments with isolated etioplasts.

4.2.2. The role of NADPH in microcycle II is to be ascribed to its role in $P_{694-682}$. The latter is a (chlorophyllide-NADPH-protein) complex whereas its partner in the microcycle, $P_{688-678}$, is a (chlorophyllide-NADP$^+$-protein) complex. The presence of NADP$^+$ in $P_{688-678}$ results from the coupling of nucleotide oxydation with protochlorophyllide reduction (Griffiths 1980; El Hamouri et al, 1981). We should thus admit that there is an alternation between reduced and oxidized state of the nucleotide in microcycle II : at each turn, photoreaction (b) should involve the replacement in the light of the reduced state by the oxidized state, while the reverse should occur in darkness (reaction (a)).

4.2.3. The data shown in fig. 18 are particularly conflicting. On the one hand, they confirm that a high [NADPH]/[NADP$^+$] ratio (of 10/1) is needed to produce $P_{694-682}$ from $P_{688-678}$, and on the other hand, they deny the need for NADPH in NADP$^+$ enriched ([NADPH]/[NADP$^+$] = 1/10) etioplasts, when the latter are illuminated for the first time. In this particular case $P_{688-678}$ appears to be transformed into $P_{694-682}$ in NADP$^+$ enriched, as well as in NADPH enriched etioplasts (fig. 16A; see also Bereza and Dujardin, this volume).

This contradicts our expectation unless we accept that an NADP$^+$ enriched environment becomes NADPH enriched at least locally, when etioplast membranes with protochlorophyllides are illuminated. Such a function belongs perhaps to the intermediates which originate from the photoactive protochlorophyllides, and to which several papers are devoted in this volume (see Iwai et al.; van Bockhove et al; Dujardin, this volume).

222

ACKNOWLEDGMENTS

The authors thank the "Fonds National de la Recherche Scientifique", Belgium, and the "Ministère de la Politique Scientifique, Actions Concertées, nr 80/85-1983", Belgium for financial support. B. Bereza had a grant from the "Patrimoine de l'Université de Liège". The authors thank also Mrs M. Crepin for skilful assistance.

REFERENCES

1. Bereza B, Dujardin E. 1984, this volume.
2. Brouers M, Sironval C. 1978. Photosynthetica,12, 399-405.
3. Dujardin E. 1976. Plant Science Letters, 7, 91-94.
4. Dujardin E. 1984, this volume.
5. El Hamouri B, Brouers M, Sironval C. 1981. Plant Science Letters. 21, 375-379.
6. Franck F, Inoue Y. 1984.

7. Franck F, Sironval C, Schmid G.H. 1984, this volume.
8. Griffiths W.T. 1980. Biochem. J. 186, 267-278.
9. Griffiths W.T.
10. Iwaï J, Ikeuchi M, Inoue Y, Kabayashi T. 1984. this volume.
11. Jouy M, Sironval C. 1979. Planta. 147, 127-133.
12. Jouy M. 1980. Ph. D. Thesis. Liège.
13. Oliver R.P., Griffiths W.T. 1982. Plant Physiol. 70, 1019-1025.
14. Shibata K. 1957. J. Biochem. 44, 147-173.
15. Sironval C. 1981. in Photosynthesis V. Chloroplast Development. Akoyunoglou G. ed. Balaban Intern. Science Services. Philadelphia. Pa. pp 3-14.
16. Sironval C. 1982. in Cell Function and Differentiation part B. Akoyunoglou G and al. eds. Alan R. Liss, Inc New-York. pp. 53-65.
17. Sironval C, Brouers M. 1970. Photosynthetica. 4, 38-47.
18. Sironval C, Brouers M. 1983. Phys. végétale. 21, 589-600.
19. Sironval C, Brouers M, Michel J-M, Kuiper Y. 1968. Photosynthetica 2, 268-287.
20. van Bochove A.C., Griffiths W.T., van Grondelle R. 1984. this volume.

EXTENSION OF THE EXPERIMENT BY J.H.C.SMITH ON THE ONSET OF THE
OXYGEN EVOLUTION IN ETIOLATED BEAN LEAVES.

F.FRANCK, C.SIRONVAL* and GEORG H.SCHMID**

* Université de Liège, Laboratoire de Photobiologie, Départe-
ment de Botanique, Sart-Tilman B22, B.4000 Liège, Belgium
** Universität Bielefeld, Fakultät für Biologie, Lehrstuhl
Zell-physiologie, Postfach 8640, D.4800 Bielefeld 1,W.-Germany.

1. INTRODUCTION

In 1954, James H.C. Smith (1) published experiments in which
he studied the appearence of the capacity to evolve oxygen in
etiolated barley leaves exposed to light. In these experiments,
leaves were first treated in the desired manner and then introduced into a device which measured oxygen via phosphorescence-
quenching. Some of the data. of Smith are summarized in table I.

TABLE I

pre-treatment	Chl (mg/g)	O_2 (μl/g)
10 min light	0.0045	0.0021
10 min light +110 min darkness	0.0089	0.088
5 min light +110 min darkness + 5 min light	0.0177	0.425

The highest oxygen production was observed when the etiolated
leaves received two 5 min illumination periods separated by a
110 min dark interval. These data show, provided a sufficiently sensitive detection method is used, that an oxygen production can be detected in conditions under which the total a-
mount of the leaf pigments is very low, and equal to the sum
of the protochlorophyll(ide) present in the etiolated leaf before the first illumination, plus the protochlorophyll(ide)
regenerated during the dark interval.
At the time Smith made these experiments, very little was known
about the existence and properties of the various chlorophyll(ide)-
protein complexes which are formed after the photoreduction of
protochlorophyll(ide). In our actual state of knowledge, we may
ask how photosynthetically active chlorophyll-proteins are for-

C. Sironval and M. Brouers (eds.); Protochlorophyllide Reduction and Greening. ISBN 90 247 2954 8
© 1984, Martinus Nijhoff/Dr W. Junk Publishers, The Hague/Boston/Lancaster.

med out of the chlorophyll(ide)-protein complexes which originate directly from the illumination of the protochlorophyll(ide)-
protein complexes. The sequence of light and dark periods used
by Smith in order to observe the oxygen production in whole
leaves is particularly well adapted for the study of such a
problem. We have repeated Smith's experiment and confirmed his
result in bean leaves by using a special oxygen electrode. In
an attempt to identify the pigment-protein complexes involved
in the oxygen production we have studied the intensity and the
spectral composition of the fluorescence emitted at room temperature by the leaves at various stages as the capacity to
evolve oxygen was developping.

2. MATERIALS AND METHODS

Seedlings of *Phaseolus vulgaris*, variety Commodore, were
grown on vermiculite in complete darkness. The plants used in
these experiments were 10 or 12 days old. Oxygen measurements
were performed in the three electrodes device described by
Schmid and Thibault (2). Leaf discs were directly used on the
platinum electrode in a 0.2 N KCL solution. The system was allowed to equilibrate during 15 min in darkness (this time is
included in the dark periods mentioned in the results).
Fluorescence measurements were performed in the set-up described by Sironval et al. (this symposium).

3. RESULTS

3.1. Confirmation of the experiment by Smith (1).

A primary leaf taken from a 10 day old etiolated bean
plant which was illuminated by continuous white light for 10
min and then kept in darkness for 2 h, did not evolve any detectable amount of oxygen during a second 5 min illumination
(Fig.1,upper trace). However, an oxygen gush was observed at
the beginning of a third illumination, given 5 min after the
second illumination period. (Fig. 1, lower trace). This result
confirms the result of Smith, showing that a growing leaf, initially etiolated, produces oxygen under illumination at a time
when it contains a very small amount of chlorophyll(ide).
The rise-time of the oxygen gush at the third illumination is

10 to 15 times slower than the rise-time of the gush produced
in a fully greened leaf. This gush is not followed by any detectable steady-state oxygen evolution. Its amplitude is about
50 times smaller than in a green leaf.

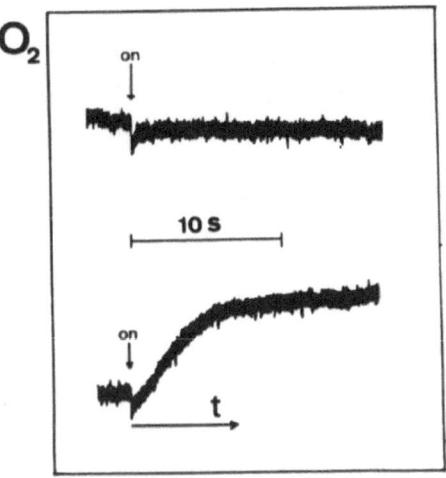

FIGURE 1. Actual oxygen evolution recordings in a single leaf
taken from an etiolated plant
which was first illuminated by
white light during 10 min and
then returned to darkness for
2 h.
Upper trace: beginning of the
                second illumination
                (5 min duration).
Lower trace: beginning of the
                third illumination
                given 5 min after
                the second one.

3.2. Fluorescence emission during the experiment of Smith

The variations of the intensity of the fluorescence emitted during the second and the third illumination are shown in
figs 2a and 3a, for 12 day old leaves.
During the second illumination, the intensity of the fluorescence first increased rapidly and reached a maximum after 1
sec; it decreased subsequently to a minimum (at time 10 sec)
and increased then slightly in the min time scale. The rapid
fluctuation was supressed by a 1 msec white flash given a short
time before the illumination; it is due to the formation of a
certain amount of chlorophyll(ide) $P_{688,678}$ from the proto-
chlorophyll(ide) which had accumulated in the dark.
Fig. 2b (full line) reproduces the room temperature emission
spectrum recorded 1 sec after the onset of the second illumination. This spectrum had a maximum around 680 nm. The fourth
derivative analysis showed that the pigments which participated
in the emission were $P_{675,668}$, $P_{684,672}$, $P_{688,678}$ and $P_{694,682}$.
The bandwidth of the spectrum narrowed during the illumination
(from time 1 sec onward) as a result of the progressive ex-

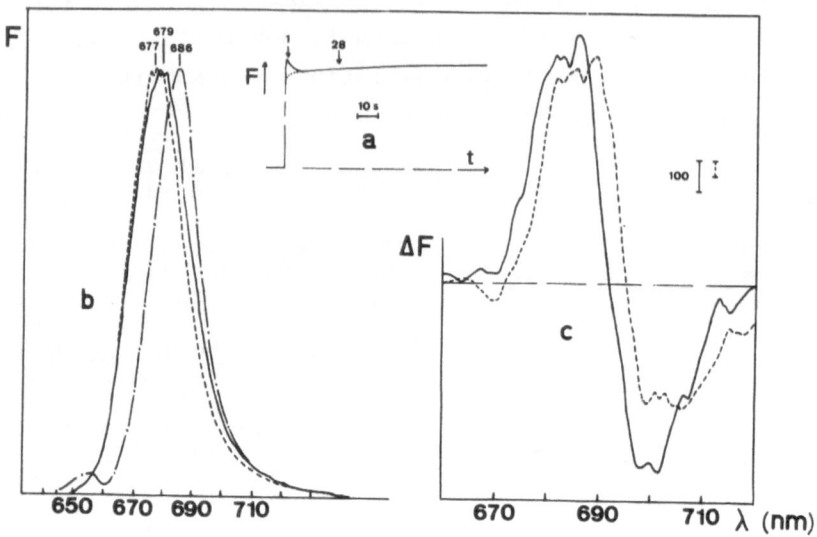

FIGURE 2. Fluorescence emitted during the second illumination in a Smith type experiment.

a. kinetics of the 690 nm fluorescence at +20°C (explanations in the text)
b. —— room temperature emission spectrum at time 1 sec
   --- room temperature emission spectrum at time 5 min
   —·— 77 K emission spectrum of the leaf frozen before the illumination (all spectra were normalized to the same maximum intensity)
c. changes in the composition of the spectrum
   —— time 28 sec minus time 1 sec
   --- time 5 min minus time 1 sec
   (the scales on the right correspond to 100 counts per channel of the ISIT target; the area of the spectra were normalized to 10^6 counts; see Sironval et al., this volume).

tinction of the emission at the long wavelength side of the spectrum, around 695 nm. At the end of the illumination (after 5 min) the spectrum in fig. 2b, dashed line, was recorded.
In spite of the variety of the pigments involved, the composition of the spectrum showed a simple change during the second illumination. Composition changes are given by differences between two emission spectra recorded at any two times taken at random, after normalization of the total area. The differences found from time 1 sec to time 5 min and from time 1 sec to time 28 sec are shown in fig. 2c. They are similar. They show that the relative contribution of the emissions at wave-

lengths higher than 690 nm decreased continuously with time, whereas the relative contribution increased continuously at shorter wavelength.

This process did not repeat itself in the third illumination. The first spectrum recorded at this illumination (time 1 sec) was identical with the last spectrum of the second illumination, and the last spectrum of the third illumination (time 5 min) was identical with the first one (Fig. 3b). This coincided with the disappearance of any significant variation of the intensity of the total fluorescence (Fig. 3a) and also with the absence of any reproducible change in the composition of the spectrum (Fig. 3c).

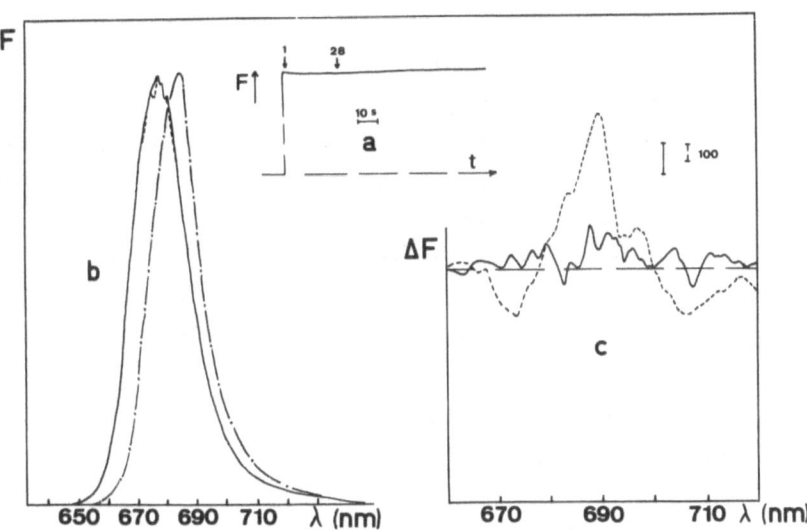

FIGURE 3. Fluorescence emitted during the third illumination in a Smith type experiment.
a,b,c: same as in Fig.2.

3.3. The effect of the two hours dark period

In order to make the oxygen gush measurable in a Smith type experiment, two continuous illuminations are required prior to the measurement, with a sufficiently long dark period in between. The need for a long dark interval becomes clear by comparing the two following situations, which differ by the duration of the dark period.

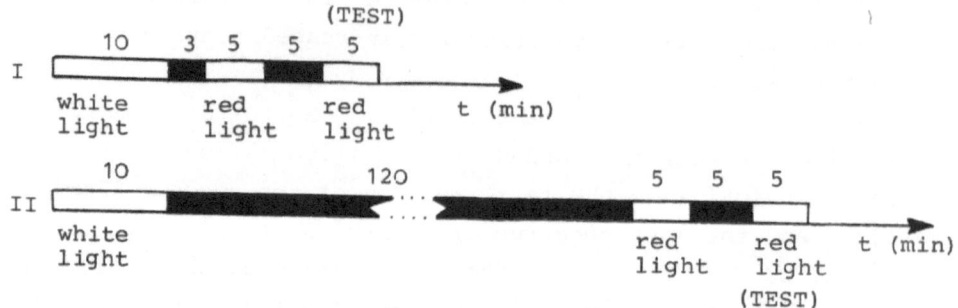

Oxygen was produced upon the third illumination in case II
but not in case I. For this reason, this third illumination
will be called "test" illumination. The spectra of fig. 4
show that, immediately before the test, essentially the chlo-
rophyll(ide) $P_{683,672}$ resulting from the Shibata shift was detected both at +20°C and at 77°K in the case where a two hours
dark period was given (broken line). The $P_{694,682}$ pigment formed as a result of the photoreduction of the regenerated pro-
tochlorophyll(ide) was not detected in this case by means of
fluorescence measurements. In contrast, $P_{694,682}$ was seen when
the dark period was 3 min only (Fig. 4, full lines)*.

When the dark period between the two first illuminations was
3 min, the variations of the fluorescence intensity during the
test as well as the changes in the composition of the emission
spectrum reflected the sequence of the two reactions belonging
to microcycle II** (Fig. 5 B,1):

$$P_{694,682} \underset{}{\overset{h\nu}{\rightleftharpoons}} P_{688,678}$$

low fluorescence   high fluorescence
yield at +20°C   yield +20°C

When the leaf was allowed to remain in darkness for two hours
these reactions could no longer be observed (Fig. 5 A,1): the
intensity of the fluorescence did not change during the test
and the composition of the spectrum did not vary significantly.

---

* In each case, freezing results in a shift of the fluorescence maximum of a few nm towards longer wavelengths.

** Concerning this microcycle, see Franck and Inoue (3) and
Sironval et al., this symposium (10).

FIGURE 4. Effect of the 2 h dark period on emission spectra.
Left: room temperature spectra at the beginning (1 sec) of
   the third illumination
Right: 77 K spectra of leaves frozen before the third illumination.
 ——— the dark interval after the first illumination is 3 min
 ---- the dark interval after the first illumination is 2 h

Thus, the main effect of the two hours dark period was that,
although a considerable amount of photoactive protochloro-
phyll(ide) had been regenerated in darkness and reduced by the
second illumination, the fluorescence associated with microcycle II could not be detected anymore.

In fig.5 A,2, the effect of the two hours dark period is shown
in the case when each of the first and second illuminations
have been substituted by a strong msec white flash. In this
case, oxygen was not emitted during the test (which was carried out during the first continuous illumination) and the fluorescence characteristics differed in two respects:

- the fluorescence kinetics showed a rapid transient, indicating that some protochlorophyll(ide) was reduced at the beginning of the illumination (Fig.5 A,2);
- the composition of the emission changed significantly during
  the test, which was in contrast with what was observed when

230

continuous illuminations had been given previously (compare figs. 5 A,2 with 5 A,1).

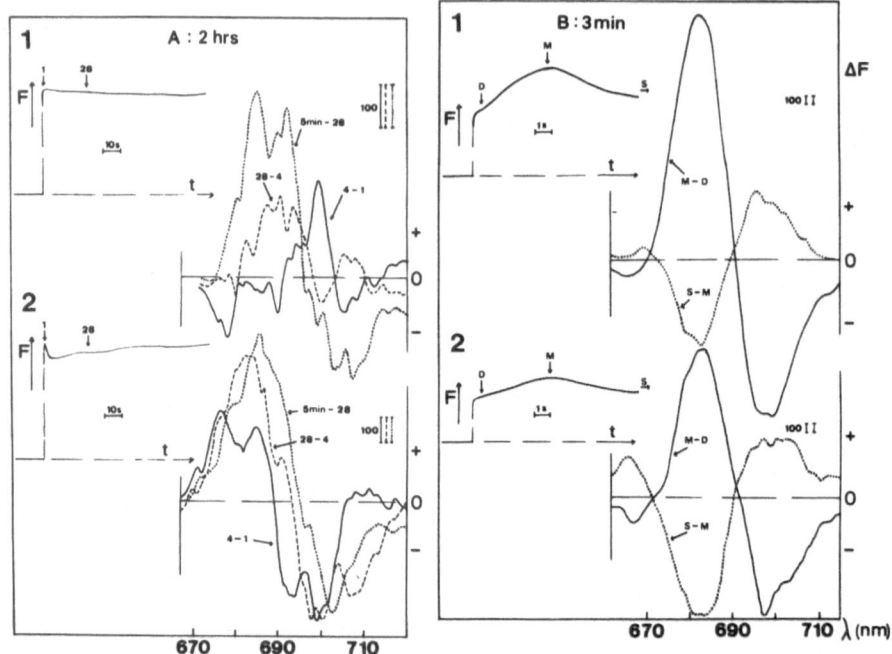

FIGURE 5. Effect of the 2 h dark period on the room temperature fluorescence variations during the third illumination.
A: the dark interval after the first illumination is 2 h
B: the dark interval after the first illumination is 3 min

1. the two first illuminations are continuous light (10 and 5 min)
2. the two first illuminations are a 1 msec white flash each.

In each case, the kinetics of the overall fluorescence and the changes in the composition of the spectrum for the indicated time intervals are shown. The scales on the right correspond to 100 counts per channel of the ISIT target; the area of the spectra were normalized to 10^6 counts (see Sironval et al., this volume, 10).

3.4. Changes in the oxygen production and in the fluorescence variations as a function of the duration of preillumination in continuous white light.

We have followed the development of the oxygen evolution capacity in 10-day old leaves as a function of the duration of a preillumination period by white light. After the preillumination period, the leaves were transferred into darkness for

30 min during which they were placed on the oxygen electrode
Oxygen was then measured during two successive 5 min illuminations in red light.

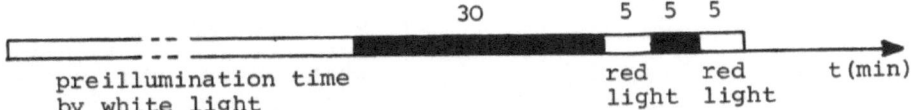

The room temperature fluorescence variations have been recorded in leaves treated in the same way, except that they received a strong msec white flash 2 min before the first illumination by red light. This flash was given in order to eliminate
the fluorescence variations due to protochlorophyll(ide)reduc-
tion (it was verified that, in fully green leaves, such a flash
did not modify the fluorescence kinetics).
An etiolated control leaf did not produce oxygen when illuminated for the first time. But, due to microcycle II, it showed
a large fluorescence fluctuation whose rise-time lasted several seconds (Fig. 6, time 0 h). When the preillumination time
was 1 h, an oxygen gush was detected upon the first illumination with red light; the same slow fluorescence fluctuation
was still observed although it was much less marked (Fig.6,
time 1 h). For longer preillumination times, the size of the
oxygen gush increased and the rise-time of the gush became
shorter. The slow fluorescence fluctuation disappeared and
was replaced by a rapid fluctuation with a rise-time of less
than 1 sec, whose fine analysis revealed the usual characteristics of the "IODP" fluorescence kinetics found in green
leaves (Fig.6, time 8 h).

Fig. 7 A represents the development of the oxygen gush at each
illumination in red light, as a function of the preillumination
time. When the preillumination was 30 min, the result was similar to the result of a Smith type experiment: no oxygen at
the first illumination, and an oxygen gush at the second one.
For preillumination times longer than 30 min, the oxygen gush
was detected at the first illumination, but was always markedly
increased at the second illumination. It must be noted that

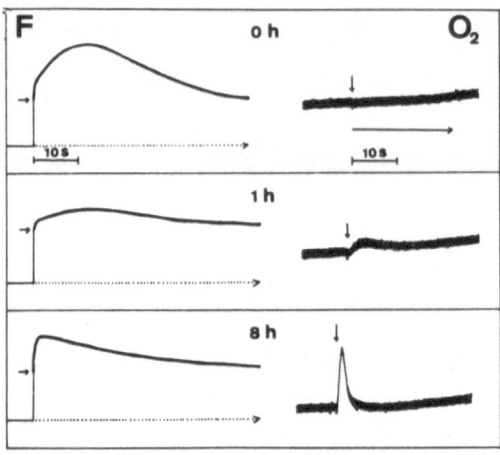

FIGURE 6. Kinetics of the 690 nm fluorescence and oxygen evolution traces obtained during the first illumination of etiolated leaves after a preillumination in white light for 0, 1 and 8 h, followed by 30 min darkness (for the fluorescence measurements, see explanations in the text).

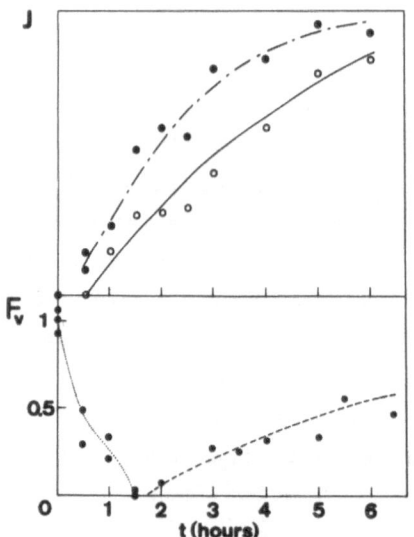

FIGURE 7. Upper part: amplitude (J) of the oxygen gush at the first (o) and second (●) 5 min illumination as a function of the preillumination time in white light; the preillumination is followed by 30 min darkness with the two measuring illuminations being separated by 5 min darkness.
Lower part: amplitude of the fluorescence variations ($F_V = F_{MAX} - F_0/F_0$) at the first illumination (see explanation in the text).

for preillumination times up to 5-6 h, the gush was the only manifestation of an oxygen evolution; for longer preillumination times only, a weak steady-state oxygen production was detected. The relative amplitude of the fluorescence fluctuations ($F_V = F_{MAX} - F_0/F_0$) is shown in fig. 7 B as a function of the preillumination time. Two phases were clearly distinguished.

For 0 to 90 min preilluminations, F_v decreased steadily. During this phase, F_v measured the amplitude of the slow fluctuation due to microcycle II. For preillumination times longer than 90 min, F_v increased steadily. In this case it measured the amplitude of the rapid fluctuation which is normally observed in a green leaf. It must be emphasized that the two fluorescence fluctuation types excluded strictly each other. A "mixed" state was never observed.

4. DISCUSSION

In the etioplasts, many of the known components of the photosynthetic electron transport chain are already present. Cytochrome f (4,5), b_6 (5) and the low potential forms of cytochrome b_{559} (6) have been detected. The high potential form of cytochrome b_{559} only appears after a few hours of greening (6); although it is thought to be associated with photosystem II, its role in the electron transport chain remains unclear (7). The main components which are needed before oxygen evolution becomes detectable are obviously the photoactive chlorophylls. In addition, a photoactivation of the donor side in photosystem II is needed (8,9) which is completed in a few min in leaves greened under flashes.

It is generally assumed that the spectral shifts which follow the photoreduction of protochlorophyll(ide) in whole leaves reflect preliminary reactions leading to the formation of free chlorophyll a. The latter would then be inserted into photosynthetic structures. In this case, the formation of the active oxygen evolving system would follow path I (see below). Ours results , together with the results of Smith 30 years ago, suggest that photosynthetically active pigments are structurally present in form of pigment-protein complexes which are formed already a short time after the photoreduction of protochlorophyll(ide) (path II). The arguments which support this suggestion may be summarized as follows:

1. In Smith's experiment, two illuminations separated by a a long (2 h) dark interval are required in order to detect an oxygen gush at the third illumination. The pigments involved

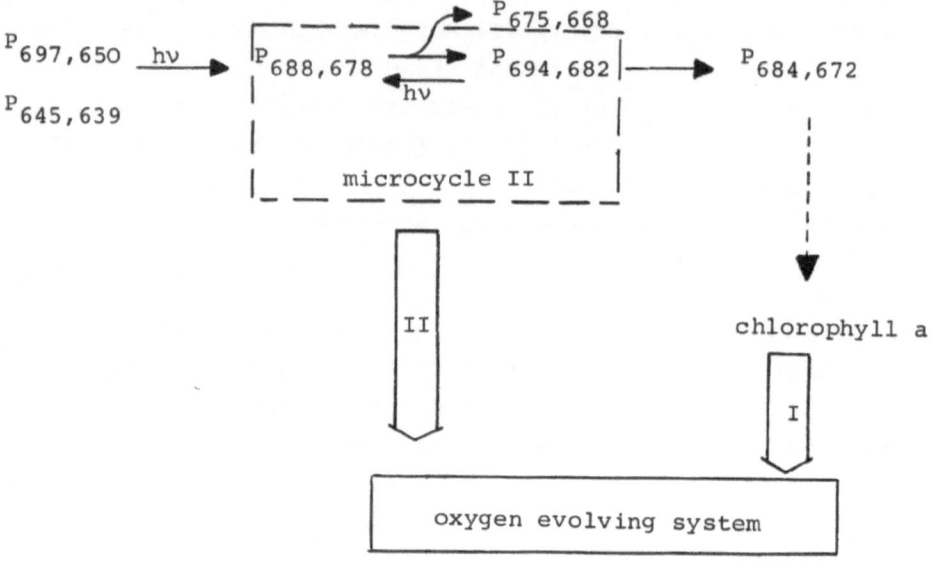

in the oxygen-evolving system in this early state of development are probably not originating from the initial protochlo-
rophyll(ide) pool of the etiolated leaf. They obviously come
from the reduction of the protochlorophyll(ide) pool regenerated during the dark interval. This point may already be deduced from the paper of Smith. Smith's data indicates that
the amount of oxygen produced during the third illumination
is roughly proportional to the amount of chlorophyll(ide) formed during the second illumination, i.e. to the amount of pro-
tochlorophyll(ide) reaccumulated during the dark period.

2. Our experiments clearly show, as this new chlorophyll(ide)
has appeared, that the oxygen gush is observed within a few
min (Fig.1). This is certainly not enough time for path I since this path should involve at least a very slow step (the
Shibata shift, from $P_{694-682}$ to $P_{684,672}$, took about 40 min
in these leaves).

3. The increase of the amplitude of the oxygen gush coincides

in time with the disappearance of the fluorescence variations linked to microcycle II described by Franck and Inoue (3) and Sironval et al.(10,this symposium). This coincidence suggests that these pigment-protein complexes, either may play a role in the system responsible for the oxygen gush, or have disappeared before it starts working.

At least two conditions are needed for the etiolated leaf pigments to become active in photosynthesis:

1. A sufficient period of time after the onset of the first illumination: we would suggest that this time is needed to allow the regeneration of some active protochlorophyll(ide) and to "prepare" the membrane structure in which the newly formed complexes become embedded.

2. An activation by continuous light: this phenomenon is probably similar to the so-called "induction process" already described by Strasser (8,9) in leaves greened under flashes. This activation occurs during the second illumination in a Smith's type experiment. It corresponds to the decrease of the contribution of the long wavelength emission to the overall fluorescence spectrum.

REFERENCES

1. Smith J H C. 1954. Plant Physiol. 29, 143-148
2. Schmid G H and Thibault P. 1979. Z.Naturforsch. 34c,414-418
3. Franck F and Inoue Y , in preparation
4. Hill R. 1954. Nature, 4428, 501-503
5. Davenport H E. 1952. Nature, 1112, 170-172
6. Whatley F R, Gregory P, Haslett B G and Bradbeer J W. 1971. Proceedings of the IInd International Congress on Photosynthesis (G Forti, M Avron and A Melandri, eds) Vol.III, 2375-2380, Dr W Junk, N.V. Publishers, The Hague
7. Bendall D S, Davenport H E and Hill R. 1971. Methods Enzymol, 327-344
8. Strasser R J. 1971. Verh. Schweiz. Naturfosch. 71, 104-106
9. Strasser R J and Sironval C. 1972. FEBS Letters 28,56-58
10.Sironval C, Franck F, Gysembergh R, Bereza B and Dujardin E, this volume.

ACKNOWLEDGMENTS

F.Franck thanks the Heinrich-Hertz-Stiftung (Minister für Wissenschaft und Forschung des Landes Nordrhein-Westfalen) for the award of a stipendium.
C.Sironval thanks the "Fonds National de la Recherche Scientifique" , Belgium, the "Ministère de la Politique Scientifi-

que" and the "Actions Concertées" nr 80/85-1983, Belgium for
financial support.
The authors thank Mr R.Gysembergh and M.Crepin for skilful
technical assistance.

THERMOLUMINESCENCE IN ETIOLATED BEAN LEAVES

F.FRANCK* and Y.INOUE**

*Universität Bielefeld, Fakultät für Biologie, Lehrstuhl Zell-
physiologie, Postfach 8640, D-4800 Bielefeld 1, W.-Germany
** Institute of Physical and Chemical Research (Rikagaku Ken-
kyusho), Laboratory of Plant Physiology, Wako-shi, Saitama
341, Japan.

1. INTRODUCTION

Since the experiments of Arnold (1) it is known that plant
material such as intact algal cells can store some of the light
energy absorbed by chlorophyll at low temperature. When the
sample is heated in darkness, the stored energy is re-emitted.
In green material illuminated at -196°C, five luminescence bands
can be observed at different temperatures during heating (2,3).
Although the assignment of each band is still a matter of controversy, four of these five bands, whose emission temperatures
are located above -20°C, are generally interpreted as the result of the recombination of electrons and holes stabilized
during the illumination at low temperature.

The fifth band, the so called Z band, is observed around -160°C,
and is not thought to be related to photosynthesis. Sane et al.
(3) observed this band also in green leaves inactivated by heating at +90°C and in dried isolated chlorophylls. They found
that its emission spectrum has a single maximum at 740 nm and
therefore concluded that the Z band results from the relaxation
of trapped electrons through the triplet state of chlorophyll,
both in vivo and in vitro. However, the nature of the pigment
state which is stabilized following excitation at -196°C remains
unclear.

In this paper, we report on some observations of the thermoluminescence Z band in etiolated bean leaves in relation to the
pigment-protein complexes contained in the leaf in the initial
stages of its development.

C. Sironval and M. Brouers (eds.); Protochlorophyllide Reduction and Greening. ISBN 90 247 2954 8
© 1984, Martinus Nijhoff/Dr W. Junk Publishers, The Hague/Boston/Lancaster.

2. MATERIALS AND METHODS

Beans (*Phaseolus vulgaris*, var.Commodore) were sown and
grown in darkness on vermiculite moistened with tap water.
Primary leaves were used after 12-14 days of growth.
Thermoluminescence experiments were carried out in a set-up
previously described by Ichikawa et al.(2). Leaf discs were
illuminated for 5 min in liquid nitrogen with a strong red light
(7 mW.cm^{-2}). Following illumination, samples were rapidly
transferred to the thermoluminescence apparatus and rewarmed
at a rate of about 1°C/sec to a temperature of +95°C. The luminescence emitted during heating was monitored and plotted
against temperature on a X-Y recorder.

3. RESULTS

A thermoluminescence profile typical of an etiolated leaf
is shown in fig.1.a. A weak band was detected around -170°C.
Its emission temperature, as well as its shape, were identical
to those of the usual Z band in a green leaf recorded in the
same experimental conditions. The profile in fig.1.b was obtained when the leaf was illuminated by a strong white flash
immediately before freezing. In this case, all the active protochlorophyllide was reduced into the $P_{688,678}$ chlorophyllide.
The Z band was then 5 to 6 times higher than in the etiolated
control leaf.
A dramatic decrease of the Z band was observed if the leaf
was allowed to remain in darkness for 2 min after the flash
at room temperature, and then was freezed (fig.1.c).
During the two min dark interval, $P_{688,678}$ was transformed into $P_{694,682}$, which has been shown to be also chlorophyllide
(4). The partial regeneration of the high Z band occured if
the leaf containing $P_{694,682}$ was illuminated for 15 sec by a
strong red light at room temperature and freezed immediately
after the end of the illumination (fig.1.d). But if the leaf
was kept in darkness for a second 2 min interval after the
15 sec illumination at room temperature, the Z band was weak
again (not shown).

In addition to the Z band, a very weak band was observed around +55°C in all profiles. At this temperature, a band is also present in green leaves (the so-called C band); its significance in green material is still unclear (2). Since this band was almost confused with the noise of the apparatus in our experiments, we were not able to study its behaviour in etiolated leaves.

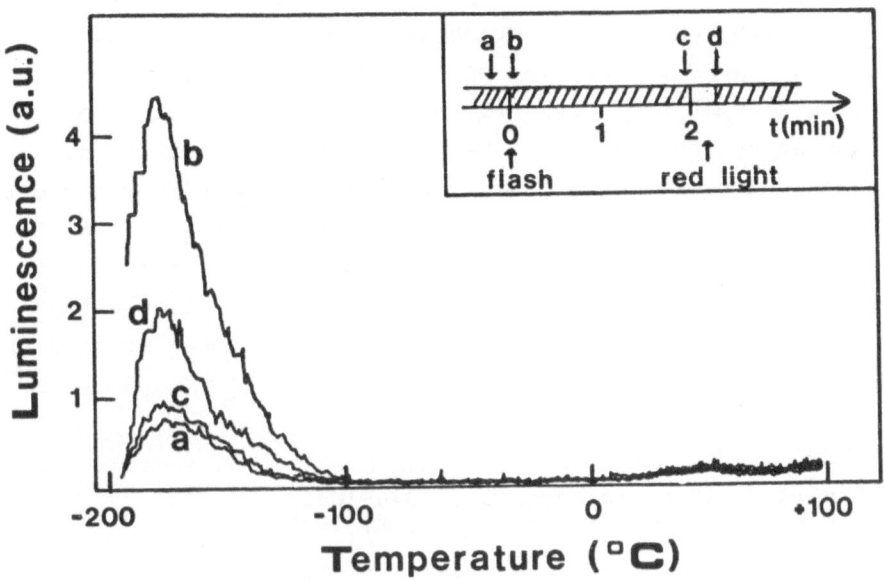

FIGURE 1. Thermoluminescence profiles of etiolated leaves illuminated in liquid nitrogen for 5 min. Pre-treatment at room temperature as indicated in the insert.

4. DISCUSSION

In previous experiments (5), we have shown that, once chlorophyllide $P_{688,678}$ has been transformed into chlorophyllide $P_{694,682}$, the latter can be converted by light back to the initial $P_{688,678}$. As this reaction is the second photoreaction observed in etiolated leaves, and as it is dark-reversible, Sironval et al. (6) referred to "microcycle II" when dealing with the couple of reactions:

$$P_{694,682} \xrightleftharpoons{h\nu} P_{688,678}$$

The results reported here show that these two chlorophyllide-protein complexes differ by their ability to emit the thermo-luminescence Z band after an illumination at -196°C. $P_{694,682}$ was the main pigment species in the leaf frozen 2 min after the initial flash, corresponding to a weak Z band (curve c). $P_{688,678}$ was present immediately after the initial flash (curve b) or immediately after the 15 sec illumination (curve d), corresponding to a higher Z band.

This suggests that the ability of pigment-protein complexes to use light in some photoreaction (such as the transformation of $P_{694,682}$ into $P_{688,678}$) may be reflected in a decrease of their ability to emit the Z type thermoluminescence. This is also suggested by the fact that the photoactive protochlorophyllides emit only a weak Z band (curve a). Such an opposite behaviour of the photoactivity and the Z type thermoluminescence also a-rises from the works of Ichikawa et al. (2) and of Inoue et al. (7), who observed a decrease of the Z band along with the activation of the water-splitting system in leaves greened un-der flashes.

REFERENCES

1. Arnold W. 1966. Science 154, 1046-1049
2. Ichikawa T, Inoue Y and Shibata K. 1975. Biochim. Biophys. Acta, 408, 228-239
3. Sane P Y, Tatake V G and Desai T S. 1974. FEBS Letters, 45 290-294
4. Ogawa T, Bovey F, Inoue Y and Shibata K. 1975. In Procee-dings of the 3rd International Congress on Photosynthesis (M Avron ed.) pp 1829-1832, Elsevier,Amsterdam
5. Franck F and Inoue Y. to be published
6. Sironval C, Franck F, Gysembergh R, Bereza B and Dujardin E., this symposium.
7. Inoue Y, Ichikawa T, Kobayashi Y and Shibata K. 1975. in Proocedings of the 3rd International Congress on Photosyn-thesis (M Avron, ed.) pp 1833-1840, Elsevier, Amsterdam.

MEMBRANE DEVELOPMENT AND PIGMENT COMPOSITION.

BIOSYNTHESIS OF THE PIGMENT-PROTEIN COMPLEXES

G. AKOYUNOGLOU

Biology Department, Nuclear Research Center "Demokritos",
Athens, Greece

1. INTRODUCTION

In dark grown plants the proplastids are transformed into etioplasts. Following the formation and growth of the etioplast we observe first the appearance of a number of perforated prothylakoids,
and later the formation of the prolamellar body. The prolamellar
body increases in size as the time in darkness increases (1).
During etioplast growth accumulation of protochlorophyllide
(PChlide), bound to the PChlide-oxidoreductase, takes place.
The rate of accumulation follows a sigmoidal curve, i.e., it is
initially low, then increases, and finally stops after reaching
a steady level (2). The PChlide is located in both the prothylakoids and the prolamellar body. Prothylakoids contain also
a number of peptides, while the prolamellar body is rich in
lipids (3,4). As the time in darkness increases further the
structure of the etioplast is slowly degraded due to senescence.

The differentiation of etioplast into chloroplast is a very
complicated and multi-step process, and in angiosperms depends
on light (5,6). It involves the biosynthesis of pigments, lipids
and proteins, and it requires the close cooperation of both,
the chloroplast and the cytoplasmic protein synthesizing systems.
Most of the proteins are synthesized on cytoplasmic ribosomes,
and then they are transported through the chloroplast envelope,
by a post-translational mechanism, processed, inserted into the
developing thylakoid, and assembled into structural and functional units (7). All of the pigments, together with lipids and
proteins, are assembled into pigment-protein complexes, which
make up the two photosystems (PS), PSI and PSII (8,9). Mild
SDS-PAGE procedures have allowed the separation of up to six ma-

C. Sironval and M. Brouers (eds.); Protochlorophyllide Reduction and Greening. ISBN 90 247 2954 8
© 1984, Martinus Nijhoff/Dr W. Junk Publishers, The Hague/Boston/Lancaster.

jor pigment-protein complexes, and three unidentified minor
ones, from SDS-solubilized thylakoids of mature chloroplasts,
the CPIa, CPI, LHCP$_1$, LHCP$_2$, CPa and LHCP$_3$ (9), CPIa and CPI
originate in PSI, and the rest in PSII. CPIa is a highly organized form containing CPI and a number of lower Mr polypeptides;
CPI is the P700-Chl$\underline{a}$-rich protein complex originating in the
core of the PSI unit; the lower Mr polypeptides are components
of the light harvesting of PSI (LHC-I), and Fe-S proteins (10).
LHCP$_1$ and LHCP$_2$ are oligomers of LHCP$_3$ and act as light harvesting antennae of PSII, while CPa originates in the core of the
PSII unit; LHCP$_1$ and LHCP$_2$ contain also a small amount of LHC-I
(Antonopoulou, et al., this volume).

2. MECHANISM OF COMPLEX ASSEMBLY AND THYLAKOID DEVELOPMENT

Recent experiments (11-13), on the reorganization of thylakoid
components, which occurs early in development (in etiolated
leaves preexposed to continuous light for a short period of time,
and then transferred to darkness), suggest a mechanism operating
in the assembly of the complexes and the development and growth
of the thylakoid. According to this mechanism, in the early
stages of etioplast development (in young etiolated plants) all
thylakoid components are synthesized, but with a relatively low
rate, except Chlorophyll $\underline{a}$ (Chl$\underline{a}$) which accumulates as PChlide $\underline{a}$.
From all thylakoid components synthesized in darkness during
etioplast development only a number of them accumulate, the rest
are digested, since in the absence of Chl$\underline{a}$ they can not be stabilized. As the age of the etiolated tissue increases a number
of the thylakoid components stops to be formed. Depending, therefore, on the age of the etiolated tissue, i.e., on the developmental stage of the etioplast, one may or may not observe the
expression of some of these components in the dark grown plants.
This can explain the conflicting data in the literature concerning the presence of mRNA of the LHCP-apoprotein in dark grown
leaves. As it is known (14,15), the mRNA of the LHCP-apoprotein
has been detected in dark grown peas, but not in dark grown barley. The difference may probably be the age of the etiolated
tissue used in each case, i.e., young etiolated peas, and rela-

tively old etiolated barley. I believe that in young etiolated leaves all thylakoid polypeptides are expressed.

Illumination of young etiolated leaves transforms the PChlide a into Chlide a, and stimulates also the synthesis of the other thylakoid components. Moreover, in old etiolated leaves, light reinduces also the synthesis of those components, which have been stopped to be synthesized in darkness. This light induced synthesis of the thylakoid components, i.e., the reinduction process, is controlled by phytochrome. This can explain our observation (16), that in young etiolated leaves, where the cells are in a growing and dividing state, and their etioplasts contain no or very small prolamellar body, there is no phytochrome effect on chloroplast development. In old etiolated leaves, however, where the cells have stopped to divide, and their etioplasts contain large prolamellar body, the phytochrome effect on chloroplast development is pronounced.

The development of thylakoid depends on the rate by which the different components are synthesized. If one, or more of the components are synthesized at a low rate, then their synthesis controls the way the thylakoid develops. The rate of Chla formation is a determining factor in thylakoid development, since most of the polypeptides formed are stabilized by forming with Chla pigment-protein complexes. If Chla is formed at a low rate, then its synthesis regulates the assembly of the pigment-protein complexes, and accordingly the development of the thylakoid. This can be seen in experiments where etiolated leaves are exposed to intermittent light (2 min light-98 min dark)(17), or to ms flashes (18), or to low intensity far-red light (19).

3. COMPLEX ASSEMBLY IN INTERMITTENT LIGHT

In the case of greening in intermittent light, the Chla formed during the 2 min illumination of every cycle is very small, and so the concentration of Chla becomes the limiting factor. There is, therefore, a competition between the different apoproteins for the small amount of Chla; it seems that the apoproteins of the reaction centers have a higher affinity for Chla than the apoproteins of the LHC-I and LHCP, so they form CPI and CPa, i.e., the core of the PSI and PSII units, while the LHC-apoproteins, not being able to be stabilized, are digested (9). It is not known whether, during the intermittent light, the LHCs are assembled first, and then in the dark they are disorganized, since their Chla is taken up to be used for the formation of CPI and CPa, and their apoproteins digested, or not. In other words, whether in this case also there is formation and reorganization of the thylakoid components, similar to the one observed in etiolated plants greened in continuous light for a short time, and then transferred to darkness (11,12). Both hypotheses, however, seem probable. This explains the absence of LHCP in the intermittent light plants (5,9), even though the mRNA of the apoprotein is present and it is translated (20,21). If during the intermittent light the relative rate of Chla formation is increased, either by shortening the dark period in the light-dark cycles (e.g., 2 min light every 60 or 20 min), or by decreasing the rate of synthesis of the reaction center apoproteins, as in the case of blue or red intermittent light, or after a long time in the intermittent light, then some Chla becomes available to bind on the LHCP-apoprotein, forming LHCP, and increasing thus the size of the PSII unit.

It is known that Chlb is formed from Chla (22), and that both Chla and Chlb are necessary for the stabilization of the LHCP-apoprotein (6,12). It is not known, however, whether,during the assembly of the LHCP,Chla binds first to the apoprotein,and then some of the Chla is transformed into Chlb, or Chlb is formed first and then both Chla and Chlb bind on the apoprotein. If the first hypothesis is correct then the LHCP-apoprotein, or an other protein close to the apoprotein, should be the enzyme which transforms Chla into Chlb. In this case, the Chlb-less mutant, which synthesizes the LHCP-apoprotein, but which does not accumulate LHCP (23), has to be deficient in the Chlb-forming enzyme, or its LHCP-apoprotein has to be different from the normal one.

Since during the intermittent illumination the Chla synthesized is used for the formation of CPI and CPa only, the plastids formed,the "protochloroplasts", have some characteristic features. They have single "primary" thylakoids, small PSI and PSII

Table 1. PSI, PSII, P700 and cytf content of plastids obtained from six-day etiolated bean leaves exposed first to intermittent light (LDC), and then transferred to continuous light (CL).

Sample	Chl(a+b) (ug/g fr wt)	Chla Chlb	Activity mg Chl.h		P700 Chl	cytf Chl
			PSI	PSII	(mmoles/mole)	
28 LDC	210	∞	485	950	6.90	7.46
57 LDC	393	26.0	480	950	7.19	7.52
86 LDC	610	14.0	475	890	6.99	7.20
28 LDC+49h CL	2765	3.0	200	170	2.52	1.55
57 LDC+49h CL	1529	3.9	225	303	2.51	2.82
86 LDC+49h CL	1214	4.7	230	384	2.56	3.58
Green Control	2800	3.0	200	170	2.50	1.56

PSII activity: umoles DCIP reduced; PSI activity: umoles oxygen consumed.

Table 2. PSI, PSII, P700 and cytf content of plastids obtained from six-day etiolated bean leaves exposed first to intermittent light (LDC), and then transferred to continuous light (CL).

Sample	Chl(a+b) (ug/g fr wt)	P700	Activity g fr wt.h		cytf nmoles/g fr wt	PSII PSI
			PSI	PSII		
28 LDC	210	1.45	102	200	1.68	2.4
57 LDC	393	2.83	188	373	3.18	2.4
86 LDC	610	4.26	289	543	4.72	2.3
28 LDC+49h CL	2765	6.96	553	470	4.70	1.0
57 LDC+49h CL	1529	3.84	344	463	4.67	1.7
86 LDC+49h CL	1214	3.10	279	466	4.73	2.0
Green Control	2800	7.00	560	476	4.70	1.0

PSII activity: umoles DCIP reduced; PSI activity: umoles oxygen consumed; The ratio PSII/PSI was estimated on the assumption that in green chloroplasts it is equal to 1.

units, i.e., high PSI and PSII activity per mg Chl and high P700/Chl ratio (see Table 1), and they are deficient in Chlb, LHCP, LHC-I and grana (5,6,17,24). The concentration of the PSI and PSII units increases as the time in intermittent light increases, and in the case of PSII, it reaches a value higher than that of the mature chloroplast, while in the case of the PSI, it reaches a much smaller value (Table 2). Throughout the inter-

mittent light exposure the PSII/PSI ratio remains constant, indicating that the core of the PSI and PSII units, i.e., CPI and
CPa, are formed with a constant rate, but the rate of assembly
of the PSII units is much higher than the rate of assembly of
the PSI units. It is not known whether intermittent light stimulates in a different degree the rate of synthesis of the reaction center apoproteins, which has an effect on the rate of
assembly of the PSI and PSII units, or if there is a different
kind of control.

All electron transport components are also synthesized throughout the intermittent light period, and incorporated in the thylakoid, so that the whole electron transport chain from H_2O to
$NADP^+$, is assembled and functioning properly. The concentration
of the electron transport chain per chloroplast, e.g. cyt/chlo-
roplast, reaches the concentration of the mature chloroplast
(Table 3). Considering the relative concentration of cytochromes and those of PSII units, PSI units and P700 (Table 2), we
can conclude that (a) the rate of cytochrome formation (i.e.,
electron transport chain/chloroplast) follows closely the rate
of PSII unit formation, and both of them reach the concentration
of the mature chloroplast; and (b) the number of PSII units operating on the same electron transport chain (PSII/cytf) remains
constant during development in intermittent light, and equal to
that of the mature chloroplast, while the number of PSI units
per electron transport chain (PSI/cytf) is much lower.

Similar thylakoid development takes also place when etiolated plants are exposed to ms flashes (one every 15 min), or
to 1 sec flashes (one every 60 min), or to far-red light. Under
the ms (or 1 sec) flashes regime Chla is mainly synthesized,
but with a low rate, while the concentration of Chlb is very
low. The Chla synthesized is mainly used for the formation of
the pigment-protein complexes, which build up the core of the
PSI and PSII units. The plastids formed have single "primary"
thylakoids with no grana stacks, small in size PSI and PSII
units, but with inactive H_2O-splitting enzymes, and they are
deficient in LHCP (11,25-28). Under prolonged greening in far-
red light Chla is initially synthesized with a low rate, and

Table 3. Cytochrome content of plastids obtained from six-day etiolated bean leaves exposed first to intermittent light (LDC), and then transferred to continuous light (CL).

Sample	Cytf	Cytb-559$_{HP}$	Cytb-559$_{LP}$	Cytb-563
			(mmoles/mole Chl)	
28 LDC	7.46(1.68)	- -	17.63(4.07)	18.30(4.16)
57 LDC	7.52(3.18)	2.40(1.02)	18.20(7.70)	18.50(7.86)
86 LDC	7.20(4.72)	5.02(3.30)	16.05(10.50)	17.50(11.57)
28 LDC+49h CL	1.55(4.70)	2.57(7.74)	3.76(11.33)	4.10(12.16)
57 LDC+49h CL	2.82(4.67)	2.50(4.13)	6.41(10.58)	7.00(11.47)
86 LDC+49h CL	3.58(4.73)	2.64(3.48)	8.93(11.70)	9.52(12.59)
Green Control	1.56(4.72)	2.58(7.78)	3.75(11.30)	4.10(12.16)

The numbers in parentheses are nmoles/g fr wt.

only after the 4th day some Chlb appears. The plastids formed have also single thylakoids with no grana stacks. *In vivo* measurements showed that these plastids have the capacity for oxygen evolution and photophosphorylation, indicating that the whole electron transport chain is assembled. The capacity for oxygen evolution, however, appears only after the first day of greening, i.e., there is a 24h lag-phase (19). No pigment-protein complexes have been isolated from these plastids, neither the size of the PSI and PSII units estimated. I expect them, however, to have small in size PSI and PSII units, and the Chla synthesized to be used for the assembly of CPI and CPa. It will be also interesting to see whether the assembly of all thylakoid components of the electron transport chain have a 24h lag-phase, or only one of them, and which one. Ogawa and Shibata (29) greened etiolated leaves under continuous illumination with white light of very low intensity. Under these conditions Chla is mainly formed at a low rate, while the accumulation of Chlb is small (high Chla/Chlb). The plastids formed have high PSI activity/mg Chl, and high P700/Chl, but no PSII activity with H_2O as electron donor. It is not known whether the PSII unit is missing, as they suggested, or it is present but with inactive H_2O-splitting enzymes. It is probable that both PSI and PSII

units, small in size, are formed, and that either the H_2O-spit-
tling enzymes are inactive, or an other electron transport component close to PSII unit is missing.

4. RATE OF COMPLEX FORMATION AND FINAL STAGE OF THYLAKOID DEVELOPMENT

The way the thylakoids develop has also an effect on the
final stage of chloroplast development. This can be seen easily
im etiolated plants exposed to intermittent light and then transferred to continuous light. In continuous light the rate of Chla
synthesis increases, Chlb and the pigment-protein complexes LHCP
and LHC-I are formed, increasing thus the size of the PSI and
PSII units, and grana stacks appear. At the same time a process
of organization of supramolecular structure is set on (9,24),
during which the monomeric forms of the complexes are transformed into oligomeric ones, so that a certain ratio of mono-
mers/oligomers is established (30,31), and the "protochloroplasts"
are differentiated into chloroplasts. The degree of differentiation, however, depends on the time of preexposure to intermittent light. The preexposure time affects the final amount of
Chla, Chlb and LHCP accumulated, the size of the PSI and PSII
units, the PSII/PSI ratio, and the structure of the chloroplast
(24,32-34). This can be seen clearly in Tables 1 and 2. In plants
transferred to continuous light after short preexposure to intermittent light the Chl accumulated, and the amount of LHCP formed,
are equal to those of the green control, while in plants transferred to continuous light after a long preexposure to intermittent light the Chl accumulated, and the amount of LHCP formed,
are much less (32,33). The final concentration of the PSII units
(PSII activity/g fr wt), after transfer to continuous light
reaches always the concentration of the green control; this
happens either by formation of new units (transfer of plants to
continuous light after a short time in intermittent light), or
by destruction of preexisting ones (transfer of plants to continuous light after prolonged exposure to intermittent light. The
final concentration of the PSI units (PSI activity/g fr wt), or
P700/g fr wt), however, decreases with time of preexposure to

intermittent light. Thus the PSII/PSI ratio drops from 2.4 (in
intermittent light) to the value of the green control (1) in
plants preexposed to intermittent light for a short time, or to
a higher value (between 1 and 2) in plants preexposed to intermittent light for a long time. The size of the PSI unit increases after transfer to continuous light, and it reaches always the size of the green control. The size of the PSII unit
increases also in continuous light, but it reaches either the
size of the green control (short preexposure to intermittent
light), or a smaller size (long preexposure to intermittent
light). In other words, prolonged preexposure to intermittent
light affects especially the growth of the PSII unit, which
reaches an intermediate stage of development.

The final concentration of cytochromes (cytf/g fr wt) after
transfer to continuous light reaches in all cases the concentration of the green control, i.e., it parallels the concentration
of the PSII units (see Tables 2 and 3). Thus the number of PSII
units, and the number of electron transport chains per chloroplast remain constant in all steady states studied, while the
number of PSI units per chloroplast changes. This indicates
that the number of PSII units per electron transport chain remains constant, but the number of PSI units per electron transport chain varies. This variation seems to depend on the size
of the PSII unit; the larger the PSII unit size the more the
PSI units per chain, the smaller the PSII unit size, the less
the PSI units per chain. This indicates that the same factor
which controls the growth of the PSII unit, affects also the
formation of the PSI units in a parallel way. The nature of
this factor is not known; it is also not known what is the mechanism which controls the termination of thylakoid growth.

Table 1 shows that the concentration of PSI units per mg Chl,
or the P700/Chl ratio remains constant in all steady states
studied, in spite of the fact that there is a great difference
in the concentration of Chl in each case. In other words, there
is a constant relationship between Chl accumulated and the PSI
units formed. This can explain the observation that the P700/
Chl ratio has almost a constant value in various plants or

mutants that have been studied (35)

Thylakoid development similar to the one observed in plants preexposed to intermittent light for different periods of time, and then transferred to continuous light (i.e., thylakoids with large PSII unit size and low PSII/PSI ratio; small PSII units and high PSII/PSI ratio),has been observed in inhabitant sun- and shade-plants. As it was found (36), the Chl/Q ratio is high in shade-plants (600-800), and low in sun-plants (200-400), while the P700/Chl ratio is almost the same in both kinds of plants. Accordingly, the PSII/PSI ratio is high in the sun- plants (small in size PSII units), and low in the shade-plants (large in size PSII units).

5. CONCLUSIONS

The formation of the pigment-protein complexes is a step- wise process, and it is regulated, directly or indirectly, by light. Light stimulates the synthesis of the different thylakoid components, their assembly into functional units, and their rel- ative concentration. This has an effect on the rate and the mode of thylakoid development. Light affects not only the rate and mode of thylakoid development, but also its final stage of development, i.e., the size of the PSI and PSII units, the PSII/ PSI ratio, the relative concentration of the electron transport components, and the differentiation of single thylakoids into stroma and grana. Depending, therefore, on the light intensity, light quality and mode of illumination we can have: (a) thyl- akoids with small or large in size PSII units; this affects the relative concentration of the PSI units, and accordingly the PSII/PSI ratio. This is the case in the sun- and shade-plants, and in plants preexposed to intermittent light for various pe- riods of time and then transferred to continuous light. (b) thylakoids with no difference in the PSII or PSI units size, but with a difference in the relative concentration of the elec- tron transport components (high-light or low-light plants); or in the relative concentration of the electron transport compo- nents and the PSII/PSI ratio as well (blue or red-light plants). Through the regulatory effect of light, therefore, the plants can adapt to their environment, and increase their photosynthetic efficiency.

REFERENCES

1. Von Wettstein, D. 1958. Brookhaven Symposia in Biology 11, 138-159.
2. Akoyunoglou, G. and Siegelman, H.W. 1968. Plant Physiol. 43, 66-68.
3. Lutz, C., et al. 1981. Eur. J. Biochem. 118, 347-358.
4. Murakami, S. and Ikeuchi, M. 1982. In: Cell Function and Differentiation, Akoyunoglou, G. et al. eds., Part B, pp. 13-23, Alan R. Liss, N.Y.
5. Argyroudi-Akoyunoglou, J.H., et al. 1971. Biochem. Biophys. Res. Comm. 45, 606-614.
6. Argyroudi-Akoyunoglou, J.H. and Akoyunoglou, G. 1973. Photochem. Photobiol. 18, 219-228.
7. Ellis, R.J. 1981. Ann. Rev. Plant Physiol. 32, 111-137.
8. Anderson, J.M., et al. 1978. FEBS Lett. 99, 227-233.
9. Argyroudi-Akoyunoglou, J.H. and Akoyunoglou, G. 1979. FEBS Lett. 104, 78-84.
10. Argyroudi-Akoyunoglou, J.H. 1982. In: Cell Function and Differentiation, Akoyunoglou, G. et al. eds., Part B, pp. 277-290, Alan R. Liss, N.Y.
11. Argyroudi-Akoyunoglou, J.H., et al. 1982. Plant Physiol. 70, 1242-1248.
12. Akoyunoglou, G. 1982. In: Cell Function and Differentiation, Akoyunoglou, G. et al. eds., Part B, pp. 171-188, Alan R. Liss, N.Y.
13. Akoyunoglou, G. 1983. In: Proc. 6th Intern. Photosyn. Congress, Sybesma, C., ed., Martinus Nijhoff/Dr. W. Junk, The Hague, The Netherlands (In Press).
14. Apel, K. and Kloppstech, K. 1978. Eur. J. Biochem. 85, 581-588.
15. Cuming, A.C. and Bennett, J. 1981. Eur. J. Biochem. 118, 71-80.
16. Akoyunoglou, G. 1980. In: Photoreceptors and Plant Development, De Greef, J. ed., pp. 269-283, Antwerpen Univ. Press, Antwerpen.
17. Argyroudi-Akoyunoglou, J.H. and Akoyunoglou, G. 1970. Plant Physiol. 46, 247-249.
18. Akoyunoglou, G., et al. 1966. Physiol. Plantarum 19, 1101-1104.
19. De Greef, J. et al. 1971. Plant Physiol. 47, 457-464.
20. Viro, M. and Kloppstech, K. 1982. Planta 154, 18-23.
21 Kloppstech, K., et al. 1982. In: Cell Function and Differentiation, Akoyunoglou, G. et al. eds., Part B, pp. 101-110, Alan R. Liss, N.Y.
22. Akoyunoglou, G., et al. 1967. Chimika Chronika 32A, 1-5.
23. Schwarz, H.P. and Kloppstech, K. 1982. Planta 155, 116-123.
24. Akoyunoglou, G. 1981. In: Photosynthesis, Akoyunoglou, G. Ed., Vol. V, pp. 353-366, Balaban Intern. Sci. Services, Philadelphia, Pa.
25. Sironval, C. 1975. In: Proc. 3rd Intern. Congress on Pho-

254

tosynthesis, Avron, M., Ed., Vol. III, pp. 2153-2170, Elsevier, Amsterdam.
26. Remy, R. 1973. FEBS Lett. 31, 308-316.
27. Strasser, R.J. and Sironval, C. 1973. FEBS Lett. 29, 286-298.
28. Akoyunoglou, G. and Argyroudi-Akoyunoglóu, J.H. 1978. In: Photosynthetic Oxygen Evolution, Metzner, H., Ed., pp. 453-488, Academic Press, N.Y.
29. Ogawa, T. and Shibata, K. 1973. Physiol. Plantarum 29, 112-117.
30. Argyroudi-Akoyunoglou, J.H. 1981. In: Photosynthesis, Akoyunoglou, G., ed., Vol. III, pp. 547-558, Balaban Intern. Sci. Services, Philadelphia, Pa.
31. Kalosakas, K., et al. 1981. In: Photosynthesis, Akoyunoglou, G., Ed., Vol. V, pp. 569-580, Balaban Intern. Sci. Services, Philadelphia, Pa.
32. Tsakiris, S. and Akoyunoglou, G. 1981. In: Photosynthesis, Akoyunoglou, G., Ed., Vol. V, pp. 513-522, Balaban Intern. Sci. Services, Philadelphia, Pa.
33. Akoyunoglou, G., et al. 1981. In: Photosynthesis, Akoyunoglou, G., Ed., Vol. V, pp. 523-532, Balaban Intern. Sci. Services, Philadelphia, Pa.
34. Akoyunoglou, G., et al. 1978. In: Chloroplast Development, Akoyunoglou, G. and Argyroudi-Akoyunoglou, J.H., Eds., pp. 843-856, Elsevier/North Holland Biomedical Press, Amterdam.
35. Boardman, N.K. and Anderson, J.M. 1978. In: Chloroplast Development, Akoyunoglou, G. and Argyroudi-Akoyunoglou, J.H., Eds., pp. 1-14, Elsevier/North Holland Biomedical Press, Amsterdam.
36. Malkin, S. and Fork, D.C. 1981. Plant Physiol. 67, 580-583.

LOW-TEMPERATURE FLUORESCENCE SPECTRA OF Phaseolus vulgaris PLASTIDS DURING
DEVELOPMENT IN PERIODIC OR CONTINUOUS LIGHT IN THE PRESENCE OF THE INTERNAL
FLUORESCENCE STANDARD PHYCOCYANIN

A. CASTORINIS, J.H. ARGYROUDI-AKOYUNOGLOU AND G. AKOYUNOGLOU
Biology Department, Nuclear Research Center "Demokritos", Athens, GREECE.

1. INTRODUCTION

Previous studies in our laboratory showed that in the early stages of
thylakoid development (during greening of etiolated leaves in intermittent
light) only CPI (the Chla-rich P_{700}-protein complex originating in Photosystem I) and CPa (the Chla rich-protein complex originating in Photosystem
II) are detected (1-3). Later during greening the monomeric form of the
light-harvesting protein appears, which gradually gives rise to oligomeric
forms; concomittantly, the more organized structure of CPI, CPIa (the complex containing in addition to CPI, 4 low molecular weight polypeptides (4,
5)) gradually increases. Thus a gradual initial increase in monomers is noticed, which is followed by their decrease and concomittant organization
of oligomeric structures (3).

Parallel studies on the room temperature Chla fluorescence yield also
showed that the Fmax/Chl initially increases and then declines to a plateau
in greening plastids of etiolated plants exposed to white, blue or red intermittent or continuous light (6,7); in addition, it was found that plastids
rich in monomeric pigment-protein complexes have a higher F685/F730 ratio
in their 77 K emission |spectrum than those containing predominately oligo-
meric structures (6). Based (a) on the observation made earlier that Mg^{++}-
induced dissociation of oligomeric pigment-protein complexes (8), isolated
by sucrose density gradient centrifugation (9), enhances the low-temperature
fluorescence ratio F685/F730 (10), and (b) on the good correlation found
between the changes in Fmax/Chl and the changes in the organization of the
pigment-protein complexes during development (6), we proposed that the Chla
fluorescence yield changes reflect changes in the organization of the pigment-protein complexes (6,10).

In the present work we study the low temperature fluorescence spectra
of plastids during greening, in the presence of the internal fluorescence
standard, phycocyanin, according to the method of Gershoni and Ohad (11).

C. Sironval and M. Brouers (eds.); Protochlorophyllide Reduction and Greening. ISBN 90 247 2954 8

Our results further support our proposal (6,10) that changes in the fluorescence parameters reflect changes in the organization of the complexes.

2. PROCEDURE

Six-day-old etiolated Phaseolus vulgaris (var. red kidney) were exposed first to periodic light-dark cycles (LDC, 2 min white light-98 min dark), and were then transferred to continuous light (CL). The growth and handling of the seedlings was done as described (12). Chloroplasts were prepared by homogenizing the leaves with 0.3 M sucrose-0.01 M KCl-0.05 M phosphate buffer, pH 7.2; after filtration through six layers of gauze, the homogenate was centrifuged at 500 x g for 2 min, and the supernatant at 1000 x g for 10 min (leaves exposed to continuous light) or first at 1000 x g for 5 min and the supernatant at 3000 x g for 10 min (leaves exposed to periodic light). The chloroplast pellet was washed once and then resuspended in the homogenization buffer.

Low-temperature fluorescence spectra were recorded in a fluorometer set-up (13) as described earlier, by immersing the samples into a glass Dewar flask, made to fit the cuvette compartment, filled with liquid nitrogen (10). Comparative values of fluorescence intensities between the various samples were obtained after adding phycocyanin (PC) as an internal standard (11). Phycocyanin (a mixture of phycoerythrin and phycocyanin) was isolated from Nostoc muscorum , as described in (11). Since the contribution of the PC fluorescence at 653 nm may deform the spectrum at 680 nm (11), we first estimated the intensity of the emission at 730 nm, in the presence of PC; this value was then corrected for the contribution at 653 nm of Protochlorophyllide, observed in samples with no PC. The net value due to PC was then multiplied by a factor to give a fluorescence intensity of 40 arbitrary units, and the corrected value for F730 was obtained by multiplying the F730 intensity (in the presence of PC) by the same factor. The F690 was then estimated from the corrected intensity of F730 and the F730/F690 ratio in samples without the internal standard.

For the low-temperature fluorescence measurements, each sample was suspended in the isolation buffer at a Chl concentration of 30 ug/ml. PC, when present, was at a PC/Chl ratio of 0.8 (w/w). Chl was estimated as in (14) and PC as in (15).

RESULTS

Figure 1 shows the changes in the F730/F690 emission ratio of plastids obtained during greening of bean plants in intermittent or continuous light.

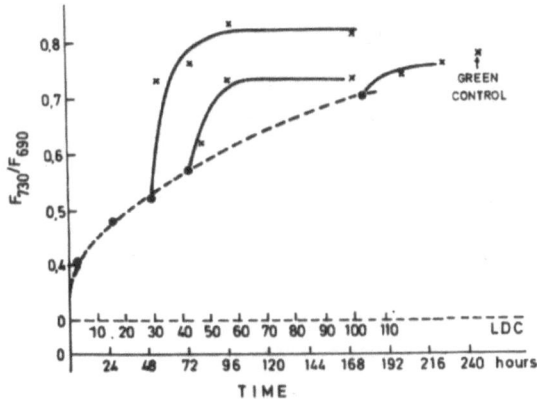

Figure 1. Changes in the F730/F690 ratio at 77 K, of plastids during greening. (- - -): plastids exposed to intermittent light; (——): plastids exposed to continuous light after preexposure to intermittent light. The F730
represents the maximum fluorescence at about 730 nm.

The ratios shown are those obtained without the use of the internal standard
phycocyanin. The increase in the ratio, which suggests that the relative
growth of F730 is faster than that of F690, is more pronounced when plants
are transferred to continuous light after a short preexposure to periodic
light, whereas after prolonged preexposure only a small increase in the ratio is noticed. In all cases the F730/F690 ratio reaches that found in the
completely green plant, used as control.

Figure 2 shows the 77 K fluorescence spectra of intermittent light plastids, in the absence of phycocyanin, normalized at 690 nm. The fluorescence
maximum at 653 nm, which is probably due to protochlorophyllide holochrome
(16) is relatively high in plastids of intermittent light plants, especially
in those of the 3 LDC, but as exposure to intermittent light decreases, the
653 nm maximum gradually decreases and appears only as a shoulder in the
emission spectrum. The maximum at 730 nm (due to the antennae Chla of PSI
(17-19)) is not present in the intermittent light plastids of 3 LDC; only
a slight shoulder is noticed, even though these plastids have considerable
activity of PSI (20). After 30 LDC, however, a clear emission band appears
at about 717 nm. This band increases faster tan that at 690 nm, and its max
is red shifted to finally reach the wavelength of 724 nm (in plastids of
103 LDC) or of 735 nm (in plastids of plants transferred to continuous light
(Figures 3 and 4).

Similar results have been reported concerning the increase of the F735
relative to F690 and its red shift, during development of Cucumis (19).

258

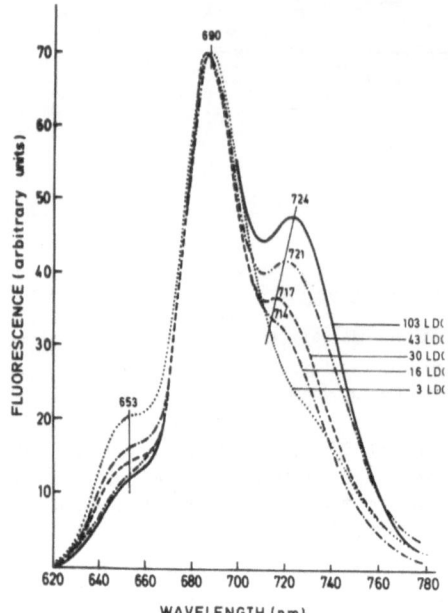

Figure 2

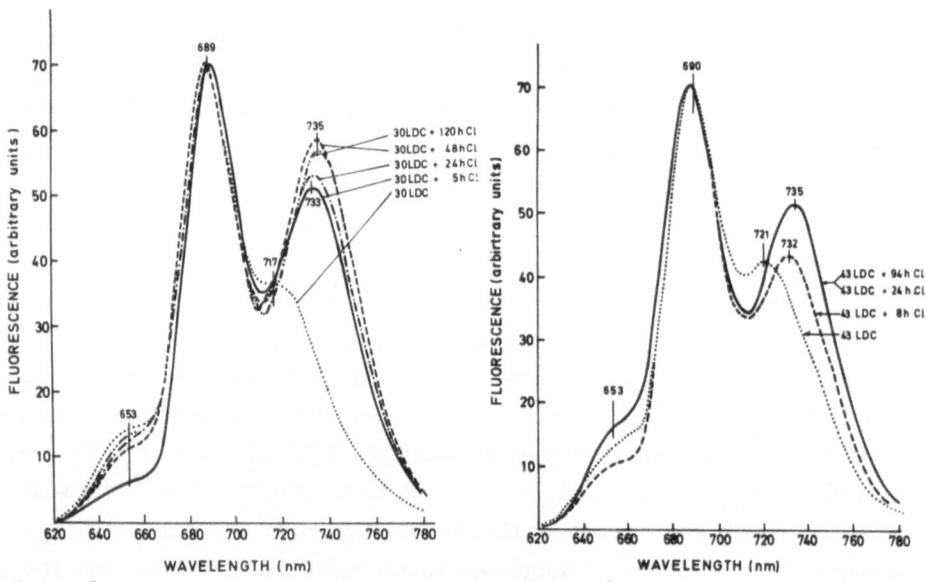

Figure 3   Figure 4

Low temperature fluorescence spectra of Phaseolus vulgaris plastids grown in intermittent light (Figure 2) and then transferred to continuous light after 30 light-dark cycles (Figure 3) or 43 light-dark cycles (Figure 4). The samples contained 30 ug Chl/ml. The normalization of the fluorescence intensity at 690 nm was done by changing the exciting light intensity. Spectra not corrected for photomultiplier sensitivity. No phycocyanin added.

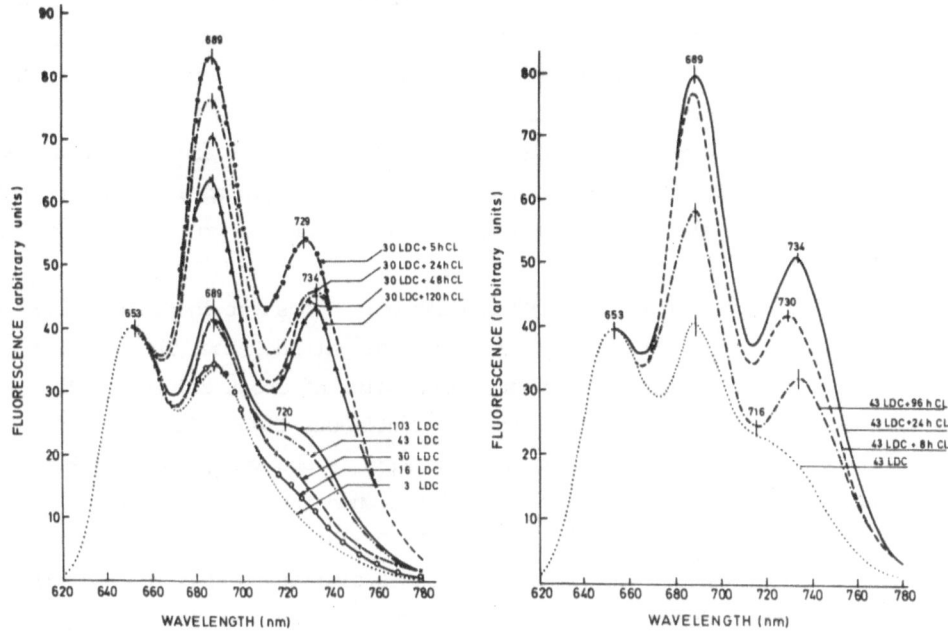

Figure 5. Low temperature fluorescence spectra, in the presence of phycocyanin, of Phaseolus vulgaris plastids, grown in intermittent light, or in
continuous light after preexposure to intermittent light for 30 light-dark
cycles (left) or 43 light-dark cycles (right). Phycocyanin was present at
a PC/Chl ratio of 0.8 (w/w). The normalization of the fluorescence intensity at 653 nm was done by changing the exciting light intensity. The samples
contained 30 ug Chl/ml.

Figure 5 shows the low temperature fluorescence emission spectra of intermittent light plastids in the presence of the internal fluorescence standard
phycocyanin. In these spectra the emission maximum at 653 nm is partly due
to protochlorophyllide holochrome (16), and mainly to phycocyanin. As the
plants green in intermittent light, an increase in both F720 and F690 is
clearly noticed; however, if we take into account the contribution of the
emission at 653 nm due to protochlorophyllide (Figure 2), then we can calculate that during greening of the plants to intermittent light the F690
emission remains constant, while the F735 one increases with parallel shift
of the maximum to longer wavelengths (Figures 2 and 5). When the intermittent light plants are transferred to continuous light we notice an initial
fast increase in both F690 and F730, and after a period of about 24 hours
in continuous light their fluorescence intensities gradually decrease (see
Figures 3 to 6).

The overall changes in F690 and F730, during the various developmental stages of bean chloroplasts are shown in Figure 6. The values shown, calculated
from the spectra in the presence of phycocyanin, have been corrected for the
653 nm contribution of protochlorophyllide: as already described, the 653 nm
contribution of protochlorophyllide is estimated from the F730 intensity in
the presence of phycocyanin, and from the F730 intensity and the one at 653
nm, in the corresponding sample without phycocyanin. To obtain then the F653
intensity due to the internal standard, the F653 due to protochlorophyllide
is subtracted.

As shown in Figure 6, the F690 remains quite constant in plants grown in
intermittent light, while the F730 shows a slight increase. Thus the ratio
F690/F730 decreases gradually from a high value of about 2.4 to a value of
1.5. When the plants are transferred to continuous light after 30 or 43 LDC,
a fast initial increase in both F690 and F730 is observed, followed by a
pronounced decrease. This is also true even when the 12% of the F690 fluorescence, reported to contribute to the F730 (18) is subtracted from the
corresponding value at 730 nm. This trend is similar to that found in the
room temperature Chl_a fluorescence yield of plastids, during greening of
plants in continuous light (6). The increase and the subsequent decrease in
the absolute values of emission after transfer of the plants to continuous
light, follows follows a more or less similar trend at both wavelengths
(690 and 730 nm). Hoever, in all cases the increase of the F730 emission
is greater that that at 690 nm, so that as greening proceeds a gradual decrease in the F690/F730 ratio occurs.

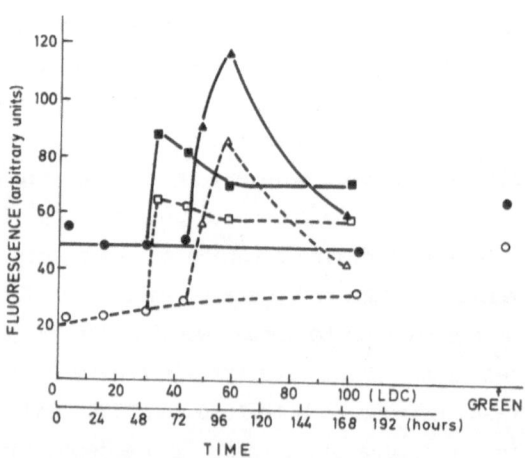

Figure 6. Changes in F690
(●, ■, ▲) and F730 (o, □, △)
in developing plastids.
Plants exposed to intermittent
light (●,o) or transferred to
continuous light after 30 LDC
(■,□) or 43 LDC (▲,△).
The F690 and F730 were calculated from the 77 K fluorescence spectra in the presence
of phycocyanin. Corrections
were as in the text. F730
represents the maximum emission at about 730 nm.

4. DISCUSSION

The results of this study show clearly that during greening in intermittent light the absolute intensity of the F690 emission remains more or less constant, but that at 720-730 nm increases steadily. Thus the ratio F690/F730 diminishes drastically from a value of about 2.2 to a value of 1.5 (after 100 LDC). During this time there is a pronounced red-shift of the long wavelength emission from about 717 nm to 725 nm. Upon transfer of the plants to continuous light, a sharp increase in both F690 and F730 occurs, followed by a gradual decline, as illumination is prolonged. The F690/F730 ratio, however, continuously declines to reach a value of about 1.2, found also in green plants. Under these illumination conditions, the long wavelength emission reaches that at 735 nm.

According to the generally accepted view, the F690 emission originates in PSII, while that at 730 nm in PSI (17-19,23). Studies on the isolated complexes show that CPa, the core complex of PSII, emits at 695 nm (23); LHCP, the light harvesting complex of PSII (24,25) at 685 nm (10, 18, 26); CPI, the core complex of PSI at 720 nm (10, 26) and the CPIa "oligomer" of CPI at 730 nm (10,26). Furthermore it has been shown that Mg^{++} (10) or H^+ (27) induce dissociation of the isolated CPI and CPIa complexes, resulting in drastic increase in the F680/F730 ratio. This suggests that the dissociated components of complexes originating in PSI may also emmit at 680 nm (10,27). We do not know whether this dissociation of the PSI complexes leads to separation of their light-harvesting antennae (which may be responsible for the 680 nm emission); however, it is very probable that PSI also contributes to the 680-690 emission depending on its microenvironment. The F720 originates in the PSI antennae closely packed around the PSI core; this is suggested from the finding that the isolated CPI complex, having a ratio of P_{700}/Chl of about 30 (9), emits at 720 nm. On the contrary, the F735 emission results from the well organized PSI complex, in which its reaction center component, CPI, is tightly connected with antennae and other polypeptides; this is suggested from the finding that the isolated CPIa "oligomer" of CPI, containing apart from CPI, 4 lower Mr polypeptides (4,5) Chlb and xanthophylls, and having a P_{700}/Chl ratio of 100 (9), emits at 730 nm (10,26).

Taking into account all these observations, our results concerning the changes observed during development, can now be attributed to the organization process of the complexes in the thylakoid: the biosynthesis and organization of components within each complex, and the subsequent gradual organization of the monomeric pigment-protein complexes into oligomeric ones within the photosynthetic units, resulting in very efficient entities for capturing light energy.

Thus we believe that early during greening in intermittent light the high F690 emission reflects on one hand the fluorescence of the PSII core complex, (CPa), and on the other the fluorescence of the less organized complexes, and their pigments not yet bound on proteins (as suggested also from the high Fo level of the room temperature fluorescence yield of these plastids (28)). The F730 emmission is very low, since no organized supramolecular structure of PSI (CPIa) is yet formed. Later in intermittent light, the CPI monomers are organized into CPIa, resulting in more pronounced emmission at 730 nm.

After transfer of the intermittent light plants to continuous light, increased biosynthesis of the complexes takes place; this is reflected by the increase in both F730 and F690 emissions . As exposure to continuous light is prolonged, however, two concurrent phenomena seem to take place: on one hand a gradual organization of the components of each monomeric form within each complex, resulting in more efficient utilization of light and thus to a decline of both emissions at 730 abd 690 nm; and on the other hand a gradual organization of monomers to aggregated supramolecular structures reflected by the transformation of the monomeric form of CPI to the highly organized structure CPIa, resulting in the F690/F730 ratio decline.

The continuous red-shift of the long wavelength fluorescence from 717 nm to 735 nm further reflects the formation of CPI in the early stages of greening, and its gradual incorporation into the highly organized CPIa form.

According to our proposal, the general concept that the fluorescence at 680-690 nm arises from the Chl a of PSII, whereas that at 725-735 nm from the Chla of PSI (7-19,23), is valid only if the PSI and PSII complexes are organized into supramolecular structures; otherwise, depending on the extent of their dissociation, PSI complexes may also fluoresce at about 680 nm (10).

This proposal explains why plastids grown in intermittent light for 3 light-dark cycles (and expected to contain less organized structures) show only the 690 nm emission maximum, eventhough they have very high PSI and low PS II activity (20). This proposal explains also, why certain preparations of PSI fluoresce at 680-690 nm (28-30), especially when high amount of detergent is used, since detergents lead to dissociation of the thylakoid pigment-protein complexes.

REFERENCES

1. Argyroudi-Akoyunoglou, J.H.,Feleki, Z. and Akoyunoglou, G. 1971. Biochem. Biophys. Res. Commun. 45, 606-613.
2. Argyroudi-Akoyunoglou, JH and Akoyunoglou, G. 1979: FEBS Lett. 104, 78-84.
3. Kalosakas, K. Argyroudi-Akoyunoglou, J.H. and Akoyunoglou, G. 1981. In: Photosynthesis, Akoyunoglou, G., ed. Vol V, pp.569-580, Balaban Int. Sci. Services, Phil. Pa.
4. Argyroudi-Akoyunoglou, J.H. 1982. In: Cell function and Differentiation Part B, pp.277-289, Alan R. Liss Inc., N.Y.
5. Anderson, J.M. 1980. Biochim. Biophys. Acta 591, 113-126.
6. Castorinis, A., Argyroudi-Akoyunoglou, J.H. and Akoyunoglou, G. 1982. Photobiochem. Photobiophys. 4, 283-291.
7. Anni, H. and Akoyunoglou, G. 1981. In: Photosynthesis, Akoyunoglou, G.,ed., Vol V, pp. 885-894, Balaban Intern. Sci. Services, Phil. Pa.
8. Argyroudi-Akoyunoglou, J.H. 1980. Photobiochem. Photobiophys. 1, 279-287.
9. Argyroudi-Akoyunogou, J.H. and Thomou, H. FEBS Lett. 135, 177-181.
10. Argyroudi-Akoyunoglou, J.H., Castorinis, A. and Akoyunoglou, G. 1982. Photobiochem. Photobiophys. 4, 201-210.
11. Gershoni, J.M. and Ohad, I. 1980. Anal. Biochem. 104, 315-320.
12. Arggyroudi-Akoyunoglou, J.H. and Akoyunoglou, G. 1970. Plant Physiol. 46, 247-249.
13. Argyroudi-Akoyunoglou, J.H. and Akoyunoglou, G. 1977. Arch. Biochem. Biophys. 179, 370-376.
14. Mackinney, G. 1941. J. Biol. Chem. 140, 315-322.
15. Bennett, A. and Bogorad, L. 1973. J.Cell Biol. 58, 419-435.
16. Kahn, A., Boardman, N.K. and Thorne, S.W. 1970. J.Mol. Biol. 48, 85-101.
17. Butler, W.L. and Kitajima, M. 1975. Biochim. Biophys. Acta 396, 72-85.
18. Strasser, R.J. and Butler, W.L. 1977. Biochim. Biophys. Acta 462, 307-313.
19. Mullet, J.E., Burke, J.J. and Arntzen, C.J. 1980. Plant Physiol. 65, 823.
20. Tsakiris, S. and Akoyunoglou, G. 1981. In: Photosynthesis, Akoyunoglou, G., ed., Vol V, pp. 513-522. Balaban Intern. Sci. Services, Phil. Pa.
21. Argyroudi-Akoyunoglou, J.H. and Tsakiris, S. 1977. Arch. Biochem. Biophys. 184, 307-315.
22. Castorinis, A. and Argyroudi-Akoyunoglou, J.H. 1981. In: Photosynthesis, Akoyunoglou, G., ed., Vol. V. pp. 501-511. Balaban Intern. Sci. Services, Phil. Pa.
23. Satoh, K. 1980. FEBS Lett. 110, 53-56.
24. Akoyunoglou, G., Tsakiris, S. and Argyroudi-Akoyunoglou, J.H. 1981. In: Photosynthesis, Akoyunoglou, G., ed., Vol V., pp. 523-532. Balaban Intern. Sci. Services, Phil. Pa.
25. Andersson, B. and Anderson, J.M. 1980. Biochim. Biophys. Acta 593, 427-440.
26. Anderson, J.M., Waldron, J.C. and Thorne, S.W. 1978. FEBS Lett. 92, 227-233.
27. Argyroudi-Akoyunoglou, J.H. and Akoyunoglou, G. 1983. Arch. Biochem. Biophys. (in press).
28. Akoyunoglou, G. 1977. Arch. Biochem. Biophys. 183, 571-580.
29. Shiozawa, J.A. 1976. Brookhaven Symposia Biol. 28,361.
30. Brown, J.S. 1977. Photochem.Photobiol. 26,519-525.

Acknowledgements. We thank Dr. G. Papageorgiou for his generous gift of N. muscorum cells.

THE COMPOSITION OF STROMA AND GRANA THYLAKOID PIGMENT-PROTEIN
COMPLEXES

P. ANTONOPOULOU, J.H. ARGYROUDI-AKOYUNOGLOU AND G. AKOYUNOGLOU
Biology Department, Nuclear Research Center "Demokritos", Athens, GREECE.

1. INTRODUCTION

The light-harvesting chlorophyll a/b Complex, is the most abundant intrinsic protein of higher plant and green algal thylakoids (1-3). LHCP contains Chlorophyll a (Chl a) and Chl b in about equal amount, carotenoids,
traces of phospholipids and certain polypeptides (3,4). According to the
generally accepted view (5-13), the presence of LHCP is necessary for the
formation of grana stacks, and its major part is located there. However,
even though the LHCP represents a minor component of stroma lamellae, one
should not disregard its presence there, nor the presence of Photosystem
II (14,15).

Thus, if indeed LHCP promotes stacking, the question arises why the
LHCP found in stroma thylakoids is incapable to form grana stacks. A possible
answer to this question might be the high content of CPI (the Chl a-rich
P_{700}-protein Complex) in stroma lamellae, which has been proposed to be ne-
gatively charged (16). This may not allow the contact and appression of thylakoids due to electrostatic repulsion. A second possibility, however, could
be that the grana LHCP may be different in composition from the stroma lamellar LHCP. This was suggested from the different behavior of stroma and
grana LHCP towards cations (17).

In an effort to understand this point, we separated stroma and grana lamellae, after French press disruption of chloroplasts (18,19); these fractions
were washed free of ATPase and RuDP-case (20), and the pigment-protein complexes of each fraction were isolated in preparative amount after separation
by sucrose density gradient centrifugation (21). The separated complexes
were then analyzed for pigment and polypeptide composition.

Our results show a pronounced difference between the LHCP of stroma and
grana thylakoids: the Chl/protein ratio in the grana LHCP is higher than
that in the stroma thylakoid LHCP; grana LHCP is composed mainly of the 25
Kd polypeptide (4), while stroma LHCP contains in addition, appreciable

C. Sironval and M. Brouers (eds.); Protochlorophyllide Reduction and Greening. ISBN 90 247 2954 8
© 1984, Martinus Nijhoff/Dr W. Junk Publishers, The Hague/Boston/Lancaster.

amount of a 21 Kd component; the ratio Chl/xan in stroma LHCP is much higher than that in grana LHCP, and so is the ratio of Chl to the individual xanthophylls.

The difference in pigment composition between stroma and grana thylakoid LHCP points to a possible involvement of xanthophylls in grana stacking, and suggests that the pigment composition of stroma LHCP may reflect the pigment composition of the PSI light harvesting component, as well, since stroma lamellae are enriched in PSI units. In addition, the presence of the 21 Kd polypeptide component in stroma thylakoid LHCP brings up the question as to whether it may play a role in the mechanism of grana stacking.

2. PROCEDURE

2.1. Material and methods

Six-day etiolated Phaseolus vulgaris plants, exposed to continuous light for 144 hours were used (2,500 lux, incandescent Sylvania 50 W and fluorescent Sylvania cool white 115 W lamps). Chloroplasts were prepared as in (19), were disrupted by French press (18,19) after two passages at 6,000 psi. Grana and stroma thylakoids were separated by differential centrifugation of the homogenate, the fraction sedimenting between 1,000 x g for 10 min and 10,000 x g for 30 min used as grana fraction, and that sedimenting between 40,000 x g for 30 min and 200,000 x g for 60 min used as stroma lamellae. The fractions were either washed in 0.05 M Tricine-NaOH, pH 7.3, or first with 10 mM Na pyrophosphate and then with 0.3 M sucrose in 2 mM Tricine-NaOH, pH 7.8 (20) and then finally with Tricine. The washed thylakoid fractions were solubilized as described earlier (21) and their pigment-protein complexes were separated by SDS-sucrose density gradient centrifugation as described earlier (21). The SDS/Chl ratio was 6.6 for grana and 8.8 for stroma thylakoids. The final SDS concentration was 0.5%; the final Chl concentration in the grana sample applied on the gradient was 770 ug/ml, and that in stroma thylakoids 628 ug/ml. The pigment-protein complexes were then separated on gradients containing 0.2% Triton X-100, 0.2% Deoxycholate and 0.1% SDS. This detergent combination allows the separation of oligomeric forms in higher amount, and permits more extensive pigment-to-protein binding in the complexes. Analysis of pigments was done by thin layer chromatography on precoated silica plates (22). Chl was determined as in (23), and carotenoids as in (24,25) in a Perkin Elmer 356 spectrophotometer. Protein was determined by the Lowry method (26). Slab gel electrophoresis was done as described by Hoober and coworkers (27) and staining was done either with coomassie brilliant blue (27) or with $AgNO_3$ (28).

3. RESULTS

The pigment composition of Phaseolus vulgaris thylakoids is shown in Table 1. The molar ratio of Chl to carotenoids in the sample not treated for removal of peripheral proteins (sample A) is about 2.5. This value aggrees with that found earlier for bean leaves exposed to continuous light for 120 hours (29). The reported value for spinach thylakoids ranges between 4.5 and 6.0 (30-32). The later ratios were considered to be relatively high (30) and was proposed that they may indicate that not all spinach leaves were exposed to light. Treatment of bean thylakoids for removal of peripheral proteins (20), however, (samples B) clearly affects the pigment binding on the membranes, so that the molar ratio Chl/car becomes twice as high. In this work we report our results obtained with subchloroplast fractions washed for removal of peripheral proteins (20). It should be noticed, however, that the absolute values shown in the Tables differ depending on previous washing of the thylakoids, since pigments are liberated to various extent. The trend and the observed differences, however, remain unaltered.

The Chl/car ratio of the stroma and grana thylakoids were found to differ slightly. This difference is mainly due to the xanthophylls, which are mainly associated with grana. Thus, the molar ratio Chl/xan in grana lamellae is lower than that in stroma lamellae (6 vs 11). Similarly, the xan/b-car ratio is much higher in the grana fraction compared to that in the stroma

Table 1. Pigment composition (moles/100 moles Chl) in thylakoids and in the separated stroma and grana thylakoid fractions of Phaseolus vulgaris.

	Total Thylakoids		grana thylakoids		Stroma thylakoids	
	A	B	A	B	A	B
Chl a	75	71	69	62 (69)	93	74 (77)
Chl b	25	29	31	38 (31)	7	26 (23)
Chl a/Chl b	2.85	2.8	2.5	1.6 (2.8)	5.3	4.6 (4.3)
b-carotene	11	7.6	14	4.0 (5.8)	16.0	8.5 (10)
lutein	20	10.4	21	11.0 (8.9)	5.0	5.4 (4.8)
Neoxanthin	7	3.8	5.0	2.2 (2.2)	2.0	1.0 (0.1)
Violaxanthin	3	1.7	4.0	3.1 (1.0)	3.0	2.6 (0.8)
Chl/car	2.44	4.3	2.27	4.9 (5.6)	3.8	5.7 (6.3)
xan/b-car	2.72	2.1	2.14	4.0 (2.1)	0.6	1.0 (0.6)
Chl/xan	3.34	6.3	3.34	6.1 (8.3)	10.0	11.0 (17.5)

All values are calculated after TLC analysis on silica plates, but the ratio Chl a/Chl b is that found in the original sample. A: fractions not washed for removal of peripheral proteins (20). B: fractions washed as in (20). The values in parenthesis are from another experiment.

(4 vs 1), reflecting the enrichment of stroma lamellae in b-carotene.

To compare the LHCP of stroma and grana lamellae we chose the second band from the top of the gradient (21) containing the oligomeric forms of LHCP (LHCP^1+LHCP^2), since the LHCP monomer band resolved at the top may be contaminated by free pigments resulting from detergent action on the complexes; in addition, the later band is also contaminated by thylakoid polypeptides of Mr lower than 25 Kd (4).

The pigment composition of grana and stroma thylakoid LHCPs is shown in Table 2. A number of differences may be noted. Earlier workers have reported for the thylakoid LHCP a Chl/car molar ratio of 3-7 (1). We have found a ratio of 3-4 for the grana LHCP and a much higher ratio for that of the stroma. This difference is mainly due to the xanthophylls. Grana LHCP is enriched in xanthophylls (Chl/xan=3.7), whereas stroma LHCP has less xan (Chl/xan=21). As far as the % of each carotenoid is concerned (expressed as molar % of total carotenoids in the complex), both grana and stroma LHCP are enriched in lutein, while the content of Neoxanthin and violaxanthin is much lower in stroma than in grana LHCP.

The molar ratio of Chl/b-car in all LHCP complexes is very high; furthermore, the Chl a/lutein and the Chl b/lutein in granal LHCP is two to three times lower than that in the stroma thylakoid LHCP. Table 2 also compares the pigment composition of the other separated complexes. CPI contains only Chl a and b-carotene. The CPI of stroma thylakoids, compared to that of grana, seems to be enriched in b-carotene, and thus the molar ratio Chl/ /b-car is higher in grana CPI. The same applies for CPIa, the higher Mr form of CPI, containing in addition lower Mr polypeptides (4); CPIa of both grana and stroma thylakoids, in addition to Chl and b-carotene, contain small amounts of lutein and violaxanthin, whereas the absence of neoxanthin must be noted. In general, comparison of the CPIa and CPI with the LHCPs of both stroma or grana thylakoids, shows that the CPIs are enriched in b-carotene, the CPIas contain in addition small amount of xanthophylls, while the LHCPs are enriched in xanthophylls. This aggrees with the results found from the analysis of the complexes obtained from total chloroplast thylakoids.

The pronounced difference between the LHCP of stroma or grana thylakoids is further established by the molar ratios shown in Table 3. There are about twice as many moles of Chl and b-carotene, and many times more xanthophylls per Mr of complex (35,000 d) in grana LHCP than in the stroma complex.

Table 2. Pigment-composition (moles/100 moles Chl) in the grana and stroma lamellar pigment-protein complexes separated by SDS-Triton X-100-Deoxycholate sucrose density gradient centrifugation from French press disrupted Phaseolus vulgaris chloroplasts.

	GRANA				STROMA			
	$LHCP^3$	$LHCP^{1+2}$	CPI	CPIa	$LHCP^3$	$LHCP^{1+2}$	CPI	CPIa
Chla	60 (63)	66 (57)	85 (79)	(78)	(68)	61 (65)	94 (90)	(82)
Chlb	40 (37)	34 (43)	15 (21)	(22)	(32)	39 (45)	6 (10)	(18)
b-car	0 (0.9)	3.9 (1.2)	2.8 (7.1)	(7.3)	(0.2)	2.6 (0.7)	10.8 (12.8)	(12.2)
lu	8.6 (9.7)	11.0 (16.6)	0 (0)	(2.5)	(5.2)	4.2 (8.3)	0 (0)	(3.0)
Nx	2.6 (3.4)	8.5 (4.8)	0 (0)	(0)	(0.8)	0.3 (1.6)	0 (0)	(0)
Vx	2.2 (9.5)	7.5 (2.6)	0 (0)	(1.8)	(1.1)	0.2 (2.2)	0 (0)	(2.0)
Chla/Chlb	1.6 (1.9)	1.4 (1.6)	10 (11)	(11)	(1.8)	1.7 (2.0)	14 (18.4)	(9.0)
Chl/car	7.4 (6.5)	3.2 (3.4)	36 (13)	(8.6)	(13.7)	13.7 (7.8)	9.2 (7.8)	(5.8)
Chl/xan	7.4 (6.8)	3.7 (4.2)	-- (--)	(23.3)	(14.1)	21.0 (8.3)	-- (--)	(20.1)

The Chla/Chlb ratio is that found prior to TLC analysis. All other values are those after TLC analysis. The values in parantheses are from a second experiment.

Table 3. Pigment to protein ratio in the pigment-protein complexes isolated by SDS-Triton X-100-Deoxycholate sucrose density gradient centrifugation from SDS-solubilized grana and stroma lamellae of Phaseolus vulgaris.

SAMPLE	Chla+Chlb	car	b-car	xan	lu	Nx	Vx
			mole / M r of complex				
G R A N A							
LHCP³	7.1 (4.7)	0.96 (0.72)	0 (0.04)	0.96 (0.68)	0.61 (0.45)	0.19 (0.16)	0.16 (0.07)
LHCP¹⁺²	7.6 (5.5)	2.36 (1.38)	0.3 (0.07)	2.06 (1.31)	0.84 (0.91)	0.65 (0.26)	0.57 (0.14)
CPI	21.5 (18.0)	0.59 (1.28)	0 (1.28)	0 (0)	0 (0)	0 (0)	0 (0)
CPIa	(23.0)	(2.67)	(1.68)	(0.99)	(0.58)	(0)	(0.41)
S T R O M A							
LHCP³	(3.71)	(0.27)	(0.01)	(0.26)	(0.19)	(0.03)	(0.044)
LHCP¹⁺²	4.1 (2.2)	0.3 (0.28)	0.11 (0.02)	0.19 (0.26)	0.17 (0.18)	0.01 (0.03)	0.01 (0.05)
CPI	36.9 (15.7)	3.99 (2.02)	3.99 (2.02)	0 (0)	0 (0)	0 (0)	0 (0)
CPIa	(16.2)	(2.79)	(1.98)	(0.81)	(0.49)	(0)	(0.32)

The molecular weights of the complexes on which the calculations are based are: 35 Kd for LHCP, 130 Kd for CPI and 190 Kd for CPIa. The values shown in parentheses are from a second experiment.

The calculations shown in Table 3, however, are based on the assumption that both stroma and grana LHCP have similar polypeptide composition and similar pigment-to-protein binding.

Dissociating SDS-PAGE on slab gels, however, shows a marked difference in polypeptide composition between the LHCP of grana and stroma thylakoids. As already known for total thylakoid LHCP, the main polypeptide component of grana LHCP is the well known 25 Kd one (some times resolved as a 25-23 Kd doublet) (4). However, the stroma LHCP contains in addition, a polypeptide of an estimated Mr of 21 Kd, which is heavily stained by AgNO$_3$. This polypeptide is present in the stroma LHCP even after washing for peripheral protein removal. It cannot be a contamination of the sucrose gradient band, since in LHCP separated by SDS-PAGE from SDS-solubilized stroma lamellae, and then disected and further analyzed on a dissociating slab gel for polypeptide composition, it is also detected. This component seems to be present as a minor component in the granal LHCP as well, as shown only after AgNO$_3$ staining. It is clear however, from Figure 1, that the LHCP monomers are free of this polypeptide. The observed difference, therefore, in the Chl to protein ratio between stroma and grana LHCP may be due to the presence of this component in stroma LHCP (on the assumption that the 21 Kd polypeptide is free of pigments).

4. DISCUSSION

Our results show that grana and granal LHCP have more carotenoids - xanthophylls - per Chlorophyll than do stroma thylakoids and their LHCP. These results aggree with those of earlier workers (30, 32) as to the prefferential association of the xanthophylls with grana. However, they do not agree with the finding that in spinach the Chl/car molar ratio is similar in both fractions(about 5). We instead find a slightly higher ratio in stroma thylakoids. This may be due to loss of carotenoids from stroma lamellae during washing. The prefferential association of xanthophylls with grana has been suggested to show a requirement for the oxygenated carotenoids for the PSII to function properly, and for the formation of grana (32). Recent results, however (14,15) show that the PS II exists in both fractions. The ratio of PSII per Chl in grana is about 3.6 times higher than that in stroma lamellae (15,33). The preferential association of xanthophylls with grana, therefore, may reflect the higher amount of PSII units in grana structures. Assuming that LHCP is a component of PSII (34), and taking into account the finding

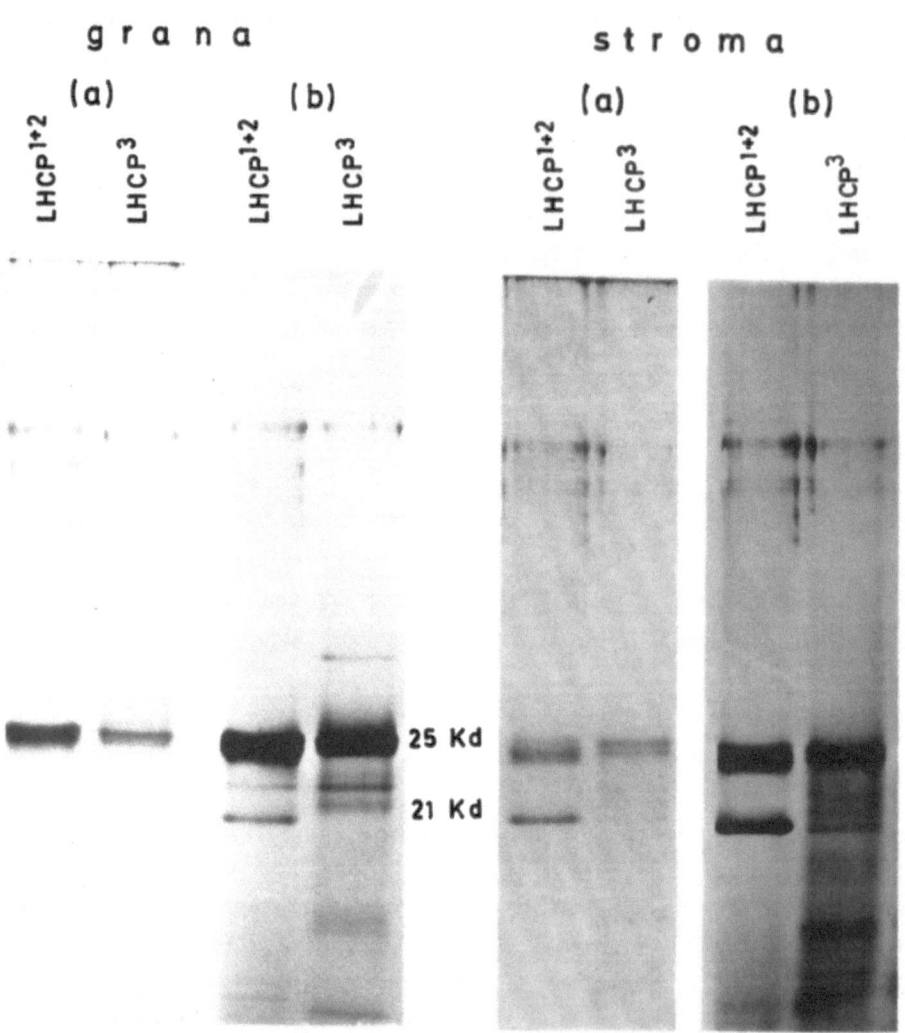

Figure 1. Electrophoretic resolution on a dissociating SDS-PAGE slab of polypeptide components of the pigment-protein complexes, LHCPs, isolated by sucrose density gradient centrifugation as earlier described (21). (a): Coomassie staining, (b): Silver Nitrate staining.

that the size of the PSII unit in stroma and grana lamellae in Phaseolus
vulgaris is similar (15), these results suggest that the size of the PSII
unit, as determined by biophysical criteria, and consequently its activity,
may not be affected by the amount of xanthophylls. This points to a structural rather than to a functional role of xanthophylls.

We can not say whether the different pigment composition in the grana
and stroma thylakoid LHCP is responsible for promoting or not grana stacking.
However, earlier workers have ¡already¡ pointed out that the hydroxyl group
of lutein may form a complex with Chl, and preferentially with Chl b (35),
which is located mainly in grana. Indeed, we have also found that the lutein
to Chl b ratio in grana and granal LHCP is twice as high as in stroma thylakoids and stroma LHCP.

As it was shown in this work, the stroma LHCP contains, in addition to
the 25 Kd polypeptide, one more polypeptide of 21 Kd. Evidence has been presented recently (36), suggesting a different composition for LHCP of stroma
and grana regions of the thylakoid. This difference was primarily attributed
to the different 25/23 Kd polypeptide ratio of heavy (grana) and light (stroma) subchloroplast fractions. In our opinion, the 25/20 Kd polypeptide ratio
shows a more stricking difference in their polypeptide patterns.

We do not know whether the 21 Kd polypeptide component found in stroma
LHCP is a contaminant or not. There is a possibility that this polypeptide
is the 20 Kd cyt-b_6 subunit of the Cyt f/b_6 complex (37), or the delta subunit of ATPase (38). Another possibility is that the 21 Kd component may
originate from the 25 Kd polypeptide due to the action of a protease, which
may be more active or abundant in stroma thylakoids. However, it should be
noticed, that this polypeptide is not present in the LHCP monomer of either
stroma or grana thylakoids; this may suggest that this component may play
a role in the organization of the oligomeric structures.

Finally, since stroma lamellae are enriched in PSI units, our finding
that stroma thylakoid LHCP differs in pigment composition from the grana
thylakoid LHCP, may reflect the difference in pigment composition between·
the LHCP complexes of the PS I and PS II units. If this is true, then our
results suggest that the LHCP of PS I contains less xanthophylls per Chl
than the LHCP of PS II.

Acknowledgments: The expert technical assistance of Spyros Daoussis is gratefully acknowledged.

REFERENCES

1. Thornber, J.P. 1975. Ann. Rev. Plant Physiol. 26, 127-155.
2. Anderson, J.M. 1975. Biochem. Biophys. Acta 416, 191-235.
3. Thornber, J.P. et al. 1977. Brookhaven Symp. Biol. 28, 132-148.
4. Argyroudi--Akoyunoglou, J.H. 1983. In: Cell Function and Differentiation Part B, pp.277-289, Alan R. Liss, N.Y.
5. Argyroudi-Akoyunoglou, J.H., Feleki, Z. and Akoyunoglou G. 1971. Biochem. Biophys. Res. Commun. 45, 606-614.
6. Argyroudi-Akoyunoglou, J.H. and Akoyunoglou, G. 1974. FEBS Lett. 42, 135.
7. Armond, P.A. et al. 1976. Arch. Biochem. Biophys. 175, 54-63.
8. Argyroudi-Akoyunoglou, J.H. 1976. Arch. Biochem. Biophys. 176, 267-274.
9. Argyroudi-Akoyunoglou, J.H. and Tsakiris, S. 1977. Arch. Biochem. Biophys. 184, 307-315.
10. Argyroudi-Akoyunoglou, J.H. and Akoyunoglou, G. 1977. Arch. Biochem. Biophys. 179, 370-377.
11. Argyroudi-Akoyunoglou, J.H. and Akoyunoglou, G. 1979. FEBS Lett. 104, 78.
12. Mullet, J.E. and C.J. Arntzen 1980. Biochim. Biophys. Acta 589, 100-117.
13. Andersson, B. and Anderson, J.M. 1980. Biochim. Biophys. Acta 593, 427.
14. Armond, P.A. and Arntzen, C.J. 1977. Plant Physiol. 59, 398-404.

15. Castorinis, A. and Argyroudi-Akoyunoglou, J.H. 1981. In: Photosynthesis, Akoyunoglou, G., ed., Vol V, pp.501-511, Balaban Intern. Sci. Services, Phil. Pa.
16. Barber J. 1980. FEBS Lett. 118, 1-10.
17. Argyroudi-Akoyunoglou, J.H., Castorinis, A. and Akoyunoglou, G. 1982. Photobiochem. Photobiophys. 4, 201-210.
18. Michel, J. and Michel-Wolwertz, M.R. 1967. Carnegie Inst. Year Book, pp. 508-514.
19. Sane, P.V., Goodchild, D.J. and Park, R.B. 1970. Biochim. Biophys. Acta 216, 162-178.
20. Strotmann, H., Hesse, H. and Edelman, K. 1973. Biochim. Biophys. Acta 314, 202-210.
21. Argyroudi-Akoyunoglou, J.H. and Thomou, H. 1981. FEBS Lett. 135, 177-181.
22. Evangelatos, G.P. and Akoyunoglou, G. 1981. In: Photosynthesis, Akoyunoglou, G., ed., Vol V, pp. 319-322, Balaban Intern. Sci. Services, Phil. Pa.
23. Mackinney, G. 1941. J.Biol. Chem. 140, 315-322.
24. Hager, A. and Meyer-Bertenrath, T. 1966. Planta, 69, 198-217.
25. Jeffrey, S.W. 1968. Biochim. Biophys. Acta 162, 271-285.
26. Lowry, O.H. et al. 1951. J. Biol. Chem. 193, 265-275.
27. Hoober, J.K., Millington, R.H. and D'Angelo, L.P. 1980. Arch. Biochem. Biophys. 202, 221-234.
28. Oakley, B.R., Kirsch, D.R. and Morris, N.R. 1980. Anal. Biochem. 105,361.
29. Tsimilli-Michael, M. and Akyunoglou, G. 1978. In: Chloroplast Development, Akoyunoglou, G. et al. eds, pp. 525-532, Elsevier/North Holland Biomedical Press, Amsterdam.
30. Lichtenthaler, H.K. and Calvin, M. 1964. Biochim. Biophys. Acta 79, 30-40.
31. Boardman, N.K. and Anderson, J.M. 1967. Biochim. Biophys. Acta 143, 187.
32. Tropster, T. and Allen, C.F. 1973. Plant Physiol. 51, 584-585.
33. Melis, A. and Brown, J.S. 1979. Carnegie Inst. Year Book 79, 172-175.
34. Akoyunoglou, G. et al. 1981. In: Photosynthesis, Akoyunoglou, G., ed., Vol V, pp.523-532, Balaban Intern. Sci. Services, Phil. Pa.
35. Sewe, K.-U. and Reich, R. 1977. Z. Naturforsch.32c, 161-171.
36. Faludi-Daniel, A., et al. 1983. Europ. J. Biochem. 131, 567-570.
37. Hurt, E. and Hauska, G. 1981. Europ. J. Biochem. 117, 591-599.
38. Suss, K.H. 1976. FEBS Lett. 70, 191-196.

EFFECT OF LIGHT AND PROTEIN CROSSLINKERS ON THE ULTRASTRUCTURE OF ISOLATED
ETIOPLAST MEMBRANES

C. SURIN and C. SIRONVAL
Laboratory of photobiology - Botanic Institute B22
University of Liège - B4000 Liège (Belgium)

1. INTRODUCTION

1.1. The growth of higher plants in darkness   - etiolation- prevents the
achievement of chlorophyll biosynthesis as well as the complete development of proplastids into chloroplasts.

In etiolated leaves, PChl(ide)-protein complexes are accumulated in
the etioplast internal membranes. There is at least three forms of those
complexes : two of them are photoreducible and absorb maximally in the
red at 650nm ($P_{657-650}$) and 639nm ($P_{-,639}$). The third form is non photoreducible; its red absorption maximum is located at 628nm ($P_{633-628}$).

The inner membranes of the etioplasts are organized into two systems
whose structures differ markedly; the prolamellar bodies (PLBs) are paracrystalline lattices which contain the most part of the membranes, while
the prothylakoids (PTs) assume the morphology of young lamellae extending
from the PLBs. The tridimensionnal architecture of the PLBs has been
described by many authors. According to Wehrmeyer (1965) and to Günning
and Steer (1975), the PLBs appear most often to be made of tetrahedrically
branched tubular units, interconnected to build hexagonal lattices.
"Ribosome-like" particles are located in the stroma spaces of the PLBs
(Günning,1965; Wellburn et al,1977).

As soon as etiolated leaves are illuminated,PChl(ide) complexes are
photoreduced to yield Chlide, and the paracrystalline configuration of

---

ABBREVIATIONS

BSA = bovin serum albumine; Chl = chlorophyll; Chlide = chlorophyllide;
DMS = dimethylsuberimidate; GA = glutaraldehyde; KD = kilodalton; PChl(ide)
= protochlorophylle and/or protochlorophyllide; PLBs = prolamellar bodies;
PTs = prothylakoids; tris = tris (hydroxymethyl) aminomethan; P_{x-y} =
pigment-protein complex, x refers to the fluorescence emission maximum at
77k and y to the red absorption maximum at 77K.

C. Sironval and M. Brouers (eds.); Protochlorophyllide Reduction and Greening. ISBN 90 247 2954 8
©1984, Martinus Nijhoff/Dr W. Junk Publishers, The Hague/Boston/Lancaster.

the PLBs is lost gradually. Several minutes later, either in darkness or in the light, Chlide is found esterified into Chl and the PLB membranes are dispersed into perforated primary thylakoids (Henningsen and Boynton, 1969; Günning and Steer, 1975).

Owing to its stable state in darkness and to its high sensitivity to light, the etiolated system has been often chosen as a starting material to study chloroplast development and the genesis of the photobiochemical activities of photosynthesis. The structure of the PLBs, their signifi- cance and the possible relationships between photoreduction and PLB dispersal have been a particular subject of studies.

1.2. The first question to be asked bears on the localization of the PChl(ide). The PChl(ide) complexes have been first thought to be located inside PLB membranes, as a red fluorescence has been reported to be limi- ted to small centres, assumed to be PLBs (Boardman and Anderson, 1964). Kahn (1968a) claimed that the PChl(ide) complexes are building parts of the PLB membranes. Physiological arguments have supported this idea Klein and Schiff (1972) correlated the growth of the PLBs during the development of the etioplasts in darkness with the accumulation of the photoreducible PChl(ide)-protein complex which absorbs at 650nm. Furthermore, the reformation of PLBs in plants growing under dim light was seen to occur in parallel to the reaccumulation of PChl(ide) (Henningsen and Boynton, 1970; Treffry, 1973). The same observation was made on plants returned to darkness after an illumination (Henningsen and Boynton, 1969; Ikeda, 1971).

The observation that the PLB dispersal occurs in the first minutes of a continuous illumination has been a strong argument in support of the idea that it is related to the photoreduction of PChl(ide). Action spectra for effect of light on PLB structure speak in favour of this conclusion (Virgin et al, 1963; Henningsen, 1967).

1.3. However data of Henningsen and Boynton (1969) show that the photo- reduction is a necessary although not a sufficient condition for PLB's dispersal. And in fact the loss of the paracrystalline structure is not so closely synchronized with the photoreduction, as the dispersal of the PLBs into primary thylakoids takes place during the dark period subsequent to a 1 min illumination. Treffry (1970) claimed that the PLB dispersal is linked to the esterification of the Chlide into Chl, rather than to the photoreduction of PChl(ide).

More recently, biochemical studies have questioned the concept of the PLB as a storage locus of substances used later in the greening process. Firstly, Kesselmeier and Budzikiewicz (1979), Kesselmeier and Ruppel (1979) have identified two saponins as the major components of PLB preparations from oat. This result was confirmed by Lütz and Klein (1979). The conclusion of Kesselmeier and co-workers, based on reaggregation experiments, was that saponins are the main building molecules in the structure of the PLBs. However, Ryberg and Sundqvist (1982a) working on wheat, Ikeuchi and Murakami (1983) working on squash could not find saponins in PLB preparations. Kesselmeier (1982), Kesselmeier and Urban (1983), Murakami et al (1983) have shown that the saponins identified in the PLB fractions came from subcellular contaminations. They would be artificially and preferentially bound to the PLBs during isolation.

The problem has been thoroughly studied by Lütz and co-workers, who separated PTs from PLBs by fractionation and have performed biochemical analysis of fractions enriched, either in PLBs or in PTs. Their results showed that the PLBs contain mainly lipids and that the amounts of proteins and PChl(ide) are low. The greatest part of PChl(ide), of the NADPH - PChlide oxidoreductase and of the proteins were found in the PTs (Lütz and Klein, 1979; Lütz and Manning, 1980; Lütz et al, 1981). Lütz and co-workers concluded that the PTs and the PLBs are morphologically, biochemically and physiologically different systems. Lütz proposed therefore a new concept for the significance of the PLBs (Lütz, 1981). This concept was later revised by Lütz and Nordmann (1983). It may be summarized as follows : during the development of the etioplast in darkness, the precursor components of the thylakoid membranes are assembled in the PT membranes. The PLBs appear later; they are secondary products of etiolation,- lipid enriched outgrowths of the assembly of the membranes. Several physiological data may be quoted to support such a concept. The presence of PLBs has been sometimes reported without any reference to PChl(ide). For instance, PLBs have been observed in dark-grown cotyledons of Pinus, which contain Chl and grana but have no detectable PChl(ide) or Chlide (Michel-Wolwertz and Bronchart, 1974). Likewise, PLBs have been shown to reappear under dim/red light in the presence of a high amount of Chlide, but in the absence of detectable PChl(ide) (Treffry, 1973). The study of the development of proplastids under continuous light has revealed that well-developped PLBs are an optional stage of the development of chloroplasts (Whatley, 1974, 1977; Boffey et al, 1980).

Although the separation of the PLBs from the PTs has introduced a new approach of the problem, the interpretation of the data is hampered by the absence of specific markers to estimate the cross contamination, and in addition the contents of the PT and PLB fractions seem to markedly depending on the plant material and on the separation procedure. If several authors reported that the PLB fraction contains a lower amount of proteins and pigments than the PT fraction , the localization of the photoreducible PChl(ide) protein complexes is controversed. Lütz et al (1981) on oat, and Gerday et al (1983) on bean have localized the main part of the photoreducible PChl(ide) and of the 36 KD peptide, characteristic of the PChlide photoreductase, in the PTs. In barley, however Ryberg and Sundqvist (1982a,b) have found the photoreducible PChl(ide) in the PLBs mainly, while the main bulk of proteins is found in the PTs. On the other hand, Ikeuchi and Murakami (1982) on squash have localized the 36 KD peptide in the PLBs fraction. It thus appears that the relationships between PLBs, PTs and photoreduction are not clarified.

1.4. This paper reports some data concerned with these controversies. Crude etioplast membranes have been isolated from bean leaves. They were essentially PLB membranes dissociated from the PTs. The structural changes due to a 1 msec flash of light and to the addition of two protein crosslinking agents (glutaraldehyde, GA, and   dimethylsuberimidate, DMS) have been studied using freeze-fracturing. Freeze-fracturing has been chosen because it avoids chemical fixation and permits in addition a rapid fixation of the structure by freezing.

The results have shown (1) that light induces *in vitro* the formation of lamellae from the PLBs, (2) that a prolonged dark treatment with GA or DMS disorganizes the regular structure of the PLBs, (3) that DMS can provoke the formation of lamellae from PLBs kept in the dark.

2. MATERIAL AND METHODS

2.1. Plant Material

Bean seedlings (*Phaseolus vulgaris* L.cvar. Commodore) were grown in complete darkness for 15 days on a mixture vermiculite-perlite. The etiolated primary leaves were harvested under a dim green safe light.

2.2. Isolation of etioplast membranes

All operations were carried at 4°C under a dim green safe light. About twenty grams of leaves were homogenized for 2 x 10 sec in 60 ml grinding medium (0.5M tris HCl buffer pH 8.0, 0.6 M sucrose, 0.005 M MgCl$_2$, 0.2% BSA, 10% glycerin) in a Braun MX 32 homogenizer. The homogenate was filtered through 4 layers of gauze and 4 layers of cleese-cloth. The filtrate was centrifuged at 120 g for 5 min. The supernatant was centrifuged for 5 min at 4,500 g. The resulting pellet was used for the experiments.

2.3. Illumination of the samples

The pellet of membranes was resuspended in the grinding buffer. For freeze fracturing, the membranes suspension was centrifuged at 4,500 g for 5 min. and the resulting pellet was divided into 3 parts. The first part was frozen directly in darkness for freeze-fracturing. The samples of the second part received a 1 msec flash of white light (photographic flash Multiblitz Report PORBA 125 J, 5,800 K) at room temperature and was frozen as soon as possible after the flash for fracturing. The third part received a 1 msec flash at room temperature and was kept 30 min in darkness at 4°C before freezing for fracturing.

2.4. Addition of crosslinking agents

The membrane pellet was resuspended in 2 ml of a 4% GA solution (0.8 M cacodylate buffer pH 8.0, 0.6 M sucrose, 0.005 M MgCl$_2$, 10% glycerin) or in 2 ml of a 0.1 M DMS solution (0.1 M triethanolamine buffer pH 8.0, 0.6 M sucrose, 0.005 M MgCl$_2$, 10% glycerin). The suspensions were incubated in complete darkness at 4°C, for 2 or 5 hours in case of GA and for 4 hours in case of DMS. At the end of the incubation, the suspensions were diluted 4 times with the buffer. Freeze-fracturing was performed on the pellet resulting from a centrifugation of the suspension at 4,500 g for 5 min.

2.5. Absorption spectrophotometry and spectrofluorometry at liquid nitrogen temperature

Red absorption spectra at 77 K were recorded using a Cary 17 spectrophotometer. 77 K fluorescence emission spectra were recorded using the apparatus described by Sironval et al (1968). The spectra were not corrected for wavelength variations of the photomultiplier response (S20).

2.6. Calculation of the inactivation index (I)

All along an incubation of etiolated membranes in GA or DMS, aliquots were taken, illuminated with a 1 msec flash and immediatly frozen.

The inactivation index at time t (I_t) was calculated from the 77 K fluorescence emission spectrum, using the following formula :

$$I_t = \frac{A_t - A_o}{A_o} \times 100 \text{ with } A = \frac{E_{690}}{E_{633} + E_{690}} \times 100 \text{ ,}$$

where A_t is the value of A at time t,

A_o is the value of A at time o.

E_{690} and E_{633} are the intensity of the emission maximum of the Chlide and of the non photoreducible PChl(ide) respectively; these intensities are corrected for wavelength variations of the photomultiplier response.

2.7. Freeze-fracturing

The samples of the membrane pellet were placed on golden holders and frozen as rapidly as possible in liquid propane (-186°C). Freeze-fracturing was performed in a Balzers apparatus (BA 360 M). Samples were fractured at -110°C and shadowed with platinum at an angle of 45°. Replicas were cleaned for 24 H in 75% H_2SO_4 and for 12 H in calcium hypochlorite. They were washed three times in twice distilled water and picked up on copper grids.

2.8. Electron microscopic observations

Observations were done using a Philips EM 101. For each experimental treatment (§ 2.3 and 2.4), at least 50 observations have been performed on at least 2 replicas of 2 different experiments.

3. RESULTS

3.1. The structure of the isolated PLB membranes

The crude etioplast membranes showed fluorescence emission and red absorption spectra at 77K similar to those of etiolated leaves (fig. 1). The 3 forms of the PChl(ide)-protein complexes, $P_{657\text{-}649}$, $P_{\text{-},639}$ and $P_{633\text{-}628}$ were present nearly in the same proportion than in the leaves.

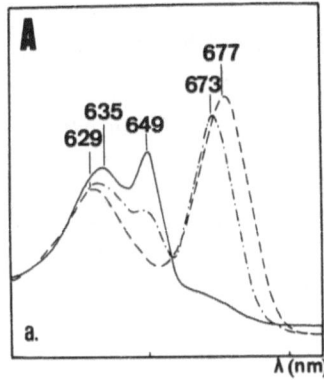

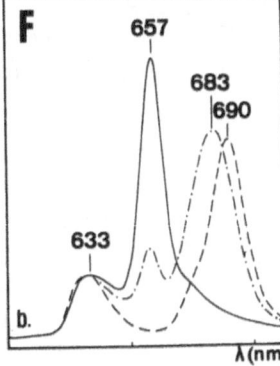

FIGURE 1. a) 77K absorption spectra and b) 77K fluorescence emission spectra of isolated etioplast membranes recorded before (—), a few seconds after (--) and 30 min after (-·-) a 1 msec flash illumination; the membranes are kept in darkness at 4°C.

The fractures showed that the etioplast envelopes were broken and lost during isolation. Furthermore, a great part of the PTs had been dissociated from the PLBs and were found dispersed into free vesicles.

Isolated membranes appeared as "PLB-like" structures, surrounded with PT vesicles in 50% of the cases only (plate 1.1). The abundance of the lamellae or vesicles has been represented by the parameter L_m which is the total length of lamellae or vesicles * per "PLB-like" structure evaluated on 30 photographs. In average, L_m was about 5 times lower in isolated membranes than in intact etioplasts (~ 5 μm).
The isolated "PLB-like structures" were variously shaped. They have been classified into three different structure types, the frequency of which was expressed as a percentage of the total number of   observations (table 1). The "paracrystalline PLBs" were found as frequently as the "inorganized PLBs". Paracrystalline PLBs are characterized by a high level of organization in a hexagonal (plate 1.2) or in a rectangular lattice (plate 1.3)..On the contrary, "inorganized PLBs" (plate 1.4) are devoided

---

* the half-perimeter of one vesicle is taken as its length.

of any obvious regularity; they are to be ascribed either to a disorganization in the PLBs, or to an irregular fracture of paracrystalline PLBs. Beside these two most frequent structure types, a few PLBs appeared as made of straight tubules pointing in different directions (plate 1.5). The length of these tubules was variable, their outer diameter being (186 ± 20) Å.

3.2. The effects of a 1 msec flash of light

Immediatly after the flash, all the photoreducible PChl(ide)s were photoreduced to Chlide absorbing at 677nm and emitting at 691nm (fig.1).

The structure of PLBs observed a few seconds after the flash were not strongly modified (table 1).

TABLE 1. Frequency of the structure types in isolated membranes before and after a 1msec flash.

	étiolated	+1msec flash	+1msec flash + 30min of darkness
PLBs $_x$ { paracrystalline (%)	46 ± 9	45 ± 5	22 ± 1
inorganized (%)	42 ± 3	30 ± 10	50 ± 0
tubular structures(%)	12 ± 9	25 ± 15	13 ± 4
lamellar structures(%)	0	0	15 4
PLBs with peripherical lamellae or vesicles (%)	55 ± 10*	50 ± 10*	97 ± 1**
L_m (μm)	0.99 ± 0.22	0.93 ± 0.14	4.40 ± 0.90

* vesicles non attached to the PLBs.

** peripherical lamellae attached to the PLBs mainly.

^{x} PLB - like structures.

PLATE 1. Structure types of isolated etiolated membranes, as shown by freeze-fracturing. (arrows indicate the direction of shadowing; bar = 0.5 μm).
    n°1 : PLB-like structure (PLB) surrounded with some prothylakoid vesicles (PTV);
    n°2 : paracrystalline PLB showing an hexagonal lattice in the fracture plane;
    n°3 : paracrystalline PLB showing a rectangular lattice in the fracture plane;
    n°4 : inorganized PLB;
    n°5 : tubular structure made of straight tubules (arrowheads).

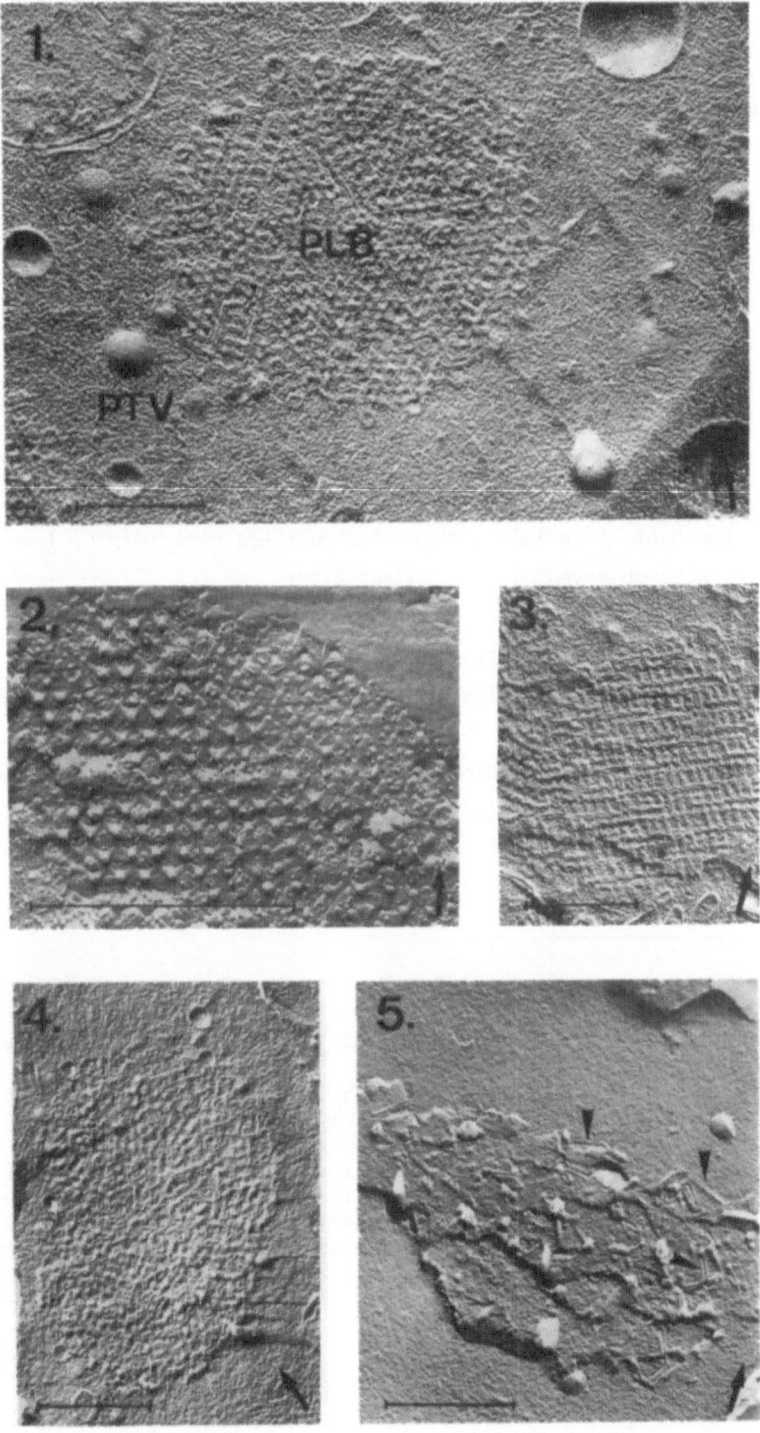

Vesicles or lamellae were found as rarely as before the flash, as shown by the value of L_m (table 1).

During the subsequent 30 min of darkness, the emission and absorption maxima of Chl(ide) shifted towards shorter wavelengths and a low amount of photoreducible PChl(ide) was reaccumulated (fig. 1).

Structural changes were observed during this dark period. The most obvious of these changes consisted in the formation of concentric lamellae around all the PLB-like structures of each type (plate 2). These concentric lamellae were attached to the PLBs in many points (plate 2.1, 2.2). They were sometimes dispersed as numerous peripheric vesicles (plate 2.3). Consequently the parameter L_m increased four times during the dark period (table 1).

The frequencies of the PLBs types themselves were also modified : paracrystalline PLBs were less often found and inorganized PLBs became predominant. A certain number of structures made only of lamellae were also observed (plate 2.4 and 2.5).

3.3. The effects of crosslinking agents

The incubation in the presence of GA as well as in the presence of DMS provoked a gradual inactivation of the PChl(ide)-protein complexes, $P_{657-649}$ and $P_{-,639}$, into the non photoreducible PChl(ide) complex $P_{633-628}$ (fig. 2 and 3). The inactivation resulted in the increase of the relative absorbance at 628nm in comparison with the absorbances at 639 and 650nm (fig. 2a and 3a). The intensity of the fluorescence emitted

---

PLATE 2. Structure types of isolated membranes, as shown by freeze-fracturing after a 30 min period in darkness at 4°C following a 1msec flash illumination (arrows indicate the direction of shadowing; bar = 0.5 μm).

    n°1 : paracrystalline PLB surrounded with concentric lamellae (L).
    n°2 : inorganized PLB surrounded with concentric lamellae (L).
    n°3 : inorganized PLB surrounded with numerous vesicles (V).
    n°4 : this structure appears exclusively made of perforated lamellae in the fracture plane; arrowhead shows a perforation.
    N°5 : fracture plane showing the connections between lamella (L) and a paracrystalline area of PLB tubules (arrowhead).

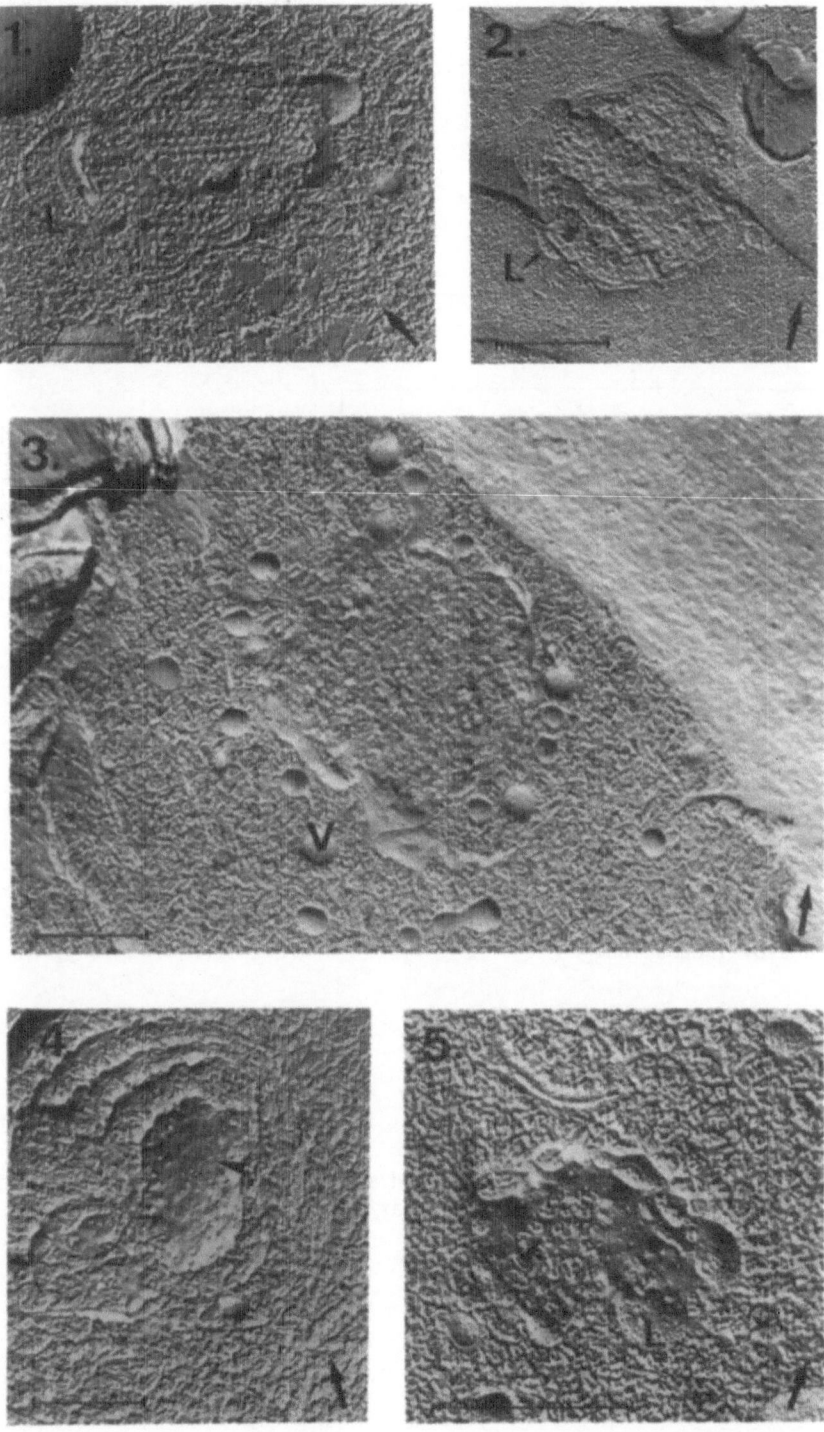

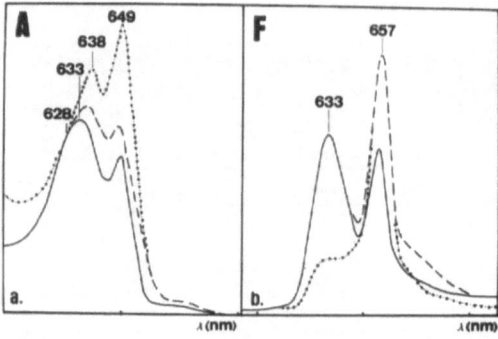

FIGURE 2. a) 77K absorption spectra and b) 77K fluorescence emission spectra of GA-treated etiolated membranes recorded after 2 hours of incubation at 4°C in the cacodylate buffer with GA (- -), after 5 hours of incubation at 4°C in the cacodylate buffer with GA (—), and after 5 hours of incubation at 4°C in the cacodylate buffer without GA (·•·).

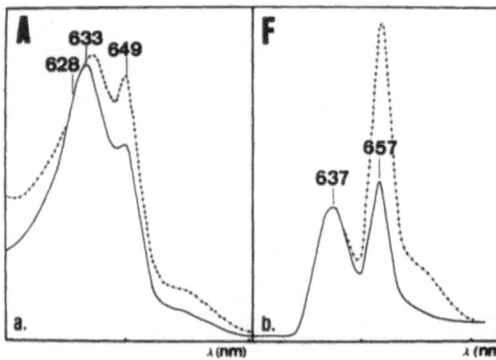

FIGURE 3. a) 77K absorption spectra and b) 77K fluorescence emission spectra of etiolated membranes incubated for 4 hours at 4°C in the triethanolamine buffer with DMS (—), or in the triethanolamine buffer without DMS (·•·).

at 633nm increased relatively to the fluorescence emitted at 657nm (fig. 2b and 3b). This inactivation has the same kinetics for both cross-linking agents as shown in fig. 4.

There is first a rapid phase, followed by a slower linear phase after 2 hours of incubation.

PLATE 3. Structure types of isolated membranes after 5 hour incubation in the cacodylate buffer with GA; (arrows indicate the direction of shadowing; bar = 0.5 μm).

  n°1 : inorganized PLB;
  n°2 : tubular structure;
  n°3 : dislocated structure made of fragments of membrane and short tubules (arrowhead).

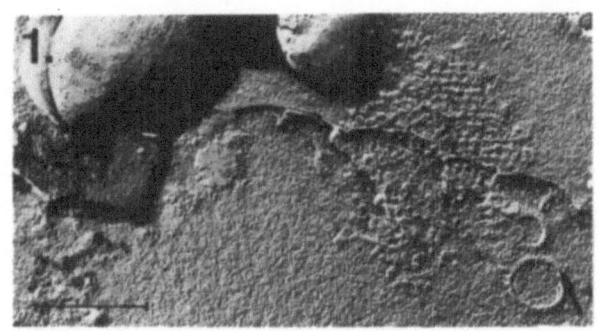

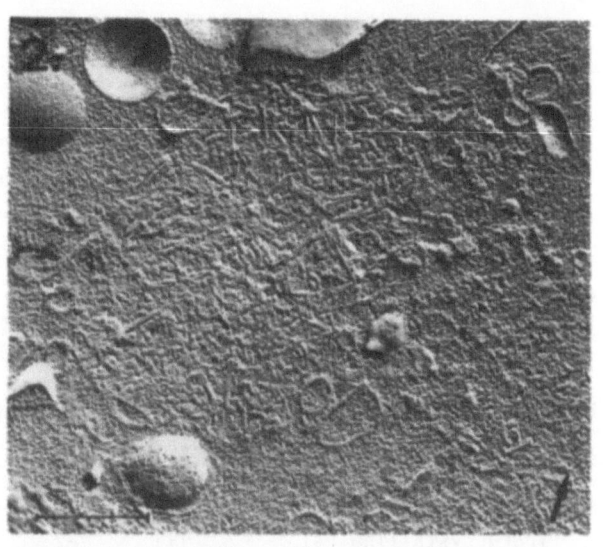

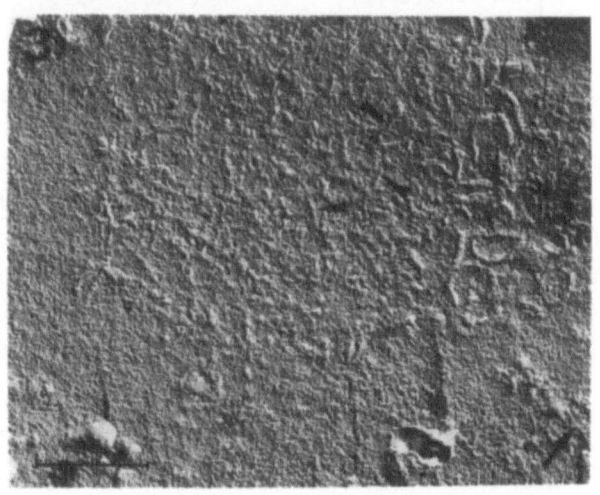

288

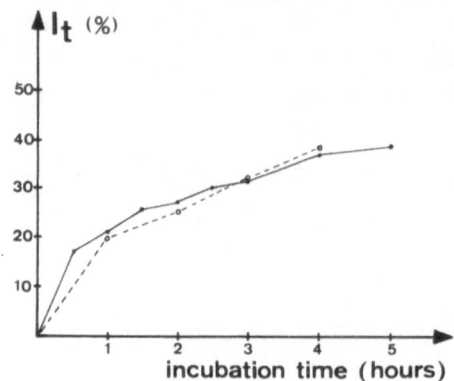

FIGURE 4 : Inactivation kinetics of the PChl(ide)-protein
complexes in isolated membranes
incubated at 4°C in the presence of GA (——) or of DMS
(---), (for the definition of
I_t, see Material and Methods,
§ 2.6)

3.3.1.Glutaraldehyde. After two hours of incubation in the presence
of GA, the PLB structures were weakly modified while the inactivation
I_t was about 27%. On the contrary, after 5 hours incubation, the
structures were completely modified although the inactivation had increased by 10% only (I_t = 37%).

TABLE 2. Frequency of the structure types in glutaraldehyde treated
membranes and the cacodylate control.

		2H GA	5H GA	2H buffer	5H buffer
PLBs	paracrystalline (%)	41±1	7±2	53±6	53±3
	inorganized (%)	27±0	46±1	39±2	27±7
	tubular structures (%)	30±3	31±3	8±8	18±2
	dislocated structures(%)	0	14±3	0	0
	lamellar structures (%)	2±2	2±2	0	2±2
PLBs with peripherical lamellae or vesicles (%)		67±10	70±18	36±9	50±10

Table 2 shows that, after 2 hours of GA incubation, paracrystalline
PLBs were found with the same frequency as at 0 hour, but tubular structures had become rather more frequent. In contrast, after 5 hours of
incubation, the paracrystalline PLBs had nearly completely disappeared.
The predominant structures were then "inorganized PLBs" (plate 3.1);
"tubular structures" were also observed but less frequently (plate 3.2).
Furthermore, a particular type of "dislocated structures" (plate 3.3),
composed of dispersed fragments of membranes or tubules, was found.

This effect is due to the presence of GA, as a same incubation in the cacodylate buffer, without GA, did not affect the structures (table 2).

We conclude that a prolonged treatment with GA provokes the disorganization and even the dislocation of the paracrystalline PLBs and that this occurs while the inactivation of the PChl(ide) does not increased markedly.

3.3.2. DMS. After 4 hours of incubation in the presence of DMS, the inactivation of the photoactive PChl(ide)-protein complexes was similar to what it was after 5 hours of GA incubation (fig.4). This 4 hour DMS incubation disorganized the paracrystalline structure of the PLBs, but the structural changes were not those seen in the presence of GA (table 3).

TABLE 3. Frequency of structure types in DMS-treated membranes and in the triethanolamine control.

		4H DMS	4H buffer
PLBs	paracrystalline (%)	0	0
	inorganized (%)	21±3	41±4
	tubular structures (%)	12±6	50±6
	dislocated in small vesicles (%)	45±1	0
	lamellar structure (%)	22±4	9±3
PLBs with peripherical lamellae or vesicles(%)		95±5**	68±13*
L_m (μm)		2.69±0.51	1.02±0.21

* vesicles non attached to the PLBs.

** peripherical lamellae attached to the PLBs, mainly.

After 4 hours, "dislocated" structures had appeared which consisted in small more or less aggregated vesicles (plate 4.1). In addition, lamellae were seen around PLBs which had lost their original paracrystalline organization (table 3). The lamellae were concentric and attached in many points to inorganized PLBs (plate 4.2) or to tubular structures (plate 4.3). Some structures were exclusively composed of concentric lamellae (plate 4.4). Paracrystalline PLBs had completely disappeared. The threefold increase of the parameter L_m (table 3) gives evidence for the formation of the lamellae. A control with triethanolamine without DMS did not provoke the formation of lamellae.

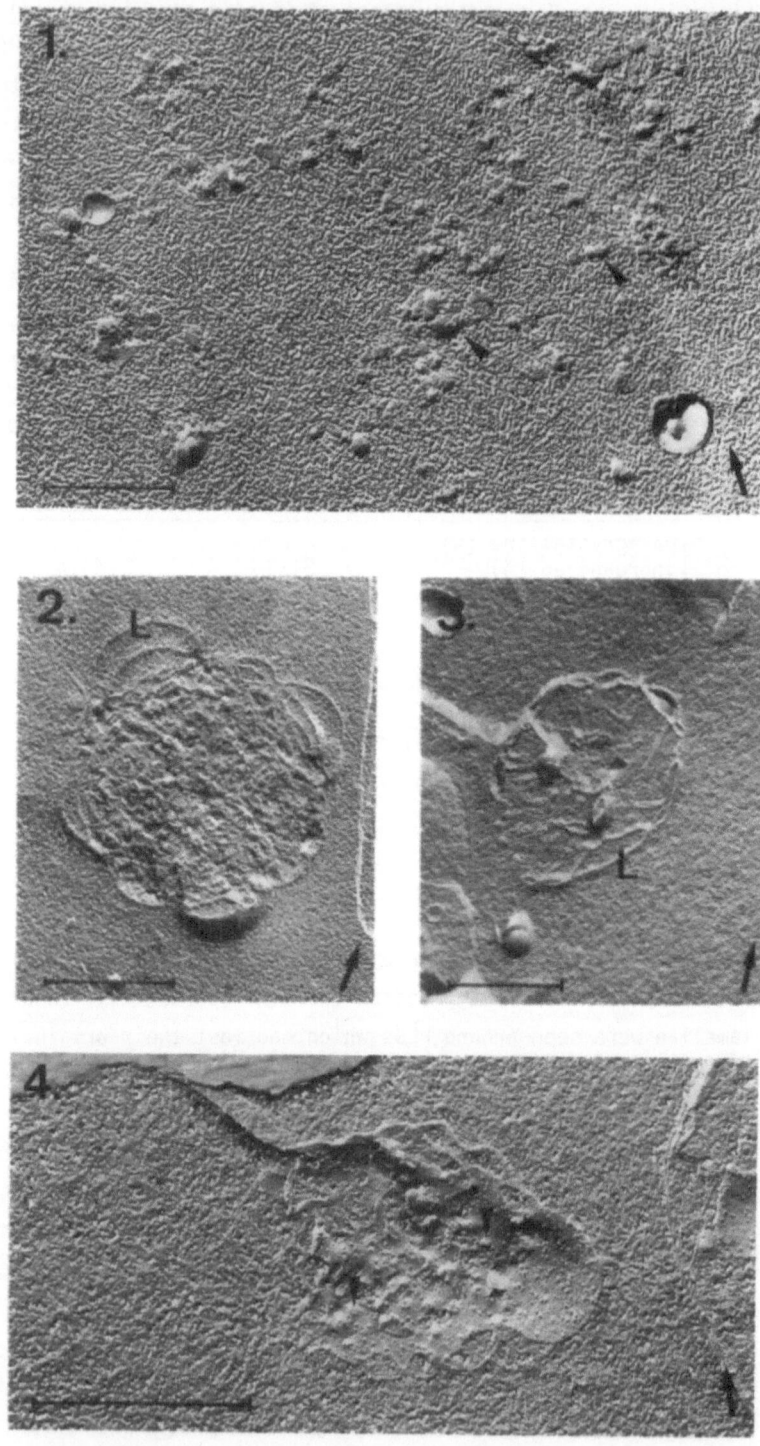

4. DISCUSSION

PLB membranes nearly completely depleted of PTs have been isolated. These membranes produce concentric lamellae or vesicles *in vitro* during a 30 min dark period subsequent to a 1msec flash, which reduces PChl(ide) into Chlide. During the same dark period, the red absorption and the fluorescence of the Chlide which have appeared shift towards shorter wavelengths and a certain amount of photoreducible PChl(ide) is reformed exactly as it occurs in entire leaves. The *in vitro* formation of lamellae around PLBs can be compared with the dispersal of PLBs into primary thylakoids, as it is observea *in vivo* (Henningsen and Boynton, 1969). Both phenomena follow the photoreduction of PChl(ide), and they occur during the subsequent dark period. So, in fact, using freeze-fracturing and a short illumination time, we have confirmed, *in vitro*, the *in vivo* observations of Henningsen and Boynton (1969).It appears that PLBs are able to perform one step of primary thylakoid development. This implies that the PLBs control this developmental step, at least partly.

The development *in vitro* of isolated, intact etioplasts has been already described by Klein and Poljakoff-Mayber (1961), Wellburn and Wellburn (1971), Wrisher (1973) and Kohn and Klein (1976). These works have established that the illumination of entire intact etioplasts induces the transformation and the dispersal of the PLBs into primary thylakoids. The behaviour of isolated PLB membranes exposed to light has been followed recently by Ryberg and Sundqvist (1983). These authors have also seen a formation of vesicles around purified PLBs 4 hours after a flash illumination.

PLATE 4. Structure types of isolated membranes after a 4 hour incubation in the triethanolamine buffer with DMS. (arrows indicate the direction of shadowing; bar = 0.5 μm).
  n°1 : dislocated structure made of small vesicles (arrowhead);
  n°2 : inorganized PLB surrounded with concentric lamellae (L);
  n°3 : tubular structure (arrowhead) surrounded with concentric lamellae (L);
  n°4 : structure which appears exclusively made of lamellae in the fracture plane; the presence of protusions on the fracture face of lamellae (arrowhead) is indicative of connections with PLB.

292

In isolated intact etioplasts, straight tubules have been often
observed (Kahn, 1968b; Wellburn and Wellburn, 1971; Wrisher, 1973;
Kohn and Klein, 1976; Simpson, 1978). We have found a few tubular structures in the isolated PLBs. The tubules are more frequent in GA-treated
PLBs. They derive from the PLBs and their formation seems to be induced
by a disturbance of the environment. A 1msec flash did not increase their
frequency significantly. The results are in agreement with the data
presented by Wrisher (1973) and Kohn and Klein (1976), showing that tubular structures are not a stage of normal development, as it was sup-
posed by Wellburn and Wellburn (1971).

The crosslinking of proteins by GA as well as by DMS inactivates
gradually the photoactive PChl(ide)-protein complexes and disorganizes
at long term the paracrystalline structure of the PLBs. DMS has a specific
effect as it provokes the formation of concentric lamellae around the
PLBs *in vitro* and in darkness. This result suggests that proteins cannot
be excluded from any role in building the paracrystalline structure of
the PLBs. They should be involved in the process of formation of lamellae
as well, since a protein crosslinking agent like DMS induces this process
in darkness.

The spectroscopy of the pigments shows further that there is no
direct relationship between the inactivation of the PChl(ide) complex
and the disorganization of the PLB structures, as GA disorganizes the
PLBs when   the inactivation has slow down. Furthermore, for a
same degree of inactivation, the structural modifications are dependent
on the crosslinking agent.

REFERENCES

1. Boardman NK and Anderson JM. 1964. Studies on the greening of dark-
   grown bean plants I Formation of chloroplasts from proplastids, Aust.
   J. Biol. Sci. 17, 86-92.
2. Boffey SA, Sellden G and Leech RM. 1980. Influence of cell age on
   chlorophyll formation in light-grown and etiolated wheat seedlings.
   Plant Physiol. 65, 680-684.
3. Gerday C, Lütz C, Michel-Wolwertz M-R and Brouers M. 1983. Some properties of purified etioplast membranes. Characterization of proto-
   chlorophyll(ide) protein complexes in PT-PLB enriched fractions of
   bean etioplasts. Proc VIth Photosynth. Congress |Brussels, in press.
4. Günning BES. 1965. The greening process in plastids I The structure
   of the prolamellar body. Protoplasma 60, 111-129.
5. Günning BES and Steer MW. 1975. Ultrastructure and the biology of
   plant cells. Edwards Arnold, London.

6. Henningsen KW. 1967. An action|spectrum for vesicle dispersal in bean plastids, in "Biochemistry of chloroplasts" (Ed TW Goodwin), vol.2, pp 453-457. Academic Press, New York.
7. Henningsen KW and Boynton JE. 1969. Macromolecular physiology of plastids VII The effect of a brief illumination on plastids of dark-grown barley leaves, J. Cell Sci. 5, 757-793.
8. Henningsen KW and Boynton JE. 1970. Macromolecular physiology of plastids VIII Pigment and Membrane formation in plastids of barley greening under low light intensity, J. Cell Biol. 44, 290-304.
9. Ikeda T. 1971. Prolamellar body formation under different light and temperature conditions, Bot. Mag. Tokyo 84, 363-375.
10. Ikeuchi M and Murakami S. 1983. Separation and characterization of prolamellar bodies and prothylakoids from squash etioplasts, Plant Cell Physiol. 24 (1), 71-80.
11. Kahn A. 1968a. Developmental Physiology of bean leaf plastids II Negative contrast electron microscopy of tubular membranes in prola-mellar bodies, Plant Physiol. 43, 1769-1780.
12. Kahn A. 1968b. Developmental physiology of bean leaf plastids III Tube transformation and protochlorophyllide photoconversion by flash illumination, Plant Physiol. 43, 1781-1785.
13. Kesselmeier J. 1982. Stereoidal saponins in etiolated greening and green leaves and in isolated etioplasts and chloroplasts of *Avena sativa*, Protoplasma 112, 127-132.
14. Kesselmeier J and Budzikiewicz H. 1979. Identification of saponins as structural building units in isolated prolamellar bodies from etioplasts of *Avena sativa* L., Z. Pflanzenphysiol. 91, 333-344.
15. Kesselmeier J and Ruppel HG. 1979. Relation between saponin concentra-tion and prolamellar body structure in etioplasts of *Avena sativa* during greening and re-etiolating and in etioplasts of *Hordeum vulgare* and *Pisum sativum*, Z. Pflanzenphysiol. 93, 171-184.
16. Kesselmeier J and Urban B. 1983. Subcellular localization of saponins in green and etiolated leaves and green protoplasts of oat (*Avena sativa* L.), Protoplasma 114, 133-140.
17. Klein|S and Poljakoff-Mayber A. 1961. Isolation of proplastids from etiolated bean leaves, Exp. Cell Res. 24, 143-200.
18. Klein S and Schiff JA. 1972. The correlated appearance of PLBs, proto-chlorophyll(ide) species and the Shibata shift during development of bean etioplasts in the dark. Plant Physiol. 49, 619-626.
19. Kohn S and Klein S. 1976. Light induced structural changes during in-cubation of isolated maize etioplasts, Planta 132, 169-175.
20. Lütz C. 1981. On the significance of prolamellar body in membranes development of etioplasts, Protoplasma 108, 99-115.
21. Lütz C and Klein S. 1979. Biochemical and cytological observations on chloroplast development. VI Chlorophylls and saponins in prolamellar bodies and prothylakoids separated from etioplasts of etiolated *Avena sativa* leaves, Z. Pflanzenphysiol. 95, 227-237.
22. Lütz C and Manning U. 1980. Determination of chlorophylls and proteins in separated prolamellar bodies and prothylakoids of etiolated *Avena sativa*, in "Photoreceptors and plant development". (Ed. J. De Greef) pp 229-236, Antwerpen Univ. Press.
23. Lütz C and Nordmann U. 1983. The localization of saponins in prolamel-lar bodies mainly depends on the isolation of etioplasts. Z. Pflanzen-physiol. 110, 201-210.
24. Lütz C, Röper U, Beer NS and Griffiths WT. 1981. Sub-etioplast loca-lization of the enzyme NADPH : protochlorophyllide oxidoreductase,

294

Eur. J. Biochem. 118, 347-353.
25. Michel-Wolwertz M-R and Bronchart R. 1974. Formation of prolamellar bodies without correlative accumulation of protochlorophyll(ide) or chlorophyllide in pine cotyledons, Plant Science Letters 2, 45-54.
26. Murakami S, Miyao M and Ikeuchi M. 1983. Prolamellar bodies and saponins. Avenacosides are not constituents of Avena etioplasts, Plant Cell Reports (in press).
27. Ryberg M and Sundqvist C. 1982a. Characterization of prolamellar bodies and prothylakoids fractionated from wheat etioplasts, Physiol. Plant. 56 : 125-132.
28. Ryberg M and Sundqvist C. 1982b. Spectral forms of protochlorophyllide in prolamellar bodies and prothylakoids fractionated from wheat etioplats, Physiol. Plant. 56, 133-138.
29. Ryberg M and Sundqvist C. 1983. Spectroscopical and ultrastructural characteristics of irradiated isolated prolamellar bodies and prothylakoids in the presence and in the absence of NADPH, Proc. VIth Photosynth. Congress, Brussels (in press).
30. Simpson D. 1978. Freeze-fracture studies on barley plastid membranes I Wild-type etioplast, Carlsberg Res. Commun. 43, 145-170.
31. Sironval C, Brouers M, Michel J-M and Kuiper Y. 1968. The reduction of protochlorophyllide into chlorophyllide. I. The kinetics of P657-647 - P688-678 phototransformation, Photosynthetica 2, 268-287.
32. Treffry T. 1970. Phytylation of chlorophyllide and prolamellar body transformation in etiolated peas, Planta 91, 279-284.
33. Treffry T. 1973. Chloroplast development in etiolated peas : reformation of prolamellar bodies in red light without accumulation of protochlorophyllide, J. Exp. Bot. 24, 185-195.
34. Virgin HI, Kahn A and von Wettstein D. 1963. The physiology of chlorophyll formation in relation to structural changes in chloroplasts, Photochem. Photobiol. 2, 83-91.
35. Wehrmeyer W. 1965. Zur Kristallgitterstruktur der sogenannten Prolamellarkörper in Proplastiden etiolierter Bohnen, Z. Naturf. 20b, 1270-1296.
36. Wellburn AR, Quail PH and Gunning BES. 1977. Examination of ribosome-like particles in isolated prolamellar bodies. Planta 134, 45-52.
37. Wellburn FAM and Wellburn AR. 1971. Developmental changes occuring in isolated intact etioplasts, J. Cell. Sci. 9, 271-287.
38. Whatley JM. 1974. Chloroplast development in primary leaves of *Phaseolus vulgaris*, New phytol. 73, 1097-1110.
39. Whatley JM. 1977. Variations in the basic pathway of chloroplast development, New phytol. 78, 407-420.
40. Wrisher M. 1973. Ultrastructural changes in isolated plastids I. Etioplasts, Protoplasma 78, 291-303.

ACKNOWLEDGMENTS

This work was supported by the Institut pour l'Encouragement de la Recherche Scientifique dans l'Industrie et l'Agriculture and by the Fonds National de la Recherche Scientifique, Belgium.

ENERGY MIGRATION.

THE ENERGY MIGRATION IN PIGMENT ASSEMBLY IN RELATION TO THE CHLOROPHYLL BIOSYNTHESIS

A.A.SHLYK, L.I.FRADKIN, I.N.DOMANSKAYA, E.R.NETKACHEVA
Institute of Photobiology.Acadamy Of Sciences of the BSSR,
Academicheskaya 27, 220733 Minsk, USSR.

The energetic interactions between pigment molecules are characteristic for functioning the photosynthetic apparatus. Therefore a study of the heterogeneous and homogeneous energy transfer is an important tool to learn more about the localization of different pigment molecules and the development of the pigment-pigment interactions in greening and ontogenesis. This approach allowed us to discover the energy transfer to Chl a even from the very first molecules of Chl b in shortly illuminated and darkened etiolated leaves (1). As Chl b arises from Chl a (2,3), their neighbouring position indicates that primary Chl a molecules were located in groups. In this way the hypothesis of group organization of Chl biosynthesis was forwarded.

In fact, a group localization of pigment molecules was revealed even at a stage of Pchlide accumulation. The active Pchlide is aggregated (4-6) and an energy transfer was observed between its various spectral forms as well as from Pchlide to the products of its phototransformation (7-9).

Pchlide reduction is followed by Shibata hypsochromic spectral shift and the energy transfer from Pchlide to Chl(ide) gradually fades due to Chl(ide) movement in the process of membrane development (8). This may be the cause for the absence of the energy transfer from Pchlide, reaccumulated upon repeated darkening, to the previously synthesized Chl (8). The observation of the Pchlide fluorescence in green leaves (10)

Abbreviations: Chl - chlorophyll; Chlide - chlorophyllide; Pchlide - protochlorophyllide.

C. Sironval and M. Brouers (eds.); Protochlorophyllide Reduction and Greening. ISBN 90 247 2954 8
© 1984, Martinus Nijhoff/Dr W. Junk Publishers, The Hague/Boston/Lancaster.

was also explained (11) by a spatial uncoupling Pchlide from Chl in chloroplasts.

Nevertheless, no kind of fractionation of the photosynthetic membranes by using various detergents, ultrasonication, French press, and grinding in combination with centrifugation or electrophoresis led to complete separation of Pchlide from Chl (12,13). The examination of pigment-protein complexes of various ranks of organization was especially impressive in this respect (14-17). Such joint localization of Pchlide and Chl in chloroplast membranes, which persisted at any procedure of membrane disintegration implied that in green leaves an energetic interaction between Chl and Pchlide, resynthesized upon darkening, is still possible. The special search really revealed an energy transfer from Pchlide to Chl in green plant (18). In the following this is described alongside with the development of such a transfer in greening etiolated leaves.

The leaves of barley of wild type and its Chl b-less mutants 2807 and 3613 from collection of H.Sagromsky (the authors are deeply thankful for this kind gift) were used. Three-day-old light-grown leaves of 4 to 5 cm length or 6-day-old leaves of 9 to 10 cm length were cut along the central vein and a solution of digitonin in 5 mM phosphate buffer, pH 7.2, was vacuum-infiltrated in one moiety while the other either remained intact or was infiltrated with the buffer only. In 20 minutes their fluorescence (excitation at 440 nm) and excitation (monitored at 730 nm) spectra were recorded.

It was reported earlier (18) that the infiltration of digitonin into green leaves causes a partial rearrangement of the Chl molecules out of its longwave forms into the shorter-wave ones. The treatment with 4% digitonin results in (Fig. 1a) a decrease in the longwave form alongside with its shift from 740 or 735 nm (before or after 5-h darkening) to 725 nm and a growth of emission at 692-695 nm. Such changes could be ascribed not only to Chl rearrangement out of the longerwave forms but also to a failure in the energy transfer to them. By comparing the excitation spectra (curves 3 and 4 in Fig. 1b) one can see the actual decrease of the band attributed to the energy transfer

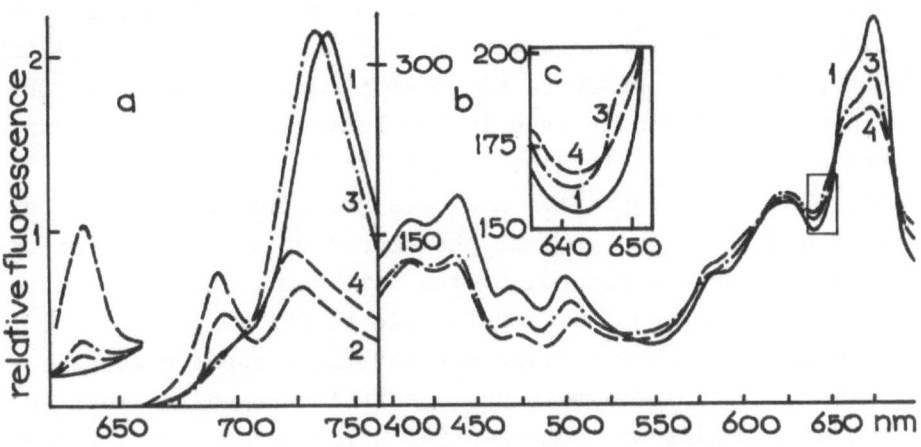

FIGURE 1. The fluorescence (a) and fluorescence excitation
(b,c) spectra of 3-day-old green leaves of the Chl b-less
mutant of barley, illuminated (1,2) or darkened for 5 h (3,4).
Phosphate buffer (1,3) or 4% digitonin (2,4) were infiltrated
in the leaves. The fluorescence spectra 1 and 3 are normalized
on the longwave band height. The excitation spectra are monitored at 730 nm and normalized at 620 nm. The part of the
excitation spectra in rectangle is scaled up in insert (c).

from carotenoids to Chl. Digitonin impaired this transfer
presumably by separating the molecules of the two pigments.

Upon darkening the green leaves accumulate Pchlide with
fluorescence bands at 635 and 655 nm (10). We observed in this
case a growing band at 648 nm in the excitation spectrum of
Chl fluorescence in both normal and mutant leaves. This excitation band was found even in the leaves in which no fluorescence band at 655 nm was detected (Fig. 1). It seems that the
pigment with the absorption maximum at 648 nm transfers the
energy to Chl a and this energy-transfer band grows when the
leaves are kept in darkness. In Chl b-less mutant it may only
be Pchlide which accumulates in the dark. In fact, when digitonin was infiltrated the impairment of the energy transfer
led to a disappearance of the Pchlide band at 648 nm in the
excitation spectrum of the Chl fluorescence alongside with
an increase in the proper emission of Pchlide. In the leaves

treated with digitonin this emission occurs only at 632-635 nm
without the fluorescence band at 655 nm. It is essential that
such Pchlide fluorescence at 632-635 nm was detected in the
presence of digitonin even in the leaves kept under illumination
when without the detergent no Pchlide fluorescence was notice-
able. The both separation and disaggregation of Pchlide correspond well to the general phenomenon of uncoupling various
pigment molecules induced by the detergent, and the impairment
of the energy transfer from Pchlide to Chl is just one of the
effects, the most important one for the present discussion.

In Fig. 2 the fluorescence spectral changes in green and
etiolated leaves are compared for a large range of the digitonin concentration. As concerns the intensity of the Pchlide
fluorescence they are of the same type in the both kinds of
the leaves, however the saturation is reached at 0.4% digitonin
in etiolated and at 1% digitonin in green leaves.

In etiolated leaves digitonin distorts the spectrum of
Pchlide fluorescence in a way that the band at 655 nm basically
disappears while the band at 633 nm grows. The integral fluorescence of Pchlide, measured as the area under the spectrum from
615 to 665 nm, is 1.54 $\pm$ 0.06 times as large at 0.4-4% digitonin
as in the control without digitonin. In green leaves 4.4-fold
increase in this area was observed when 1-4% digitonin was applied. If instead of digitonin sodium dodecylsulfate or organic

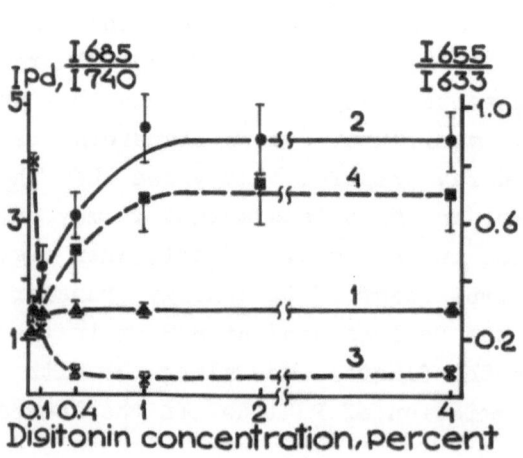

FIGURE 2. Influence of
digitonin in different concentrations on some optical
characteristics of 6-day-old
barley leaves. Ipd - intensity of protochlorophyllide
fluorescence, measured as
the area under spectrum from
615 to 655 nm, of etiolated
(1) and green (2) leaves.
I655/I633 - the ratio of the
fluorescence intensities of
protochlorophyllide in etiolated leaves (3). I685/I740 -
the ratio of the fluorescence
intensities of chlorophyll
in green leaves (4). In each
case the value for intact
leaves is taken for unity.

solvents were infiltrated there was again a much greater in-
tensification of the Pchlide emission in green leaves than
in similarly treated etiolated leaves. For instance, in case
when 40% acetone was infiltrated, the difference between both
kinds of the leaves was as high as 3.5 times. The much greater
changes in green leaves means that the effects observed could
not be fully attributed to a change in the yield of the Pchlide
fluorescence done by an alteration of the pigment arrangement.
On the other hand, just such results should be obtained due
to a separation of Pchlide and Chl molecules with breaking
the energy transfer from Pchlide to Chl and increasing the
proper Pchlide emission.

A parallel course of the curves 2 and 4 in Fig. 2 is also
impressive. It shows a similarity in the action of digitonin
on the energy transfer from Pchlide to Chl and on the state
of Chl molecules monitored by the ratio of the Chl shortwave
to longwave fluorescence bands. The both may reflect the
general destabilization of the chloroplast membranes under a
loosing action of the detergent.

An estimation of the efficiency of the energy transfer from
Pchlide to Chl was based on the assumption that at the saturated concentration of digitonin the transfer was blocked
completely. In this case the final intensity of the Pchlide
fluorescence corresponded to the energy absorbed by itself,
though this intensity had to be reduced by a factor of 1.54
(see above). Then the energy which in the intact leaves is
transfered from Pchlide to Chl might be determined as the
difference between the Pchlide fluorescence intensities in
the digitonin-treated and untreated leaves. An uncertainty
was brought in, however, because the factor of 1.54 had to be
taken from the measurement made for etiolated leaves and applied to the green ones. It was found that the efficiency of
the energy transfer from Pchlide to Chl varies the duration
of darkening the leaves (18). The most careful examinations
were performed with 6-h darkening when the efficiency was
averaged around 60 $\pm$ 7% with occasional values from 24 to 85%.
The origin of such scattering was not studied as yet.

302

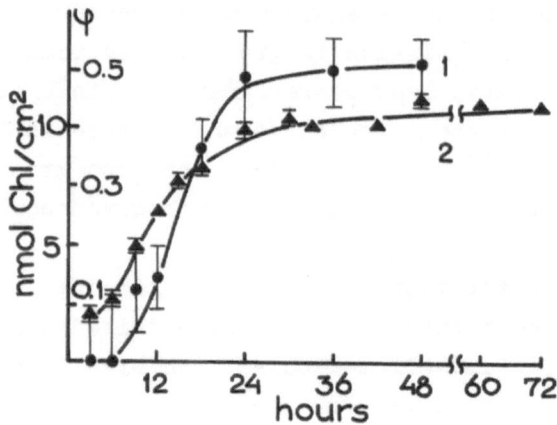

FIGURE 3. Kinetics of the evolution of energy transfer efficiency (1) and chlorophyll accumulation (2) during illumination of 6-day-old etiolated leaves of barley

The efficiency of the energy transfer from Pchlide to Chl depends upon a stage of the greening process. This was learned from the experiments with the etiolated leaves of barley which were illuminated for various periods of time and darkened for 6 h in order to have enough Pchlide accumulated (Fig. 3). In shortly illuminated leaves there was no energy transfer from Pchlide to the first molecules of Chl. In such leaves digitonin created the same increase in the Pchlide fluorescence as in unilluminated leaves. This corresponds to the previous conclusion by Thorne (8) that Pchlide resynthesized in the postetiolated leaves of bean does not transfer energy to Chl. The curve 1 in Fig. 3 goes up only if the leaves were illuminated longer than for 6 h when the Chl synthesis accelerates. The both curves reach saturation to about 24 h, thereafter Chl accumulates only very slowly and the energy transfer seems to remain constant at the level similar to that in green leaves.

Hence, three stages can be distinguished in the evolution of the energy transfer in correlation with changing situation in the work of the centers of Chl biosynthesis as producers of multimolecular groups of the pigment (5,12,19,20). At the first stage all elements of the future structural units of the photosynthetic apparatus are still to be developed and contain not enough Chl. These conditions favor molecular translocations which follow Pchlide photoconversion. So the first Chl molecules

easily leave the sites of their initial origin and are bound
rather far from the molecules of Pchlide, those which possibly
remained and those which are synthesized to substitute the
phototransformed ones. The second stage is intermediate on the
way to the third, when an opposite situation comes. It may be
suggested that in the process of building the pigment systems,
when more and more pigment molecules fill the vacant places
of their destination and the growing assemblies become larger,
the distance between Chl accumulated and Pchlide-forming sites
shortens. Then, the shorter the distance, the better conditions
for the energy transfer. In this way the Chl accumulation may
be an important (though not the sole) prerequisite for the
evolution of the energy transfer under discussion.

It is generally accepted nowadays that Pchlide accumulation
up to a certain limit stops its further production at the early
step of the biosynthetic pathway where 5-aminolevulinic acid
is created (21,22). The whole mechanism may be visualized in
frames of the concept of the centers of Chl biosynthesis more
precisely. Each center has a definite capacity for a group of
Pchlide molecules. When this capacity is filled up, the center
is switched off. In light Pchlide undergoes photoreduction and
molecules of the arising Chl leave the center and open the
capacity for new Pchlide to be synthesized. Under illumination
more and more Chl molecules can be harvested from the same site
of Pchlide production and they are piling up in the growing
unit of the photosynthetic apparatus. In its turn, the piling
also may not continue indefinitely and is limited by construc-
tion of the corresponding pigment carrier. It seems logical
that the growing assembly of the pigment molecules may finally
become so proximate to the Pchlide-producing site as to prevent
any translocation of a newly formed molecule. In these case
even after photoreduction of Pchlide the molecule could not
leave the site of its origin, the capacity would remain filled
and the center would remain switched off. It could turned out
that in the green leaves Chl molecule would play the same role
for the feed-back regulatory mechanism as Pchlide usually does
in the etiolated ones. Since then the center could work only

304

for Chl turnover, i.e. for producing new pigment molecules to substitute those which degrade and are out.

One more point should be mentioned. As the chloroplast membrane system is metabolically heterogeneous and contains not only mature but also developing structural units (2,12, 13,17,20), the conditions for the interaction of Pchlide and Chl there vary. The efficiency of the energy transfer from Pchlide to Chl given above is therefore averaged from very low in the units which just started their development to higher ones in those units which are completed.

REFERENCES

1. Fradkin LI, Shlyk AA, Kolyago VM. 1966. Dark biosynthesis of chlorophyll b in shortly illuminated etiolated seedlings, Dokl. Akad. Nauk SSSR 171, 222-225.
2. Shlyk AA. 1965. Chlorophyll metabolism in green plants. Minsk, Nauka i tekhnika Publ.
3. Shlyk AA. 1971. Biosynthesis of chlorophyll b, Annual Review Plant Physiol. 22, 169-184.
4. Butler WL, Briggs WR. 1966. The relation between structure and pigments during the first stage of proplastid greening, Biochim. Biophys. Acta 112, 45-53.
5. Fradkin LI, Shlyk AA, Kalinina LM, Fáludi-Daniel Á. 1969. Fluorescence studies on the reaction centers of chlorophyll biosynthesis at the early stages of greening, Photosynthetica 3, 326-337.
6. Schultz A, Sauer K. 1972. Circular dichroism and fluorescence changes accompanying the protochlorophyllide to chlorophyllide transformation in greening leaves and holochrome preparations, Biochim. Biophys. Acta 267, 320-340.
7. Kahn A, Boardman NK, Thorne SW. 1970. Energy transfer between protochlorophyll(ide) molecules: protochlorophyll(ide)-protein complex in vivo and in vitro, J. Mol. Biol. 48, 85-101.
8. Thorne SW. 1971. The greening of etiolated bean leaves. I. The initial photoconversion process, Biochim. Biophys. Acta 226, 113-127.
9. Litvin FF, Belyaeva OB. 1971. Characteristics of individual reactions and general scheme of the biosynthesis of native forms of chlorophyll in etiolated leaves of plants, Biokhimiya 36, 615-622.
10. Litvin FF, Krasnovskii AA, Rikhireva GT. 1959. Formation and transformation of protochlorophyll in green leaves of plants, Dokl. Akad. Nauk SSSR 127, 699-701.
11. Fradkin LI, Shlyk AA. 1967. On energy transfer between carotenoids and green pigments. In Vasil'ev RF, Chumakova RI, eds. Bioenergetics and spectrophotometry, pp. 135-140. Moskva, Nauka.

12. Shlyk AA. 1975. Progress in the studies of metabolic heterogeneity of photosynthetic membranes. Review of the initial steps. In Shlyk AA, ed. Biosynthesis and state of chlorophyll in plants, pp. 104-160. Minsk, Nauka i tekhnika Publ.
13. Shlyk AA. 1980. Current concept of organization of chlorophyll biosynthesis. In Mazliak P, ed. Biogenesis and function of plant lipids, pp. 311-320. Amsterdam, Elsevier/North Holland.
14. Fradkin LI, Kolyago VM, Shlyk AA. 1972. Fractionation of digitonin-treated chloroplasts of barley by a gel-electrophoretic method, Dokl. Akad. Nauk SSSR 207, 453-456.
15. Chaika MT, Savchenko GE. 1974. Fractionation of the chloroplasts pigment pool by a gel-electrophoretic method. In Shlyk AA, ed. Chlorophyll, pp. 269-279. Minsk, Nauka i tekhnika Publ.
16. Shlyk AA, Fradkin LI, Shalygo NV, Averina NG. 1981. Localization of magnesium protoporphyrin IX monomethyl ester in submembrane particles of barley chloroplasts, Biofizika 26, 1102-1104.
17. Fradkin LI, Chkanikova RA, Shlyk AA. 1981. Coupling of chlorophyll metabolism with submembrane chloroplast particles, isolated with digitonin and gel electrophoresis, Plant Physiol. 67, 555-559.
18. Fradkin LI, Domanskaya IN, Shlyk AA. 1981. Energy transfer from protochlorophyllide to chlorophyll in green plant, Dokl. Akad. Nauk SSSR 261, 220-223.
19. Shlyk AA, Prudnikova IV, Fradkin LI, Nikolaeva GN, Savchenko GE. 1969. Relationship between chlorophyll metabolism and heterogeneity of pigment apparatus. In Metzner H, ed. Progress in photosynthesis research, v. 2, pp. 572-591. Tübingen, Laupp.
20. Shlyk AA, Fradkin LI, Rudoi AB, Prudnikova IV, Savchenko GE. 1978. Group mechanism of formation of pigment assembly in centers of chlorophyll biosynthesis. In Akoyunoglou G, ed. Chloroplast development, pp. 119-130. Amsterdam, Elsevier/North Holland.
21. Lascelles J. 1964. Tetrapirrole biosynthesis and its regulation. N-Y, Benjamin.
22. Kaler VL. 1976. Autoregulation of the chlorophyll formation in higher plants. Minsk, Nauka i tekhnika Publ.

THE BIOENERGETIC DESCRIPTION OF LIGHT ENERGY MIGRATION IN PHOTOACTIVE MEMBRANES; EQUIVALENCE BETWEEN THE THEORY OF THE ENERGY FLUXES AND THE THEORY OF THE PROPORTION OF PIGMENTS FORMS TO TOTAL PIGMENTS.

C. SIRONVAL[*], R. STRASSER[**], M. BROUERS[*].
[*]Lab. of Photobiology University of Liège B.22 4000 Liège,Belgium
[**]Dept. of Bioenergetics University of Stuttgart, 7000 Stuttgart FRG

1. INTRODUCTION

The energy exchanges between pigments in biological membranes irradiated with light can be described in a number of ways. It is shown in this paper that the theory of the interrelationships between pigment species, as expressed by their proportions to total pigment, which has been used to describe the exchanges between protochlorophyllide and chlorophyllide in irradiated etioplast membranes (Brouers and Sironval, 1978), is equivalent to the theory of the energy exchange fluxes developed by Strasser (1978). Yet, the first of these theories was derived from the analysis of events observed when etiolated leaves are irradiated for the first time, while the second was developed to account for energy transfers from light harvesting pigments to active centres in fully green leaves. The exchanges in green and in etiolated leaves will be shown to have common features, and the connection between the methods of accounting for them will appear.

2. FUNDAMENTALS

2.1. Consider a coloured biological membrane irradiated at temperature T by a stationary light flux of intensity I (joules.cm^{-2};sec^{-1}). This membrane exchanges energy and matter with its surroundings and it may perform biochemical work. The molecules of S which absorb light are, by definition, the pigments. At the stationary state the energy of the light which these pigments absorb per cm^2 of membrane during a given time is equal to the energy they dissipate. Dissipation appears as heat conduction, performed work or light reemission. The ratio of the reemitted light to the amount of absorbed light :

C. Sironval and M. Brouers (eds.); Protochlorophyllide Reduction and Greening. ISBN 90 247 2954 8
© 1984, Martinus Nijhoff/Dr W. Junk Publishers, The Hague/Boston/Lancaster.

$$\phi = \frac{\text{amount of reemitted light}}{\text{amount of absorbed light}} \tag{1}$$

defines the yield of light reemission.

Let the membrane contain N_x pigment molecules of a certain pigment kind X per cm^2. When exposed to light these molecules absorb :

$$E_x \ (joules.cm^{-2}.sec^{-1}) = \alpha_x.I.N_x \tag{2}$$

where α_x ($cm^2.molecule^{-1}$) is the absorption coefficient of the X's. The product $\alpha_x.I$ is the amount of light absorbed per X and per second, in joules.

Denoting ε_x ($joules.molecule^{-1}.sec^{-1}$) the amount of energy reemitted as light by each of the X's, the N_x pigment emit :

$$F_x \ (joules.cm^{-2}.sec^{-1}) = \varepsilon_x N_x \tag{3}$$

One finds an expression for (1) by dividing (3) by (2) :

$$\phi_x = \frac{F_x}{E_x} = \frac{\varepsilon_x}{\alpha_x I} \quad \text{implying} \quad \frac{\varepsilon_x}{\alpha_x} = \phi_x I \tag{4}$$

which applies to a stationary flux of light.

2.2. Let us suppose that, among the works performed by the membrane in the radiant flux I, there is the irreversible, chemical transformation of pigment P into pigment C.
Then at any time t :

$$N_p + N_c = N; \ N \text{ constant.}$$

The ratios :

$$\frac{N_c}{N} = c \quad \text{and} \quad \frac{N_p}{N} = p = 1-c$$

vary as the transformation proceeds.

Assuming the P's to differ from the C's in the spectrum of the light they absorb or emit, one can write using (2), (3) and (4) :

$$F_c = \varepsilon_c.N_c = \alpha_c.I.N_c.\phi_c = E_c.\phi_c \tag{5.1}$$

$$F_p = \varepsilon_p.N_p = \alpha_p.I.N_p.\phi_p = E_p.\phi_p \tag{5.2}$$

These formulae express the reemitted light as the ϕ_pth (or ϕ_cth) part of the absorbed light. They imply that the light is absorbed, and emitted by a same pigment.

2.3. (5.1) and (5.2) may be rewritten as follows :

$$F_c = \alpha_c (\tfrac{I}{N} N_c) N \; \phi_c = \alpha_c \cdot I_c \cdot N \cdot \phi_c = E_c \phi_c \qquad (6.1)$$

$$F_p = \alpha_p (\tfrac{I}{N} N_p) N \; \phi_p = \alpha_p \cdot I_p \cdot N \cdot \phi_p = E_p \phi_p \qquad (6.2)$$

in which : $I_c = \tfrac{I}{N} N_c$

$I_p = \tfrac{I}{N} N_p$

I_c and I_p are the parts of the incident flux which reach the C's and the P's respectively :

$$I = I_c + I_p.$$

If the membrane was to contain N pigments P (or N pigments C), the incident flux should be I_p (or I_c) for these pigments to absorb energy E_p (or E_c) and to emit energy F_p (or F_c).

3. ENERGY TRANSFER

The C's will emit energy originally absorbed by the P's in case of energy transfer from the P's to the C's. We will consider this case only, the reverse being symetrical.

Then, the flux transferred from the P's adds to the incident light to compose the total flux which reaches the C's :

$$I'_c = I_c + I_p \, e_{pc} \qquad (7.1)$$

in which e_{pc} is the part of the flux I_p which is transferred. The flux I'_p is then left to the P's :

$$I'_p = I_p - I_p \, e_{pc} \qquad (7.2)$$

Replacing I_c by I'_c in (6.1) and I_p by I'_p in (6.2), the energies emitted in case of transfer are :

$$F'_c = \alpha_c [I_c + I_p \, e_{pc}] N\phi_c = E'_c \phi_c$$

$$F'_p = \alpha_p [I_p - I_p \, e_{pc}] N\phi_p = E'_p \phi_p$$

Introducing the definitions of I_c and I_p (§ 2.3) and rearranging, one obtains :

$$F'_c = \alpha_c \cdot I.N [\tfrac{Nc}{N} + \tfrac{Np}{N} e_{pc}] \phi_c = E'_c \phi_c \qquad (8.1)$$

$$F'_p = \alpha_p \cdot I.N [\tfrac{Np}{N} - \tfrac{Np}{N} e_{pc}] \phi_p = E'_p \phi_p \qquad (8.2)$$

(8.1) and (8.2) give the partition between pigments of the energy emitted at the stationary state when the P's transfer to the C's. In the absence of this transfer (e_{pc} = 0) they are reduced to (5.1) and (5.2). The factor $(\frac{Nc \text{ or } Np}{N} \pm \frac{Np.epc}{N})$ is the partition factor. The part e_{pc} of the flux I_p, which defines the migrating energy, is identified in this factor with the proportion to total pigments of the P's which transfer energy.

4. ENERGY TRANSFER EXPRESSED AS A RELATIONSHIP BETWEEN PIGMENT PROPORTIONS TO TOTAL PIGMENTS (Brouers and Sironval, 1978).

4.1. The formulae (8.1) and (8.2) apply at any time during the transformation of the P's into the C's. If one adopts the hypothesis that this transformation is irreversible, N_p approaches zero and N_c approaches N after a sufficiently long period. Then the transfer from P to C vanishes and (8.1) becomes :

$$(F'_c)_{t \to \infty} = \alpha_c . I . N . \phi_c \qquad (9)$$

Dividing (8.1) by (9) one finds :

$$\frac{(F'_c)_t}{(F'_c)_{t \to \infty}} = \frac{N_c}{N} + \frac{N_p}{N} e_{pc} = c + (1-c)e_{pc} = i_c \qquad (10)$$

This relationship deals with the set of pigments which contribute energy to the emission of the C's. It states that the proportion i_c of these pigments to total pigments comprises (I) the proportion c of the C's to total pigments, and (II) the proportion to total pigments of the P's which transfer energy $(1-c)e_{pc}$. It shows further that i_c is measured at any time t by the ratio of the energy emitted by the C's at this time, to the energy emitted by the C's when the P→C transformation has been completed, i.e. for $t \to \infty$. Any simultaneous measurement of c and i_c permits to calculate e_{pc}.

4.2. An energy transfer from one pigment to another does not necessarily exclude the reverse transfer. However, in some cases, the transfer is in one direction only, as for instance during the photoreduction of protochlorophyllide into chlorophyllide in higher plants. This particular reaction fits the situation described in the previous sections :

a. The protochlorophyllides (the P's) and the chlorophyllides
   (the C's) are in biological membranes (the prothylakoïds
   and prolamellar bodies);
b. the P's and the C's absorb and reemit light, and part of
   the light is used for work;
c. light transforms irreversibly the P's into the C's;
d. the P's transfer energy to the C's and the reverse tranfer
   has not been demonstrated.

The theory comes to the conclusion that, the proportion
to total pigments of the pigments which contribute energy to
the C's is given by (10). The facts prove that this conclusion is correct provided e_{pc} has the meaning which is given
below.

Sironval and Kuiper (1972) have measured i_c at a series
of times chosen during the phototransformation of protochlorophyllide into chlorophyllide in illuminated, etiolated
leaves. They have measured c at the same times. These measurements have shown that c is linked to i_c by the time independent, empirical relationship :

$$(A-i_c)c = Ki_c \tag{11}$$

in which A and K are constants. The hyperbolic form of this
expression implies A-K = 1.

On the other hand, Brouers and Sironval (1978) have deduced
the empirical result by defining e_{pc} as the probability that
1 protochlorophyllide transfers energy to 1 chlorophyllide.
They have expressed this probability as :

$$\dot{e}_{pc} = \frac{c}{K+c} \tag{12}$$

They have found (11) by substituting (12) for e_{pc} in (10),
i.e. by prolonging the theory.

4.3. It appears in this way that, if one accepts (12), the
theory which gives (10) is verified by the experimental demonstration of (11). This theory is a theory of the relationships between proportions to total pigments of pigments of
different kinds, as it deals with c, p, i_c and e_{pc} only.
Vaughan and Sauer (1974) have shown experimentally that one
has to accept (12), and they have calculated e_{pc}. Other reasons

to accept (12) are found in Thorn and Boardman (1972) and
Nielsen and Kahn (1973).

5. THE THEORY OF THE ENERGY FLUXES (Strasser, 1978).

The theory of the energy fluxes postulates that any complex
arrangement of interconnected pigment systems can be expressed in term of basic equations making use of the following
variables :

a) the excitation energy influx E_i into pigment system i
   (number of excitations.cm^{-2}.sec^{-1});

b) the deexcitation energy outflux E_{ij} due to energy migration
   from pigment i to pigment j (number of deexcitations.cm^{-2}.
   sec^{-1}):

c) the enrgy content P_i^* of pigment system i (number of excited i.cm^{-2}).

(notations and units are taken from Strasser, 1978).

These variables are defined by the three following basic
equations :

$$E_i = J_i + \Sigma_h E_{hi} \tag{13}$$

$$E_{ij} = E_i \cdot P_{ij} \tag{14}$$

$$P_i^* = E_i \cdot \tau_i \tag{15}$$

(13) gives the excitation influx E_i into the set of pigments
i. E_i comprises the influx J_i of the light absorbed directly
by the i's and all energy transfer fluxes E_{hi} from neighbour
pigments h to the i's.

(14) defines the energy transfer flux leaving the i's.
This flux is E_i times the probability p_{ij} that an excitation
which leaves the i's is transferred to the j's. p_{ij} is measured by the ratio of the energy transfer flux E_{ij} to the
total sum of all fluxes leaving the i's, $\Sigma_h E_{hi}$ (heat conduction, chemical work, fluorescence emission, energy transfer,
etc...). At the stationary state this total outflux is equal
to the total influx E_i, and each individual outflux can be
represented by its rate constant k_{ih} (which includes the concentration of the energy acceptor if necessary).

Therefore :

$$p_{ij} = \frac{E_{ij}}{\Sigma_h \, E_{ih}} = \frac{E_{ij}}{E_i} = \frac{k_{ij}}{\Sigma_h \, k_{ih}} \tag{16}$$

(15) defines the energy content of the i's as being the energy influx E_i multiplied by the half-time τ_i of the excited state of the i's, -with τ_i equal to :

$$\tau_i = \frac{1}{\Sigma_h \, k_{ih}} \tag{17}$$

It follows from (15), (16) and (17) that the transfer flux E_{ij} may be expressed indifferently by one of the three following equations :

$$E_{ij} = E_i \cdot p_{ij} \tag{14}$$

$$E_{ij} = E_i \cdot \overbrace{\tau_i \cdot k_{ij}} \tag{18}$$

$$E_{ij} = p_i^* \cdot k_{ij} \tag{19}$$

The theory of the energy fluxes is able to represent any pattern of energy exchanges, -whatever its complexity, using the set of equations (13 to 15), (18, 14 and 19).

6. EQUIVALENCES

6.1. In order to discuss the connection of the flux theory with the theory of the relationships between pigment proportions to total pigments, (8.1) is rewritten :

$$\frac{F_c'}{\phi_c} = \alpha_c \cdot I . N_c + \alpha_c \cdot I . N_p \cdot e_{pc} \tag{20}$$

The first term of (20) is the amount of energy absorbed by the C's in the absence of any transfer from the P's. It is the influx directly absorbed by the C's. This term is noted J_i in Strasser's formula (13) :

$$\alpha_c \cdot I . N_c = J_c \tag{21}$$

6.2. The second term of (20) refers to the transfer of the energy. In this term, the P's are considered as the sole source and the C's as the sole target of the transfer. The C's are implied both in the fraction e_{pc} of the P's which donate energy (formula 12); and in the proportionality factor α_c.

The latter is the cross-section of the acceptor molecule for the energy flux.

The transfer is described as a flux by considering a sum of individual transfer acts, each of which consists in an energy quantum q jumping from 1P to 1C. The excited P's, -the P*'s-, contain q during a mean time τ_p (the mean life-time). At the stationary state the deexcitation of 1P* is counterbalanced by the excitation of 1P, and the total energy flux passing through the P's (the "transit" flux) is :

$$E_p = \frac{N_{p*} \cdot q}{\tau_p} \tag{22}$$

A fraction e_{pc} of this flux is transferred to the C's :

$$E_p\, e_{pc} = \frac{N_{p*} \cdot q}{\tau_p}\, e_{pc}$$

This implies the following equivalence :

$$\alpha_c \cdot I \cdot N_p \cdot e_{pc} = \frac{N_{p*} \cdot q}{\tau_p}\, e_{pc} = E_p \cdot e_{pc} \tag{23}$$

(23) applies (14) to the energy transfer from the P's to the C's.

6.3. The relationship :

$$E_p \cdot \tau_p = N_{p*} \cdot q = [P^*] \tag{24}$$

is easily identified to (15). It gives the energy content of the P's.

It must be noticed that (22), which defines E_p, does not specify the origin of the excitation of the P's; the variety and the nature of the sources of this excitation are not identified. Generalizing (22) one can write :

$$\frac{N_{x*} \cdot q_x}{\tau_x} = E_x$$

This relation furnishes a way for expressing (20) as a total excitation flux.

Taking (21) and (23) into account, (20) becomes :

$$E_c = J_c + E_p \cdot e_{pc} \tag{25}$$

which applies (13) to the total excitation of the C's.

7. CONCLUSIONS

The formula (11) and (25) use distinct languages for expressing the same facts.

In the theory of the relationships between pigment proportions to total pigments, each term designates a collection of molecular objects with defined properties. If additional collections of objects have to intervene new terms have to be added to (10), each of them remaining individual. In the theory of the energy fluxes, several flux contributions may be mixed in a single flux.

We intend to show in a later stage how the theory of the energy fluxes describes the general case of reciprocal transfer, i.e. when a transfer from the C's to the P's is superimposed on the transfer from the P's to the C's. The theory of the relationships between pigment proportions to total pigments can also adapt to this general case.

The theory of the relationships between pigment proportions to total pigments postulates "morphological" differences between interacting molecular "species" and it manages to preserve the individuality of these "species". The theory of the energy fluxes refers to the concept of "grouping" in a set of subunits (Strasser, 1980). It is also able to distinguish between "species" with specific "morphologies" in the photosynthetic apparatus. It remains to decide in each particular case to what extent the "species" postulated by one theory are identical to the "species" postulated by the other.

This paper is a shortened version of a paper published in French by the authors under the title : "Equivalence entre la théorie des flux et la théorie des relations entre proportions de pigments pour la description de la répartition de l'énergie lumineuse absorbée par les membranes photoactives", Bull. Acad. Roy. de Bel., Classe Sciences, 67, 248-259, 1981.

REFERENCES
1. Brouers M, Sironval C. 1978. Photosynthetica, 12, 399-405.
2. Nielsen O.F., Kahn A. 1973. Biochim. Biophys. Acta, 292, 117-129.

3. Sironval C, Kuiper Y. 1972. Photosynthetica, 6, 254-275.

4. Strasser R.J. 1978. in Chloroplast Development (Akoyunoglou et al. eds) pp. 513-524, Elsevier, Amsterdam.

5. Strasser R.J. 1981. in Proc. Fifth International Congress Photosynthesis. G. Akoyunoglou ed. III 727-737. Balaban Int. Science Services Philadelphia.

6. Thorne S.W., Boardman N.K. 1972. Biochim. Biophys. Acta 267, 104-110.

7. Vaughan G.D., Sauer K. 1974. Biochim. Biophys. Acta, 374, 393-394.

THE DYNAMICS OF THE PHOTOREDUCTION OF PROTOCHLOROPHYLL(IDE) INTO CHLOROPHYLL(IDE)

RETO J. STRASSER

Department of Bioenergetics, Inst. of Biology, University of Stuttgart, Ulmerstrasse 227, 7000 Stuttgart 60, FRG

1. THE MODEL

The active protochlorophyllide-protein-complex is converted in the light to chlorophyllide. This reaction has been analyzed by several authors. A serie of publications by Sironval, Brouers, Kuiper (1968-1980) report data which support a model for the protochlorophyllide photoreduction. The model proposes:

1) Protochlorophyllide P_1 is reduced in the light to chlorophyllide P_2. The electron donor is not limiting.
2) The excited protochlorophyllide transfers a part of its excitation energy to chlorophyllide.

The authors developed for the description of this model a theory of the relations between proportions. Another theory describing the energy fluxes of any kind of photobiological system was developed by Strasser (1978). It has been shown elsewhere (Sironval, Strasser, Brouers 1981) that both theories lead to the same equations when applied to the model for the protochlorophyllide photoreduction as proposed by Sironval and Brouers. The energy flux theory can be applied for any complexity of a model. Therefore the model will be extended the following way and calculated by this theory. Any other model can be calculated accordingly. The extended model proposes in addition:

3) Energy transfer from chlorophyllide P_2 to protochlorophyllide P_1.
4) Energy transfer from one pool of protochlorophyllide to another pool of protochlorophyllide.
5) Energy transfer from a pool of chlorophyllide to another pool of chlorophyllide.

The extended model is shown in Fig.1 and the reactions of this model are as follows:

C. Sironval and M. Brouers (eds.); Protochlorophyllide Reduction and Greening. ISBN 90 247 2954 8
© 1984, Martinus Nijhoff/Dr W. Junk Publishers, The Hague/Boston/Lancaster.

$$P_1^* + NADPH \xrightarrow{k_{1P}} P_2 + NADP \qquad \text{Photoreduction}$$

$$P_1^* + P_2 \xrightarrow{k_{12}} P_1 + P_2^* \qquad \text{transfer } P_1 \text{ to } P_2$$

$$P_1^* \xrightarrow{k_{1F}} P_1 + F_1 \qquad \text{fluorescence of } P_1$$

$$P_2^* \xrightarrow{k_{2F}} P_2 + F_2 \qquad \text{fluorescence of } P_2$$

MODEL by Sironval and Brouers

$$P_1 + P_2^* \xrightarrow{k_{21}} P_1^* + P_2 \qquad \text{transfer } P_2 \text{ to } P_1$$

$$P_1^* + P_1 \xrightarrow{k_{11}} P_1 + P_1^* \qquad \text{transfer } P_1 \text{ to } P_1$$

$$P_2^* + P_2 \xrightarrow{k_{22}} P_2 + P_2^* \qquad \text{transfer } P_2 \text{ to } P_2$$

EXTENTION

Fig. 1

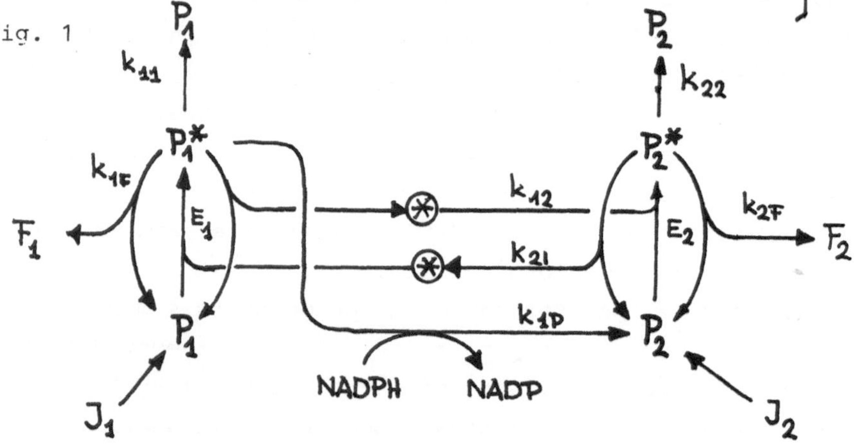

2. THE ENERGY FLUX THEORY APPLIED TO THE MODEL

The theory of the energy fluxes postulates that any type of complex arrangement of interconnected pigment systems can be expressed in terms of basic equations for the following variables:

a) <u>Excitation energy in flux</u> E_i or excitation rate of the pigment i (number of excitations.cm^{-2}.sec^{-1}).

b) <u>De-excitation energy outflux</u> E_{ij} migrating from the pigment system i to a surrounding pigment system j (number of de-excitation. cm^{-2}.sec^{-1}).

c) <u>Energy content</u> P_i^* of the whole pigment system (number of excited pigments i. cm^{-2}).

The variables are defined in the three basic equations:

Excitation rate $\quad E_i = J_i + \Sigma E_{hi}$ $\qquad$ Eq.1

De-excitation rate $E_{ij} = E_i \cdot P_{ij}$ $\qquad$ Eq.2

Exciton density $\quad P_i^* = E_i \cdot \tau_i$ $\qquad$ Eq.3

Eq.1 describes the excitation of pigment system i. And E_i is the sum of the light flux J_i absorbed by the pigments i and of all energy transfer fluxes from surrounding pigment systems h towards i. Eq.2 refers to de-excitation of pigments i by energy transfer towards j. And p_{ij} is the probability that an exciton is de-excited by transfer from i to j; it is defined by the ratio between energy transfer flux E_{ij} (from i towards j) and the sum of all types of de-excitations like fluorescence emission, heat dissipation, photochemistry and energy transfer. Each individual de-excitation flux is characterized by its rate constant k_{ij} (for second order de-excitations by the product of this rate constant and the concentration of the energy acceptor). Hence

$$p_{ij} = \frac{E_{ij}}{\Sigma E_{ih}} = \frac{k_{ij}}{\Sigma k_{ih}} = \frac{E_{ij}}{E_i} \qquad \text{Eq.4}$$

Eq.3 defines the energy content of pigment system i as the product of the excitation energy influx of pigment system i and the lifetime τ_i of the excited state of the pigment i.

$$\tau_i = \frac{1}{\Sigma k_{ih}} \qquad \text{Eq.5}$$

Thus each de-excitation flux from i towards j may be expressed differently by the following equations. The transferred energy being used either for photochemistry, heat dissipation or still re-emitted as fluorescence.

$$E_{ij} = E_i \cdot P_{ij} \qquad \text{Eq.6}$$
$$E_{ij} = E_i \cdot \tau_i \cdot k_{ij} \qquad \text{Eq.7}$$
$$E_{ij} = P_i^* \cdot k_{ij} \qquad \text{Eq.8}$$

All models can be expressed by the relations Eq.1 to 8.

3. THE EQUATIONS OF THE EXTENDED MODEL

The total excitation rate E_i of the pigment pool of protochlorophyllide can be written according to Eq.1 as the sum of all energy fluxes converging in this pool. These fluxes are: the light
absorption flux J_1, the energy transfer flux from a neighbour protochlorophyllide pool to another protochlorophyllide pool E_{11} and
the energy transfer flux E_{21} from excited chlorophyllide back to
the protochlorophyllide pool. According to Eq.2 we can write:

$$E_1 = J_1 + E_1 \cdot P_{11} + E_2 \cdot P_{21} = (J_1 + E_2 \cdot P_{21})(1 - P_{11})^{-1} \qquad \text{Eq.9}$$

The total excitation rate of chlorophyllide is accordingly:

$$E_2 = J_2 + E_2 \cdot P_{22} + E_1 \cdot P_{12} = (J_2 + E_1 \cdot P_{12})(1 - P_{22})^{-1} \qquad \text{Eq.10}$$

The equations 9 and 10 can be solved in such a way that the total
excitation rates of protochlorophyllide E_1 and that of chlorophyllide E_2 can be expressed in terms of the absorption fluxes and
the probabilities of different energy transfer fluxes.

$$E_1 = \{J_1(1 - P_{22}) + J_2 P_{21}\}\{(1 - P_{22})(1 - P_{11}) - P_{12}P_{21}\}^{-1} \quad \text{Eq.11}$$

$$E_2 = \{J_2(1 - P_{11}) + J_1 P_{12}\}\{(1 - P_{22})(1 - P_{11}) - P_{12}P_{21}\}^{-1} \quad \text{Eq.12}$$

$$\text{absorption term} \quad . \quad \text{excitation gain factor}$$

4. THE PROBABILITIES OF ENERGY TRANSFER

The probability that an exciton flows from a location i to another
location j is defined as the fraction of the energy flux which goes
from i to j compared to all de-excitation energy fluxes leaving
the excited pigment pool i. As the pigment concentrations are changing and the energy transfer fluxes depend on the energy donor and
energy acceptor concentrations, the energy transfer probabilities
become a function of the actual pigment concentration. For the
energy transfer probability that an exciton goes from an excited
protochlorophyllide molecule to a chlorophyllide molecule we can
write according to Eq.4:

$$P_{12} = \frac{E_{12}}{E_{12} + E_{11} + E_{1P} + E_{1F}}$$ (in the following the term for heat dissipation is always included in the term for fluorescence emission)

According to the reactions of the extended model follows:

$E_{12} = k_{12} \cdot P_2 \cdot P_1^*$ Energy transfer from P_1 to P_2

$E_{11} = k_{11} \cdot P_1 \cdot P_1^*$ Energy transfer from P_1 to P_1

$E_{1P} = k_{1P} \cdot P_1^*$ Rate of photochemistry

$E_{1F} = k_{1F} \cdot P_1^*\quad = F_1$ Fluorescence emission of P_1

We write for the relative concentration of the pigment P_1 the term c_1 and for the concentration of P_2 the term c_2 so that $c_1 + c_2 = 1$. The energy transfer probabilities are therefore: Eq.13

$$P_{12} = \frac{k_{12} \cdot c_2}{k_{1P} + k_{1F} + k_{11} + (k_{12} - k_{11}) \cdot c_2} ; \quad P_{21} = \frac{k_{21} \cdot c_1}{k_{2F} + k_{22} + (k_{12} - k_{11}) \cdot c_1}$$

$$P_{11} = \frac{k_{11} \cdot c_1}{k_{1P} + k_{1F} + k_{12} - (k_{12} - k_{11}) \cdot c_1} ; \quad P_{22} = \frac{k_{22} \cdot c_2}{k_{2F} + k_{21} - (k_{12} - k_{11}) \cdot c_2}$$

$$P_{1F} = \frac{k_{1F}}{k_{1P} + k_{1F} + k_{11} + (k_{12} - k_{11}) \cdot c_2} ; \quad P_{2F} = \frac{k_{2F}}{k_{2F} + k_{21} - (k_{12} - k_{11}) \cdot c_2}$$

The probability that an exciton in P_1 is used for photochemistry:

$$P_{1P} = \frac{k_{1P}}{k_{1P} + k_{1F} + k_{12} - (k_{12} - k_{11}) \cdot c_1} = \frac{k_{1P}}{k_{1P} + k_{1F} + k_{11} + (k_{12} - k_{11}) \cdot c_2}$$

The dynamics of these energy transfer probabilities during the transformation of protochlorophyllide into chlorophyllide is presented in Fig.2.

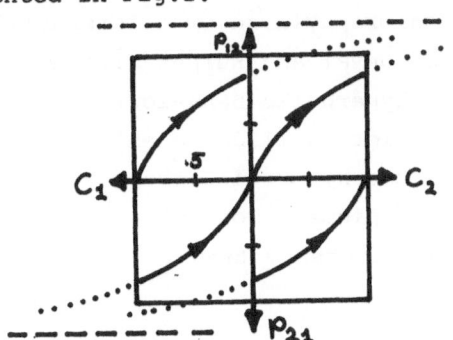
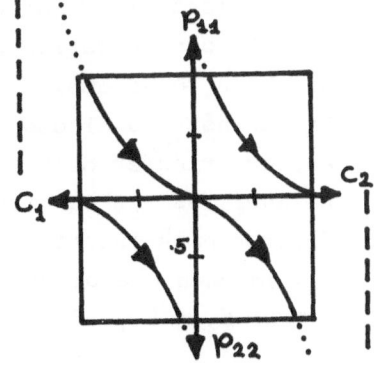

5. THE ENERGY FLUXES

The light energy flux J_i absorbed by the pigment pool i depends on the concentration of that pigment, its absorption coefficient, the spectrum and intensity of the exciting light source. The light energy flux J_i is therefore proportional to one minus the transmission of the sample. That means for low pigment concentrations J_i is proportional to the pigment concentration c_i. According to Eq.2 and 11 we can write for the experimental fluorescence signal of chlorophyllide:

$$F_2 = \{c_2 \cdot (\frac{E_{1F} + E_{1P}}{\Sigma E_{1h}}) + \frac{E_{12}}{\Sigma E_{1h}}\}\frac{E_{2F}}{\Sigma E_{2h}}\{\frac{(\Sigma E_{2h} - E_{22})(\Sigma E_{1h} - E_{11}) - E_{12}E_{21}}{\Sigma E_{2h} \cdot \Sigma E_{1h}}\}^{-1}$$

If we replace the energy fluxes E_{ij} by their terms for rate constants k_{ij} and pigment concentrations (according to Eq.8), the experimental signal for the fluorescence of protochlorophyllide F_1 or of chlorophyllide F_2 can be expressed as a function of the actual pigment concentration. (Eq.14 and 15)

$$F_2 = \frac{(1 + Z + X) \cdot c_2}{(1 + Z)(1 + Y) + \{X - Y(1 + Z)\} \cdot c_2} = \frac{K \cdot c_2}{(K - 1) + c_2}$$

$$F_1 = \frac{(1 + Y) \cdot c_1}{(1 + Z)(1 + Y) + \{X - Y(1 + Z)\} \cdot c_1} = \frac{1}{1 + Z} \cdot \frac{(K - 1) \cdot c_1}{K - c_1}$$

where: $K = (1 + Z + X)/\{X - Y(1 + Z)\}$ and $X = k_{12}/k_{1F}$; $Y = k_{21}/k_{2F}$; $Z = k_{1P}/k_{1F}$

This hyperbolic relationship between the fluorescence emission and the concentration of protochlorophyllide or chlorophyllide has been found empirically by Sironval and Kuiper already in 1972. The authors concluded that this hyperbolic behaviour indicates energy transfer only from protochlorophyllide to chlorophyllide. However, as the equations show, the hyperbolic character of the fluorescence functions versus the pigment concentration is always valid when one or several types of energy transfer fluxes $(E_{12}; E_{21}; E_{11}; E_{22})$ occur.

Whenever the expression:

$$\frac{k_{12}}{k_{1F}} - \frac{k_{21}}{k_{2F}} \cdot (1 + \frac{k_{1P}}{k_{1F}}) \quad > \quad 0 \quad \text{then } F_2 \text{ vs } c_2 \text{ is a horizontal hyperbolic}$$

function, or if

$$< \quad 0 \quad \text{then } F_2 \text{ vs } c_2 \text{ is a vertical hyperbolic}$$

function, or if

$$= \quad 0 \quad \text{then } F_2 \text{ is proportional to } c_2. \text{ That means}$$

no energy transfer E_{21} and E_{12} at all or

$$k_{12}/k_{1F} = (1 + k_{1P}/k_{1F})k_{21}/k_{2F}$$

The data reported by Sironval and Kuiper (1972) show that F_2 vs c_2 is a horizontal hyperbolic function. That means that energetic communication between protochlorophyllide and chlorophyllide does exist. However, it cannot be evaluated by these experiments if an energy transfer flux from chlorophyllide to protochlorophyllide does exist or not, because:

$$k_{12}/k_{1F} > (1 + k_{1P}/k_{1F})k_{21}/k_{2F} \quad \geq \quad 0 \quad \text{The equations show, that the}$$

terms K and k_{1P}/k_{1F} can be experimentally determined if we are able to measure the concentrations c_1 or c_2 and the value of the ratio F_1/F_2.

Fig.3 shows the dynamics of the fluorescence signals and the concentrations of proto- or chlorophyllide during the photoreduction. The mathematical expressions of the experimental kinetics F_1; F_2; c_1; c_2; versus time can be found by integration of the function F_2 vs c_2. Fig.3

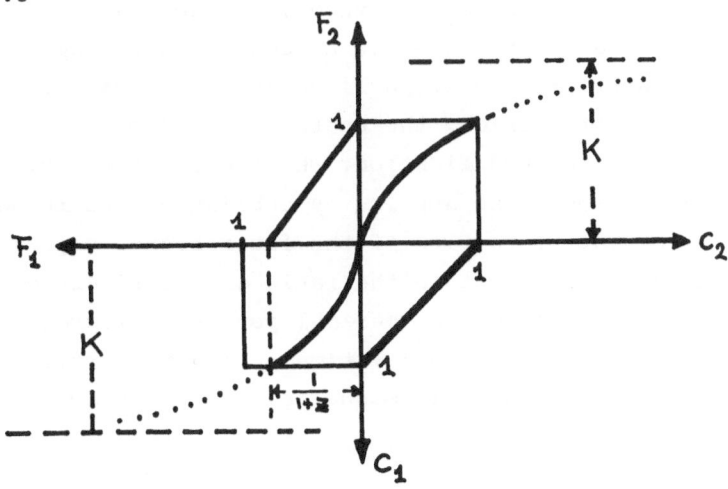

6. THE NORMALIZED FLUORESCENCE FLUXES

Very often it is not possible to measure the absolute fluorescence
values of each sample. In this situation normalized fluorescence
values will be reported. We define the relative fluorescence signal
$F_{1(rel)}$ or $F_{2(rel)}$ as the fluorescence emission of proto- or chlorophyllide at the time t divided by the maximal available fluorescence emission. $F_{1(rel)} = F_1/F_{1M}$ and $F_{2(rel)} = F_2/F_{2M}$

F_{1M} is measured when $c_1 = 1$ and $c_2 = 0$
F_{2M} is measured when $c_1 = 0$ and $c_2 = 1$. Combining these definitions
with the equations for the fluorescence signals (Eq.14 and 15)
leads to the following expressions: Eq.16

$$F_{1(rel)} = \frac{(K - 1) \cdot c_1}{K - c_1} \quad \text{and} \quad F_{2(rel)} = \frac{K \cdot c_2}{(K - 1) + c_2}$$

therefore
$$F_{2(rel)} = 1 - F_{1(rel)} \qquad \text{Eq. 17}$$

7. WHAT'S WRONG WITH THE PROPORTIONS ?

In an earlier paper (Sironval, Strasser, Brouers 1981) we showed
that the same equations for the photoreduction of protochlorophyllide to chlorophyllide according to the model by Sironval and
Brouers (where only energy transfer from Protochlorophyllide to
chlorophyllide is assumed) can be derived whether the theory of
the proportions or the theory of the energy fluxes is used. If the
model has to be extended as done in this paper by assuming energy
transfer from chlorophyllide back to protochlorophyllide (E_{21}) or
any other version, the energy flux theory can still be used
without any modification. The theory of the proportions however
can no longer be applied as published. The reasons are the following: The proportion of the fluorescence signal of chlorophyllide
at a given time t is the ratio of the fluorescence signal at time t
and the hypothetical maximal possible fluorescence signal under
the given conditions at time t. The proportions of F_2 defined by
Sironval et al as i_c means in physical terms the following:

$$i_c = \frac{F_{2(t)}}{F_{2(max)(t)}} = \frac{E_{2(t)}}{E_{2(max)(t)}} \cdot \frac{p_{2F(t)}}{p_{2F(t)}}$$

Empirically however the proportion i_c is measured (for technical reasons) as the ratio of the fluorescence signal at the time t and the fluorescence signal at the time t = ∞. This experimental ratio expresses the proper fluorescence proportion i_c only when $p_{2F(t)} = p_{2F(\infty)}$. The model by Sironval and Brouers assumes no other energy transfer than that from protochlorophyllide to chlorophyllide. In such a case and only in that case p_{2F} is a constant and the experimental value $F_{2(t)}/F_{2(\infty)}$ is indeed the theoretically defined proportion i_c. In all other situations the fluorescence probability of chlorophyllide (p_{2F}) becomes a function of the pigment concentration and the expression $F_{2(t)}/F_{2(\infty)}$ is no longer a proportion but a normalization on the signal at t = ∞ .

The difference between proportions and normalization seems very subtle but the distinction is crucial when one or the other theory is applied. Of course every extended model can be theoretically expressed by the theory of the proportions as described by Sironval and Brouers, however the link to the experimental measurements is no longer given by the ratio $F_{2(t)}/F_{2(\infty)}$.

8. MULTI-PARAMETER-ANALYSIS OF THE PHOTOREDUCTION

The rate of photochemistry can be described according to Eq.8 as follows:

$$-\frac{dc_1}{dt} = E_{1P} = P_1^* \cdot k_{1P} \quad \text{(if the electron donor is not limiting)}$$

$$F_1 = P_1^* \cdot k_{1F} \quad \text{(according to Eq.3) therefore}$$

$$-\frac{dc_1}{dt} = \frac{dc_2}{dt} = F_1 \cdot \frac{k_{1P}}{k_{1F}} \quad \text{or after integration}$$

$$c_2 = \frac{k_{1P}}{k_{1F}} \cdot \int_0^t F_1 \, dt$$

The proportion of chlorophyllide c_2 can now be expressed by the normalized area-growth curve S_1 of the fluorescence signal F_1. This expression remains unchanged if instead of the absolute fluorescence signals the relative (normalized) fluorescence signals are used. Using the Eq.17 leads to:

$$c_2 = \int_0^t F_1\, dt / \int_0^\infty F_1\, dt = \int_0^t F_{1(rel)}\, dt / \int_0^\infty F_{1(rel)}\, dt = S_1 = S_2$$

it follows Eq.18

$$C_2 = \text{area-growth of } F_{1(rel)} = S_1 = \text{area-growth of } (1- F_{2(rel)}) = S_2$$

==

The full dynamics of the protochlorophyllide phototransformation
can now be analyzed by the experimental fluorescence kinetics of
chlorophyllide as shown in Fig.4. The multi-parameter-analysis
(Strasser 1978,1981) is a tool to analyze such a problem. The experimentally accessible function $F_{2(rel)}$ versus S_2 can now be taken as a measure for the physical equation $P^*_{2(t)}/P^*_{2(\infty)}$ vs c_2.

$$F_{2(rel)} = \frac{K \cdot S_2}{(K-1)+S_2} \equiv \frac{K \cdot c_2}{(K-1)+c_2} \equiv K-(K-1)\cdot\frac{F_{2(rel)}}{S_2}$$

This experimental treatment of data is shown in Fig.4.
The physical meaning of the measurable constant K is given above.

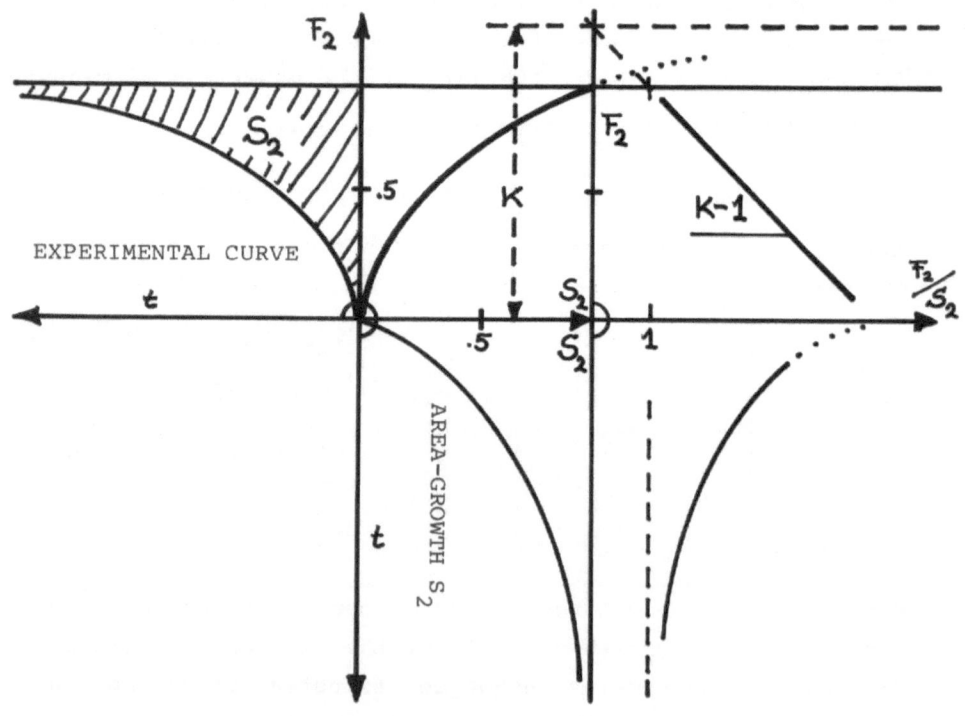

9. CONCLUSION

In this paper it is shown how a complex model can be described by the energy flux theory. Any other model can be analyzed accordingly. Often only the overall reaction of protochlorophyllide to chlorophyllide is considered. In these cases the model by Sironval and Brouers proves to be very useful. If several intermediate forms (eg. free pigments, pigment enzyme complex etc) and energetic communication between such forms are assumed, then the dynamics of the system becomes immediately very complex. The energy flux theory enables the description of such systems on every level of complexity and it makes the models accessible to computer simulations. The figures shown in this paper visualize the derived equations when specific values for the rate constants are assumed. The theoretical predictions of the shape of these functions are in full agreement with the published data of other authors. The physical meaning of the empirically found constants are given by the equations and their quantitative values may be measured under different environmental conditions in the future.

10. REFERENCES

The Reduction of Protochlorophyllide into Chlorophyllide
1) C. Sironval, M. Brouers, J.M. Michel, Y. Kuiper
                        Photosynthetica 2 (4), 268-287, (1968)
2) C. Sironval, M. Brouers; Photosynthetica 4 (1), 38- 47, (1970)
3) E. Dujardin, C. Sironval;Photosynthetica 4 (2), 129-138, (1970)
4) C. Sironval, Y. Kuiper; Photosynthetica 6 (3), 254-275, (1972)
5) M. Brouers, Y. Kuiper, C. Sironval
                        Photosynthetica 6 (2), 169-176, (1972)
6) C. Sironval   Photosynthetica 6 (4), 375-380, (1972)
7) M. Brouers, C. Sironval; Photosynthetica 12(4), 399-405, (1978)
8) C. Sironval, M. Brouers; Photosynthetica 14(2), 213-221, (1980)
The energy flux theory and the multi-parameter-analysis
    R. J. Strasser; in Chloroplast Development, G.Akoyunoglou et al
                        (ed) Elsevier Holland, 513-524, (1978)
    R. J. Strasser; in Photosynthesis III, G.Akoyunoglou (ed)
                        Balaban Int.Sc.Serv.PA;727-737, (1981)
    R. J. Strasser, H. Greppin; as above   717-727, (1981)
Comparison of both theories
    C. Sironval, R. J. Strasser, M. Brouers;Bulletin de l'Academie
                        Royale de Belgique 5,(67)248-259(1981)
    C. Sironval, R.J. Strasser, M.Brouers, 1984, this volume.

M O D E L S.

PHOTOCHEMICAL REDUCTION OF PROTOCHLOROPHYLLIDE IN DETERGENT MICELLES WITH THE FINAL CHLOROPHYLLIDE FORMATION

A.A. KRASNOVSKY, M.I. BYSTROVA, I.A. SAFRONOVA

Institute of Biochemistry, USSR Academy of Sciences, Moscow

The photoreduction of protochlorophyllide is the final photochemical stage of chlorophyll formation in plants. This stage is dependent on the specific native state of proto-chlorophyllide, which is associated with the lipoprotein mo-lecules in etioplast membranes.

The modelling of the in vivo state and photoconversions of the chlorophyll precursor is an approach to the study of its structural and functional organization in plants. In or-der to model the natural state of the chlorophyll precursor the different systems as solutions of protochlorophyll pig-ments in nonpolar solvents (1-3), solid films of protochlo-rophyll (4-6), its solutions in detergent micelles (7), and the others were used succesfully. We expected aqueous solu-tions of protochlorophyllide in detergent micelles would be most suitable for modelling of photoconversions of this pig-ment.

In general, aqueous micellar solutions of photosynthetic pigments in detergents as the systems modelling properties and the state of native pigment-protein-lipid complexes ha-ve been always attractive.

The present study describes the photochemical reduction of isolated protochlorophyllide in the aqueous solution of Triton X-100. The final photohydrogenation of protochloro-phyllide occurs at the site of the semiisolated double bond in the pyrrol ring IV. The yield of chlorophyllide obtained was up to 20% of the total amount of the initial pigment, being quite effective as compared to the reduction of proto-chlorophyll in the same conditions of the experiment.

C. Sironval and M. Brouers (eds.); Protochlorophyllide Reduction and Greening. ISBN 90 247 2954 8
© 1984, Martinus Nijhoff/Dr W. Junk Publishers, The Hague/Boston/Lancaster.

Protochlorophyllide was isolated from 10-12-d-old etiolated leaves of wheat (var. "Iubileinaya") in the following way. Leaves were extracted by 80% cooled acetone. The extract was repeatedly washed with hexane and heptane successively to remove protochlorophyll and carotenoids. These stages were rather near to the technique described previously (8). Then we transferred protochlorophyllide into diethyl ether. Beforehand, the pH of water-acetone solution had been adjusted to pH 7 by saturated solution of KH_2PO_4. Thus, without purification by chromatography, we obtained the pigment which was identical to protochlorophyll by absorption and luminescence spectra of its solutions in polar organic solvents.

Protochlorophyll was isolated from inner coats of pumpkin seeds by the method used in our laboratory. Protopheophorbide was prepared via protochlorophyllide pheophytinization in 1% aqueous solution of Triton X-100 at pH 3. Such acidity was obtained by adding of ascorbic acid up to 10^{-2}M. Chlorophyllide was prepared from chlorophyll a by the procedure described earlier (9).

Photoreduction was carried out after the vacuum pumping of samples in Thunberg tubes ajusted to install them into the cell compartment of Perkin-Elmer MPF-44B Spectrofluorometer. Reaction systems were illuminated with incandenscent lamp through the colored glass filter OS-13 which transmits light in the range 600-800 nm ($5 \cdot 10^4$ergs/cm^2. sec). All runs were made at 18-20°C. Progress of the photoconversion was controlled by fluorescence spectra recorded in the course of illumination. It took about 2 minutes to record a fluorescence spectrum, and so the illumination of samples was interrupted in all our experiments. When the photoprocess was over, air was admitted into the system, and spectra were recorded during the back dark reaction of reduced intermediates.

Photoreduction of protochlorophyllide was carried out in 1% solution of Triton X-100 with the admixture of 5% pyridine and 10^{-2}M of ascorbic acid. Under the action of light

the decline of the initial luminescence of protochlorophyllide rapidly occured. After 2-5 minutes of illumination a new
maximum appeared at 672-675 nm, which increased during the
subsequent illumination, while protochlorophyllide luminescence continued to decline but more slowly than in the be-
gining of the process. Usually, in 7-8 minutes of the interrupted illumination, light saturation took place, and a
luminescence peak near 674 nm reached its maximum. Very peculiar is that in the dark the intensity of the newly formed
longwave maximum increased significantly after the admitting
of air into the tube and usually reached up to 10-15% of the
initial value of the main protochlorophyllide luminescence
maximum as the result of the reversible oxidation of intermediate photoreduced forms. The course of a typical protochlorophyllide photoreduction sequence under described conditions is traced in Fig. 1.

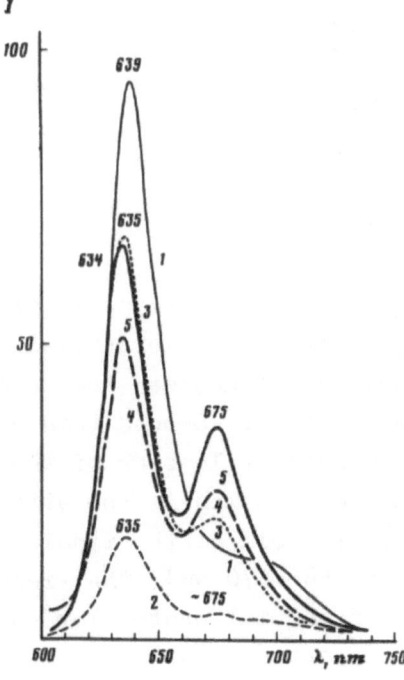

FIGURE 1. Photoreduction of
protochlorophyllide in 1%
Triton X-100 aqueous solution
in the presence of 10^{-2} M ascorbic acid and 5% of pyridi-
ne. Fluorescence spectra ($\lambda_{ex.}$
440 nm): 1, initial; 2-4,
during photoreduction, after
2-12 min. illumination; 5,
after the back reaction in the
presence of air. (Spectra 1, 2
were recorded at the sensitivity 1 /in arbitrary units of
measurement/; the sensitivity
for 3, 4 spectra is x10; and
the sensitivity for spectrum
5 is x3).

The maximum of 674 nm is assigned to the quite stable product which can remain in the reaction system for a long time after photoconversion. The fluorescence spectra of the system after photoreaction have been compared with the fluorescence spectra of an artifical mixture, taken before and after illumination. This mixture was composed of the equal amounts $(0.2 \cdot 10^{-6}$ M) of protochlorophyllide and chlorophyllide a in the reaction medium (Fig. 2). This allowed to estimated the yield of photoproduct with the maximum at 674 nm as about 20% of protochlorophyllide initial content and identify the product as chlorophyllide a.

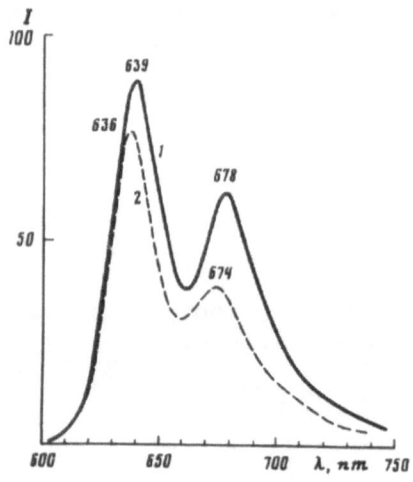

FIGURE 2. Fluorescence spectra of protochlorophyllide (0.2· 10^{-6} M)+chlorophyllide a (0.2· 10^{-6} M) mixture in 1% Triton X-100 aqueous solution in the presence of 10^{-2} M ascorbic acid and 5% of pyridine: 1, initial; 2, after 10 min. illumination.

Thus, in the micellar detergent solution the semiisolated double bond in the pyrrole ring IV of protochlorophyllide is reduced on illumination. It should be emphasized that protochlorophyll is much less reduced at the 7, 8-bond than protochlorophyllide under the same conditions. In case of the protochlorophyll reduction the photodestruction is too strongly pronounced to create the sufficient yield of the final long-wave product (with fluorescence at 668-670 nm); the yield of this product does not exceed 5% of initial protochlorophyll.

Under the same conditions of experiments with protopheophorbide (magnesiumless derivate of protochlorophyllide) no stable reduced product at the site of semiisolated double bonds appears. The photoreduction of protopheophorbide results in a reversible formation of a photoproduct having fluorescence maximum about 690 nm (Fig. 3).

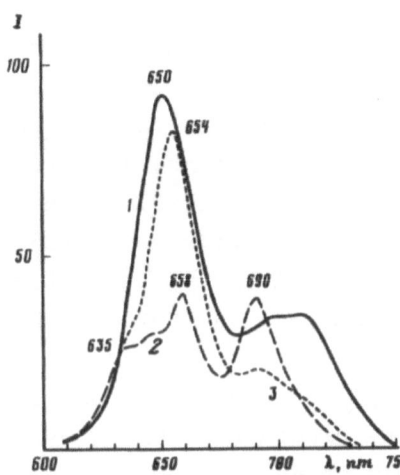

FIGURE 3. Photoreduction of protopheophorbide in 1% Triton X-100 aqueos solution in the presence of $2 \cdot 10^{-2}$ M ascorbic acid and 5% of pyridine. Fluorescence spectra ($\lambda_{ex.}$ 420 nm): 1, initial; 2, after 9 min. illumination; 3, after the back reaction in the presence of air.

These data indicate that such features of protochlorophyllide structure as the presence of a central magnesium atom and a free propionic group in the 7^{-th} C-atom are essential for the effective reduction of semiisolated bonds in the micellar detergent solution as in etiolated leaves (10).

Kinetics of photoprocess indicates that the chlorophyllide formation is not a single reaction; it proceeds stepwise and includes several stages. Among them there are both the actually photochemical and dark reactions; it reminds of the situation in etiolated leaves to a certain extent too. In detergent micelles the initial photochemical reaction leads to formation of nonfluorescence intermediates, it is another analogy with that in leaves. The mentioned intermediates are the photoproducts which have maxima near 480 nm and 530 nm as far as we can judge from absorption spectra. They appear, probably, from initial pigment successively as a result of the primary reduction at the site of the system of conjugated double bonds /P480/ (11), and then at the site of the 7, 8-double bond /P530/ in protochlorophyllide molecules. The oxidation converts P530 to chlorophyllide. The simplified scheme is given below.

$$PChde639 \xrightarrow[D]{h\nu} \text{reduced products} \dashrightarrow Chde675$$

(PChde, protochlorophyllide; Chde, chlorophyllide; D, electron /hydrogen/ donor; the numerials denote the positions of the main fluorescence maxima).

Probably, this is the principal pathway of the chlorophyllide formation in micellar detergent solutions. It corresponds to the phototransformation of protochlorophyll into chlorophyll in pyridine, investigated in our previous studies (12-14), and in the study (15), as well as to the photoreduction of Zn-derivatives of porphyrins and Zn-protopheophytin (16-18).

The highest yield of chlorophyllide (up to 20% of the initial pigment) was observed in the presence of 5% of pyridine. The clorophyllide formation diminished with the decrease of the pyridine content up to 1%, and became negligible small in the absence of pyridine. However, the rate of the reaction became slower and the yield of chlorophyllide fell, when we used pyridine without detergent as the reaction medium. Using ethanol-pyridine and water-pyridine solutions one could observed no formation of chlorophyllide under prolonged illumination. Thus, protochlorophyllide shows most completely its capasity for reduction of semiisolated double bonds in the micellar detergent solution. When we used in our experiments phosphate buffers in the pH range from 4 to 9, the effectiveness of the chlorophyllide formation greatly decreased. The increase to 5-10 fold in the amount of ascorbic acid in the system had analogous effect. Therefore, one may assume that in the detergent micelles protochlorophyllide posesses a peculiar molecular organization which is associated with the state of the detergent micelles and may determine protochlorophyllide photoreactivity. Indeed, in the aqueous solution of Triton X-100 (0.1-1%) at 20° C, the absorption and

luminescence spectra of protochlorophyllide differ from protochlorophyll spectra in displacement of maxima by 4-6 nm
to the long-wave side (Fig. 4). The difference between fluorescence spectra of protochlorophyllide and protochlorophyll
remains at -196° C, but at the same time the freezing causes
a short-wave shift of the both pigment maxima.

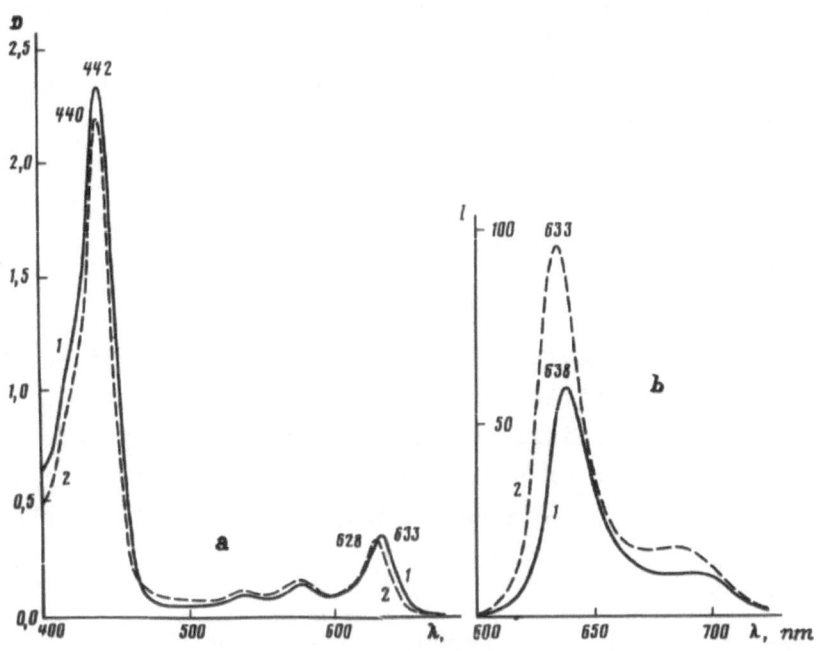

FIGURE 4. Absorption (a) and fluorescence (b) spectra of
pigments in 1% Triton X-100 aqueous solution: 1, protochlorophyllide; 2, protochlorophyll.

The comparative study of photochemical destructive oxidation of pigments in Triton X-100 by oxygen revealed a faster
rate of the protochlorophyllide oxidation in comparison with
that of protochlorophyll.

The study of pheophytinization in micellar solution of
Trition X-100 showed that in the case of protochlorophyllide
the replacement of magnesium by hydrogen ions occurs much
more rapidly than in protochlorophyll; halflife-time for
conversion into pheophytin takes 30-40 min. for protochlo-

rophyllide and 8.5-9 h. for protochlorophyll.

The resultes obtained may indicated the existence of a specific interaction between protochlorophyllide molecules and detergent micelles. This interaction is important for the manifestation of protochlorophyllide photochemical activity. At the same time a coordination binding of pigment molecules to nitrogen base molecules appears to be very significant for the photoreduction of protochlorophyllide into chlorophyllide a.

The system studied may be considered as a simple functional model of protochlorophyllide-holochrome.

Summary

Under the action of light protochlorophyllide is reduced in 1% solution of Triton X-100 in the presence of pyridine (5%) and ascorbic acid (10^{-2} M) with the final formation of the photoproduct having fluorescence maximum at 675 nm. This product was indentified as chlorophyllide a, its yield reached up to 20% of the initial amount of protochlorophyllide. Photochemical behaviour of protochlorophyllide differs from that of its analogs such as protochlorophyll and protopheophorbide.

REFERENCES

1. Seliskar C.J., Ke B. 1968. Protochlorophyllide aggregation in solution and associated spectral changes. Biochim. Biophys. Acta, 153, 685-691.
2. Brouers M. 1972. Optical properties of in vitro aggregates of protochlorophyllide in non-polar solvents. 1. Visible absorption and fluorescence spectra. Photosynthetica, 6 (4), 415-423.
3. Rasquain A., Houssier C., Sironval C. 1977. The dimerization of protochlorophyll pigments in non-polar solvents. Biochim. Biophys. Acta, 462, 622-641.
4. Krasnovsky A.A., Bystrova M.I., Lang F. 1971. The modelling of the different forms of protochlorophyll pigments in solid films and etiolated leaves treated by δ-ALA. Dokl. Akad. Nauk. USSR, 201, 1485-1488.
5. Böddi B., Soós J., Láng F. 1980. Protochlorophyll forms with different molecular arrangements. Biochim. Biophys. Acta, 593, 158-165.

6. Bystrova M.I., Safronova I.A., Krasnovsky A.A. 1982. The study of molecular arrangement of protochlorophyll aggregated forms in solid films. Mol. Biol. USSR, 16, 291-301.

7. Böddi B., Kovács K., Láng F. 1983. Spectroscopic properties of protochlorophyll forms in Triton X-100 detergent micelles. Biochim. Biophys. Acta, 722, 320-326.

8. Bombart P., Dujardin E. 1980. A new method for the purification of protochlorophyllide. Abstracts of 5th International Congress on Photosynthesis, Greece, p. 76.

9. Mapleston R.E., Griffiths W.T. 1980. Light modulation of the activity of protochlorophyllide reductase. Biochem. J., 189, 125-133.

10. Griffiths W.T. 1980. Substrate-spesifity studies of protochlorophyllide reductase in barley (Hordeum vulgare) ethioplast membranes. Biochem. J., 186, 267-278.

11. Krasnovsky A.A., Voinovskaya K.K. 1949. Photochemical characteristics of protochlorophyll. Dokl. Akad. Nauk. USSR, 66, 663-666.

12. Krasnovsky A.A., Bystrova M.I. 1962. Spectral and photochemical properties of protochlorophyll pigments in model systems. Biochimiya. USSR, 27, 958-968.

13. Bystrova M.I., Umrikhina A.V., Krasnovsky A.A. 1966. Photoreduction of protochlorophyll and protopheophytin. Biochimiya. USSR, 31, 83-92.

14. Krasnovsky A.A., Bystrova M.I., Lang F. 1970. The study of photoreduction of protochlorophyll to chlorophyll in the solution. Dokl. Akad. Nauk. USSR, 194, 1441-1444.

15. Suboch V.P., Losev A.P., Gurinovitch G.P. 1974. Photoreduction of protochlorophyll and its derivatives. Photochem. Photobiol., 20, 183-189.

16. Seely G.R., Talmadge K. 1964. Photoreduction of zinc porphin by ascorbic acid. Photochem. Photobiol., 3, 195-206.

17. Sidorov A.N. 1964. Spectroscopic investigation of Zn-TPP photoreduction. Dokl. Akad. Nauk. USSR, 158, 973-976.

18. Suboch V.P., Losev A.P., Gurinovitch G.P., Sevchenko A.N. 1970. Photochemical hydrogenation of porphyrin metallocomplexes, protochlorophyll derivatives. Dokl. Akad. Nauk. USSR, 194, 723-726.

MOLECULAR STRUCTURE OF PROTOCHLOROPHYLL FORMS

B. BÖDDI and F. LÁNG
Department of Plant Physiology, Eötvös University
Budapest, Hungary

SUMMARY

Aggregation and spectroscopic properties of protochlorophyll were investigated in solid films and in Triton X-100 detergent micelles. The protochlorophyll form with absorption maximum at 650 nm appeared in solid films treated with vapour of acetone or water as well in detergent micelles. In addition to this form, several long-wavelength forms were also found especielly in water vapour treated solid films. While the 650 nm and the 676 nm forms had intense CD signals - with positive Cotton effects around 650 nm - another long-wavelength forms were optically inactive. The structural reasons of the differences of spectroscopic properties of protochlorophyll forms are discussed.

INTRODUCTION

It is well established that the in vivo state of protochlorophyll/ide /PChl/ide/ - similarly to other chlorophyll pigments - differs from that in polar solvents diethyl ether, acetone etc. /which compound are used for the extraction of these pigments/ /1,2/. Different variations of the in vivo state of PChl/ide were observed with spectroscopic methods /3,4/; the main red absorption and/or fluorescence maxima are used for their characterization. The species described this way are called "forms". In different plant materials the following PChl/ide forms were described /indexes are the absorption maxima/: $PChl_{628-630}$, $PChl_{636}$, $PChl_{640}$, $PChl_{650}$ in etiolated leaves /5,6/ and in pumpkin seed coats /7,8/. Computer analysis of absorption spectra of etiolated leaf homogenates at -196 °C showed bands at 629, 638, 650, 657, 669, 675, 685, 697 and

C. Sironval and M. Brouers (eds.); Protochlorophyllide Reduction and Greening. ISBN 90 247 2954 8
© 1984, Martinus Nijhoff/Dr W. Junk Publishers, The Hague/Boston/Lancaster.

711 nm /9/. $PChl_{640}$ and $PChl_{671}$ forms were found in Cyclanthera seed coats /10/. Corresponding to the above described absorbance measurements also fluorescence spectroscopy showed the multiplicity of PChl/ide forms, however, not all absorption maxima can be corresponded to fluorescence emission bands in vivo /2/ and there are bands in the fluorescence emission spectra of etiolated leaves of which molecular origin has not been satisfactory discussed. In these spectra, bands at 633-635, 655-657, 675, 686, 712 and 726 nm were observed /6,11/. Computer analysis showed maxima at 657, 669, 682, 692, and 712 nm /9/. The fluorescence emission spectra of seed coats contain long-wavelength /above 655 nm/ bands with high intensity at 675 and 695 nm in case of pumpkin seed coats and at 691 nm in case of Cyclanthera seed coats /8,10/

To investigate the structural origin of the differences in spectroscopic characteristics of PChl/ide forms, model systems of different nature are often used. On the basis of these works the following conclusions can be drawn:
- Polar solvents and other non-chromophore compounds having electron donor or acceptor groups can connect to PChl molecules which interaction cause the perturbation of the electron system of PChl molecules. This perturbation - depending on the refractive index and on the dielectric constant of the solvent results in small /several nm/ shifts of the absorption maxima of PChl /12,13/. These solutions of PChl show the intrinsic CD signals of PChl due to the presence of asymmetrically substitued carbon atom in the porphyrin ring /14/.
- In apolar solvents /14-19/ or in apolar microenvironment /i.e. inside of detergent micelles/ /20/ PChl or PChlide molecules connect to each other; their aggregation takes place. This type of interaction results in the splitting of the absorption and fluorescence bands which causes the complexity of the red region of spectra of these samples /21/. These systems usually have intense CD signals which are due to the exciton splitting and the asymmetric structure of the aggregates /21,22/.
- Ligands with two /or more/ functional groups - water, dioxane - may build into the aggregates connecting Chl or PChl molecules

to each other. The species formed this way have particularly
red-shifted absorption maxima, and in case of crystallic structures, large CD signals /22,23/.

Despite the great number of data about spectroscopic properties
of Chl and PChl/ide forms, there are important questions to
be answered in this topic: concerning the size and the molecular
arrangement of individual forms.

MATERIAL AND METHODS

PChl was extracted from pumpkin seed coat and purified by
column chromatography /12/.

Solid films were prepared on glass surface from diethyl ether
solution of PChl. The method of preparation and the treatment of
films with acetone or water vapour are described elsewhere /24,25/.

Micellar solutions of the detergent Triton X-100 /LOBA/
were prepared by dissolving the detergent in phosphate buffer
/ 0.05 M, pH 7.2/. The concentration of Triton X-100 was 7×10^{-4} M.
For preparation of micellar solutions of PChl, different amounts
of diethyl ether solution of PChl were added to the above
described solution /20/.

Absorption spectra were measured with a Unicam SP 1800 and
with a SF 18 spectrophotometer. Fluorescence emission spectra
were registered with a Perkin-Elmer MPF 44B type spectrofluorometer. The excitation wavelength was 480 nm. The CD spectra were
measured with a CNRS-Roussel-Jouan-Dichrograph III type spectro-
polarimeter with samples directly in front of the photomultiplier.

RESULTS

PChl solid films after preparation usually had the main red
absorption maximum at 632-635 nm /24/. These films had no CD
signals. Treatments of these films with vapours of acetone or
water resulted in the transformation of films by slackening the
films and giving this way a possibility for the aggregation of
PChl or by building into the aggregates. As a result of this
transformation of films the main absorption maximum appeared
at 650 nm in the case of acetone vapour treatment for 1-40 min
and in the case of water vapour treatment for 24 hours at 40 °C.

At 635 nm there was a shoulder and the shape of the absorption
spectra of these films was very similar to that of spectra of
etiolated leaves and pumpkin seed coats. /Figure 1A and B/
After 6 days treatment of films with water vapour at 25 °C
a band at 676 nm dominated in the absorption spectrum.
/Figure 1C/

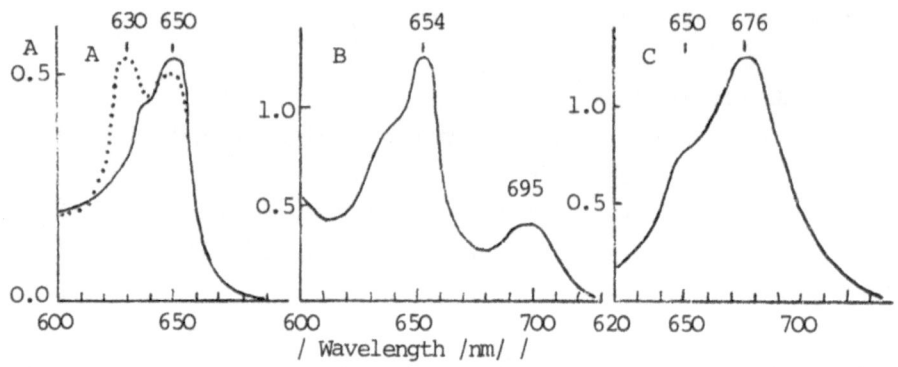

Figure 1. Absorption spectra of protochlorophyll solid films:
A: treated with acetone vapour for 1 min /dotted line/ and
for 40 min /continuous line/, B: treated with water vapour
for 24 hours at 40 °C, C: treated with water vapour for 6 days
at 25 °C.

The shapes, positions and intensities of the CD signals of these
films refer to the presence of aggregates of PChl: after 1 min
acetone vapour treatment a large negative band appeared at 650 nm.
Further treatment resulted in the decrease of this band and the
appearance of a positive CD signal around 650 nm. /Figure 2A/
This positive CD signal around 650 nm was also found in the CD
spectra of water vapour treated films / 24 hours, 40 °C /:
a positive Cotton effect was found with a negative band at 644 nm,
a zero-crossing at 650 nm and a positive band at 658 nm. This
positive band gradually increased in time of the treatment.
/Figure 2B/. Positive Cotton effect was found also in the CD
spectrum of films treated with water vapour for 6 days at 25 °C:
negative signal at 642 nm, zero crossing at 650 nm and positive
band at 667 nm. /Figure 2C/

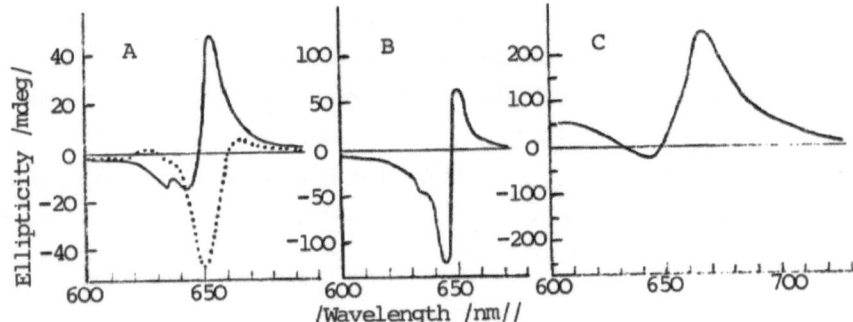

Figure 2. Circular dichroism spectra of protochlorophyll solid films: A: treated with acetone vapour for 1 min /dotted line/ and for 40 min /continuous line/, B: treated with water vapour for 24 hours at 40°C, C: treated with water vapour for 6 days at 25 °C.

PChl forms prepared in Triton X-100 detergent micelles had similar spectroscopic properties to those in solid films. In paralell with the increasing concentration of PChl, a band around 650 nm increased. At small concentration /1.6 x 10^{-5}M/ the main band was at 632 nm, at 1.8 x 10^{-4} M concentration of PChl the main band appeared at 652 nm and the spectrum was very similar to spectra of etiolated leaves and solid films treated with acetone or water vapour. /Figure 3A/ The CD spectra of these solutions had positive signals and positive Cotton effects around 650 nm. /Figure 3B/

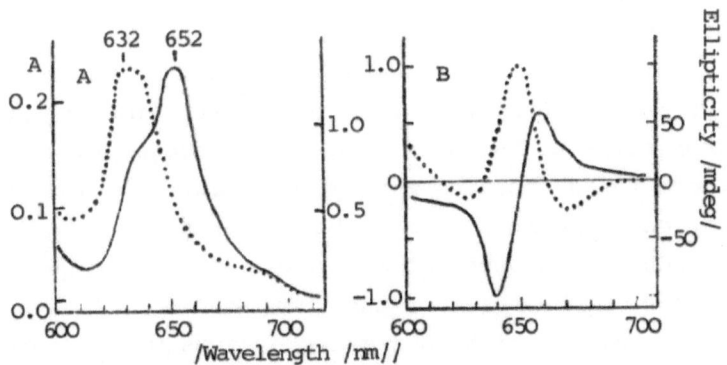

Figure 3. Absorption /A/ and circular dichroism /B/ spectra of micellar solutions of PChl. Triton X-100 concentration: 7 x 10^{-4}M, PChl concentrations: 1.6 x 10^{-5}M /dotted line/ and 1.8 x 10^{-4}M /continuous line/

Similarly to water vapour treated films also micellar solutions
of PChl contained long-wavelength forms of PChl. This was supported
by the fluorescence emission spectra of these samples which
showed maxima at 693 and 707 nm in the case of solid films
and at 678 and 696 nm in the case of detergent micelles.
Despite the small amount of these long-wavelength forms in the
micellar solutions, their bands are dominating in the fluorescence emission spectra because of an effective energy migration
existing in this system. /Figure 4 A and B/

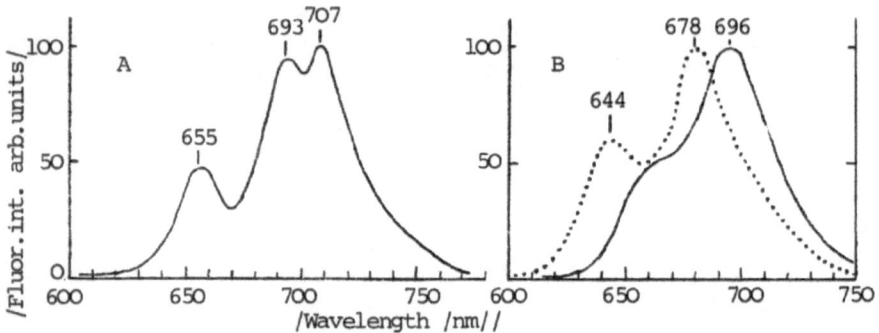

Figure 4. Fluorescence emission spectra of protochlorophyll solid
film treated with water vapour for 6 days at 25 °C /A/ and of
micellar solutions of protochlorophyll. Triton X-100 concentration
7 x 10^{-4}M. PChl concentrations: 1.6 x 10^{-5}M /dotted line/ and
1.8 x 10^{-4}M /continuous line/

DISCUSSION

   Solid films and micellar solutions proved to be suitable
systems for preparing PChl forms with similar spectroscopic
properties to those in vivo. As in these protein-free systems
these forms appeared a very important role must have the aggregation of PChl in forming these forms.

   PChl forms appeared in these systems can be ranged into two
groups on the basis of the CD spectra:
l, Aggregates with CD signals - forms with absorption maxima at
640, 650 and 676 nm may be classed among forms of this group.
These aggregates must have a definite geometry which is due to
interactions of definite molecular groups /infra red data showed
interactions of keto C=O groups with central Mg atoms /18/ /.

In different circumstances different molecular arrangements
can be favourable: PChl forms with similar absorption and
fluorescence properties can have different arrangements and
consequently, different CD signals /24/. In this paper the
650 nm form of PChl is demonstrated with different CD signals
/see Figure 1A and Figure 2A/. The data presented here show
that in stabile state the structures with positive CD signals
or positive Cotton effects are favourable: such signals were
found in stabile state of solid films, in detergent micelles
and in pumpkin seed coat, too /8/

In systems containing water /water vapour treated films and
detergent micelles/ also water molecules must be built into the
aggregates and this way also water molecules have a role in
forming the molecular arrangement and spectroscopic properties
of aggregates. Long-wavelength forms with absorption maxima
above 650 nm can be thought to have such structure, as these
forms appeared in great amount in water containing systems.
/Similarly , "watered" aggregates of Chl-a and Chlide-a had
strongly red-shifted maxima at 700 and 740 nm /26,27//.
2, Long-wavelength forms without CD signals were also found
in solid films and in detergent micelles /see Figure 1B and
Figure 2B; Figure 3B and Figure 4B/. Cyclanthera seed coat
had the main absorption maximum in this region, but it had
intense CD signals /10/. On the basis of the position of the
absorption and fluorescence bands belonging to these forms one
can conclude a high aggregation degree for these forms. In
case of definite geometries these forms should have intense
CD signals. Consequently, the long-wavelength forms found in
our systems must have indefinite or random geometry. In solid
films and in micelles PChl molecules or aggregates of small
aggregation degree can come so close to each other that their
π-electron systems overlap. Because of small steric circumstances molecular groups cannot interact and there is no directing
rule for the geometry of aggregation and form only molecular
"stacks" with random arrangement.

Several short-wavelength forms were present in the examined
systems - with absorption maximum at 630-632 nm -. These forms

are probably due to interactions of PChl with solvent or Triton
X-100 molecules.

The results of this work show that Pchl - similarly to other
chlorophyll pigments - have a great number of different spectral
forms. As several long-wavelength forms in solid films and in
detergent micelles had absorption and fluorescence maxima in
similar positions to that in spectra in etiolated leaves, it
cannot be excluded, that etiolated leaves contain these forms,
too. In the spectra of etiolated leaves these bands have small
intensities probably because of the small amount of these forms
or a less effective energy migration can there exist.

The presented data support that the term "form" denoted
with an absorption or fluorescence maximum cannot be satisfying
for exact characterization of species existing in Chl containing
systems: aggregates with different arrangements or with different
types of interactions can have similar absorption and fluorescence
properties.

REFERENCES

1. Boardman N.K. 1966. in: The Chlorophylls ed. L.P.Vernon _and
   G.R.Seely pp. 437-480 Academic Press New York/London
2. Virgin H. 1981. Annu.Rev.Plant Physiol. 32, 451-463
3. Shibata. K. 1957. J.Biochem. /Japan/ 44, 147-173
4. Brouers M. and Sironval C. 1975. Plant Sci Letters 4, 175-181
5. Virgin H. 1975. Photosynthetica 9, 84-92
6. Brouers M. and Sironval C. 1974. Plant Sci.Letters 2, 67-72
7. Inada Y. and Shibata K. 1960. Plant Cell Physiol. 1, 311-316
8. Böddi B. Soós J. and Láng F. 1979. Plant Sci.Letters
   16, 75-79
9. Litvin F.F. and Stanitsuk I.N. 1980. Fiz.Rast. 27, 1024-1030
   /in Russian/
10. Sundqvist C. Ryberg H. Böddi B and Láng F. 1980. Physiol.Plant.
    48, 297-301
11. Láng F. Vorobyeva L.M. and Krasnovsky A.A. 1972. in:
    Photosynthesis, Two Centuries after its discovery by Joseph
    Pristley ed. Forti G. Avron M. and Melandry A. Vol. 3.,
    pp. 2309-2317 Dr. W. Junk, The Hague
12. Houssier C. and Sauer K. 1969.Biochim.Biophys.Acta 172, 476-491
13. Seely G.R. and Jensen R.G. 1965. Spectrochim.Acta 21, 1835-1845
14. Houssier C. and Sauer K. 1970. J.Am.Chem.Soc. 92, 779-791
15. Seliskar C.J. and Ke B. 1968. BichimBiophys.Acta 153, 685-691
16. Brouers M. 1972. Photosynthetica 6, 415-423
17. Brouers M. 1975. Photosynthetica 9, 304-310
18. Brouers M. 1979. Photosynthetica 13, 9-14
19. Rasquain A. Houssier C. Sironval C. 1977. Biochim.Biophys.
    Acta 462, 622-641

20. Böddi B. Kovács K. and Láng F. 1983. Biochim.Biophys.Acta
    722, 320-326
21. Tinoco I. 1963. Rad.Res. 20. 133-139
22. Dratz E.A. 1966. Ph.D. Thesis Univ.California, Berkeley
23. Katz J.J. 1972. in: Chemistry of Plant Pigments ed. Chichester
    C.O. pp. 103-122 Academic Press New York
24. Böddi B. Soós J. Láng F. 1980. Biochim.Biophys.Acta 593,
    158-165
25. Böddi B. Rákász É. and Láng F. 1983. Photobiochem.Photobiophys.
    5, 27-33
26. Ballschmiter K. and Katz J.J. 1972. Biochim.Biophys.Acta
    256, 307-327
27. Fong F.K. and Koester V.J. 1976. Biochim.Biophys.Acta 423,
    52-64

S P E C I A L   T O P I C S.

DARK SYNTHESIS OF CHLOROPHYLL IN VIVO AND DARK REDUCTION OF PROTOCHLORO-
PHYLLIDE IN VITRO BY PEA CHLOROPLASTS

H. ADAMSON, N. PACKER.   School of Biological Sciences, Macquarie
University, North Ryde, NSW 2113 Australia.

INTRODUCTION

There is evidence that some plants which accumulate chlorophyll in
darkness have two reductive pathways for the conversion of protochlorophyllide
to chlorophyllide operating in parallel (Castelfranco & Beale, 1981).
Griffiths and Mapleston (1978) showed that dark grown spruce (Picea)
seedlings and Chlamydomonas reinhardtii (wild type) both contain the light-
dependent enzyme, NADPH-Protochlorophyllide oxidoreductase. However the
ability of these plants to form chlorophyll in the dark indicates that they
do not rely on this enzyme for the reduction of protochlorophyllide to
chlorophyllide. The simplest explanation is that they contain a light-
independent protochlorophyllide reductase as well. Direct evidence for
the existence of such an enzyme is lacking. There are several reasons
for this. The chlorophyll already present in the membranes makes small
increases in chlorophyllide or chlorophyll in the dark difficult to detect
and a reliable method for assaying the light-independent protochlorophyllide
reductase has not yet been published. However, we believe the main reason
so little is known about the dark enzyme is because it is not present in
dark-grown angiosperm seedlings.

Almost all studies of chlorophyll biosynthesis and chloroplast
development in multicellular plants have taken dark-grown angiosperm
seedlings as their starting point. On illumination, these etiolated plants
begin to green, accumulated protochlorophyllide is converted to chlorophyllide, prolamellar bodies in the etioplasts begin to disperse, photosynthetically competent thylakoids are synthesised and autotrophic nutrition
is established. The assumption underlying all studies of chloroplast
development from etioplasts is that they reflect normal patterns of
chloroplast development. Although this assumption is sometimes questioned
(Leech, 1973; Whatley, 1974) and other systems for studying normal
chloroplast development proposed (Baker & Leech, 1977; Boffey, Sellden &
Leech, 1980; Adamson, Hiller & Vesk, 1980; Adamson & Hiller, 1981)

C. Sironval and M. Brouers (eds.); Protochlorophyllide Reduction and Greening. ISBN 90 247 2954 8
© 1984, Martinus Nijhoff/Dr W. Junk Publishers, The Hague/Boston/Lancaster.

etiolated plants remain the preferred experimental material. No doubt most workers agree with Bradbeer (1981) that "there is no evidence that investigating greening etiolated plants instead of normal plants has led to serious error". We disagree.

Our observations of green plants transferred from a normal diurnal environment to darkness (Adamson, Hiller & Vesk, 1980; Adamson & Hiller, 1981; Adamson, 1982a,b) highlight an important difference between dark and light grown angiosperms. Dark grown angiosperms require light for chlorophyll synthesis. In the absence of light they accumulate protochlorophyllide. On the other hand a variety of light grown angiosperms continue synthesising chlorophyll when they are transferred to darkness. In other words, etiolated seedlings appear to have only a single light-dependent pathway for the reduction of protochlorophyllide to chlorophyllide; normal green plants appear to have two pathways, one requiring light, the other not. This is an important difference. It immediately raises questions concerning the relative contribution of the two pathways of protochlorophyllide reduction in the light and the proportion of total chlorophyll synthesised at night. It also suggests (and we have confirmed, Adamson, 1982b) that, contrary to popular belief, chloroplasts may go on synthesising thylakoids at night.

Our earlier work, already cited, and that of Godnev, Shlyk & Rotfarb (1959), Robbelen (1956) and Popov & Dilova (1969) has clearly demonstrated that the ability to synthesise chlorophyll in the dark is not confined to lower plants and gymnosperms. It is also present in a variety of monocotyledonous and dicotyledonous angiosperms adapted to both low and high light situations and aquatic and terrestrial environments. Nor is it confined to relatively uncommon plants adapted to specialised ecological situations which have not been studied previously. Barley is one of the most investigated plants in photosynthesis research. We reasoned that if the ability of barley to make chlorophyll in the dark could be overlooked for 100 years and evidence to this effect (Popov & Dilova, 1969) ignored for over twenty, other plants commonly used in studies of chloroplast development might possess the same ability. We transferred light-grown peas (Pisum sativum L) to darkness and observed that, like barley, peas did not stop making chlorophyll when light was withheld.

This paper gives some indication of the rate of chlorophyll accumulation in the dark by shoot tips and expanding leaflets of normal light grown peas and provides evidence of the ability of pea chloroplast membranes to catalyse the in vitro conversion of protochlorophyllide to chlorophyllide in the absence of light.

MATERIALS AND METHODS

Pisum sativum L var. Greenfeast was grown for 2-3 weeks in a glasshouse before being transferred to darkness.

To measure chlorophyll accumulation in the dark by pea leaflets, plants were selected for uniformity of height and leaf expansion and placed in paper cups with their roots in water. 10 plants per sample and 5 to 25 replicates per treatment were used. Changes in the youngest, still folded leaflets to have emerged from the apex were followed. One leaflet of each pair was carefully removed at the start of the experiment to provide the initial sample and the plants immediately transferred to complete darkness. After 17-20 hours the other leaflet of the pair was removed. These leaflets formed the final dark sample. All manipulations after the plants were transferred to darkness were carried out using a dim green safelight. (Kodak Wrattan, Filter No. 54). Chlorophyll was extracted in 80% acetone and determined spectrophotometrically (Arnon, 1949). Leaf expansion was estimated from the increase in leaf fresh weight. Tritiated ALA (50uCi/ml) was supplied as a 10mM solution (pH 6.8). Esterified and unesterified pigments were separated using a phase separation technique (Treffry, 1970) . The esterified pigments were separated by HPLC. Chlorophyll a and b were collected and counted using standard procedures.

To demonstrate the activity of the light-independent protochlorophyllide reductase, chloroplasts were obtained, by the method of Anderson et al (1971), from shoot tips of 8 day old pea plants grown under glasshouse conditions. Prior to obtaining the chloroplasts the plants were placed overnight in a dark room at $25^{o}C$. All manipulations thereafter were carried out either under dim green safelight or complete darkness. The chloroplasts were lysed in a 10mM phosphate buffer (pH 7.4) containing 0.03M sorbitol, 1mM $MgCl_2$; 0.5% B.S.A. The membranes were pelleted at 10,000g, resuspended in the buffer solution, without B.S.A. and incubated in a reaction mixture containing approximately 15-20 nmoles solubilised protochlorophyllide, an NADPH regenerating system and 1mM sodium ascorbate. The procedure of Griffiths (1978) for obtaining and solubilising protochlorophyllide and generating NADPH were followed. The reaction was stopped by adding ice cold acetone, to give a final concentration of 85%. The esterified and

unesterified pigments were separated using a modification of Treffry
(1970). After standing for 30 minutes on ice the reaction mixture was
centrifuged at low speed using a bench centrifuge, the supernatant
decanted and ammonia (880) added. A final ratio of ammonia : acetone :
water of 1:44:8 gave reliable phase separation with 40-60 petroleum ether.
The ammoniacal acetone to petroleum ether ratio was 3:1. The ammoniacal
acetone phase containing the unesterified pigments was back-washed twice
with petroleum ether to remove contaminating esterified pigments. The
petroleum ether phases obtained after back washing were added to the
original phase. The amounts of protochlorophyllide and chlorophyllide
contained in the lower ammoniacal acetone phase and chlorophyll and
protochlorophyll in the upper petroleum ether phase were determined
spectrophotometrically.

RESULTS

1. Chlorophyll accumulation in the dark.

When 3 week old pea plants are transferred to darkness the chlorophyll
concentration (ug/g f.wt) of the young expanding leaflets usually falls.
The decline in chlorophyll concentration is directly related to the amount
of leaf expansion which occurs in the dark (Fig.1).

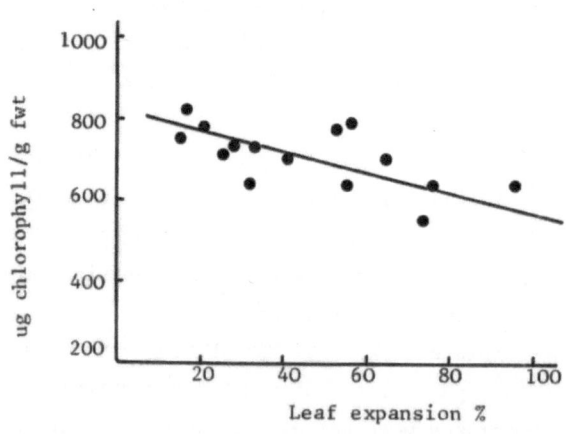

Fig.1. Chlorophyll
concentration of pea
leaflets after 17 hours
darkness as a function of
expansion (% change in
leaflet fresh weight).

In this experiment, although the leaflets were getting paler the
total amount of chlorophyll per leaflet was increasing significantly (Fig.2)
and the increase was greatest in those leaflets which expanded most. The
chlorophyll concentration fell because chlorophyll synthesis in the dark
in these plants was not able to keep pace with leaf expansion.

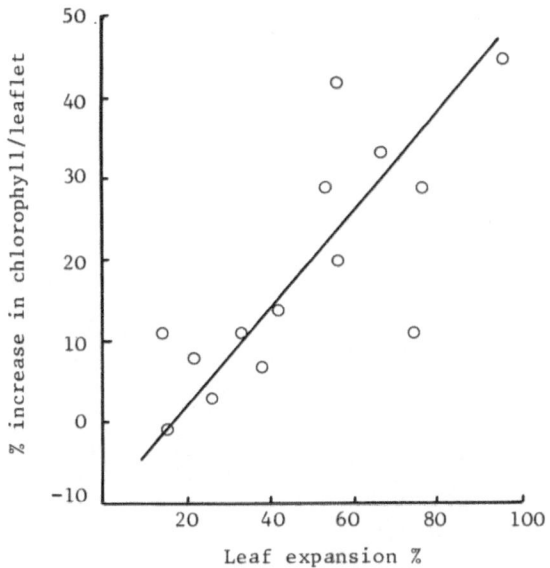

Fig.2. % increase in chlorophyll per leaflet after 17 hours darkness as a function of expansion (% change in leaflet fresh weight)

P < .01

Occasionally we have observed an increase in chlorophyll concentration in the dark. This occurs under conditions of particularly rapid chlorophyll synthesis when the percentage change in chlorophyll per leaf or tip exceeds the percentage change in their fresh weight(Fig.3).

2.   Incorporation of ^{3}H-ALA into chlorophyll in the dark.

As reported elsewhere (Adamson, Packer & Sanders, 1983) chlorophyll synthesis in the dark was confirmed by the incorporation of radioactive label from ^{3}H-ALA into both chlorophyll a and b. Tips became more highly labelled than leaves (Table 1). The ratio of cpm/mg fresh weight for esterified pigments extracted from tips and leaflets was 1.7. This is very similar to the ratio of 1.5 for % change in chlorophyll/tip and /leaflet observed in Fig.3.

TABLE 1.   Incorporation of ^{3}H-ALA by whole pea plants into esterified pigments in the dark (36hr).

	Tips	Leaflets	Ratio (T:L)
cpm/mg fr.wt.	11070	6517	1.5

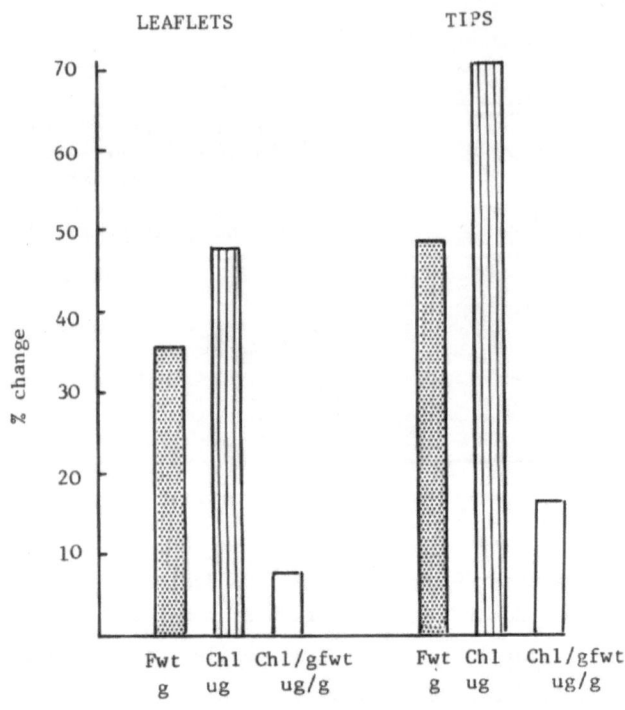

Fig. 3. % change in
fresh weight (g),
chlorophyll content (ug)
and chlorophyll concent-
ration (ug/g) of pea
leaflets and tips after
20 hours darkness.

leaflets n = 5
Chl/leaflet P < .005

3. <u>In vitro conversion of protochlorophyllide to chlorophyllide in the dark</u>

When chloroplast fragments from peas were incubated in the dark with
protochlorophyllide and NADPH, chlorophyllide was formed. Fig. 4 shows
that the reaction was substrate dependent. Chlorophyllide was not formed
in the absence of protochlorophyllide or NADPH. The rate of chlorophyllide
formation <u>in vitro</u> in this experiment was approximately twice the rate of
chlorophyll formation in the dark by intact pea shoots. (Table 2).

TABLE 2. Rates of dark formation of chlorophyllide <u>in vitro</u> and chlorophyll
<u>in vivo</u> by chloroplasts of normal green pea shoots.

Chloroplast membranes	1.3×10^{-3} mols chlide/mol chl/min
Shoot tips	0.58×10^{-3} mols chl/mol chl/min

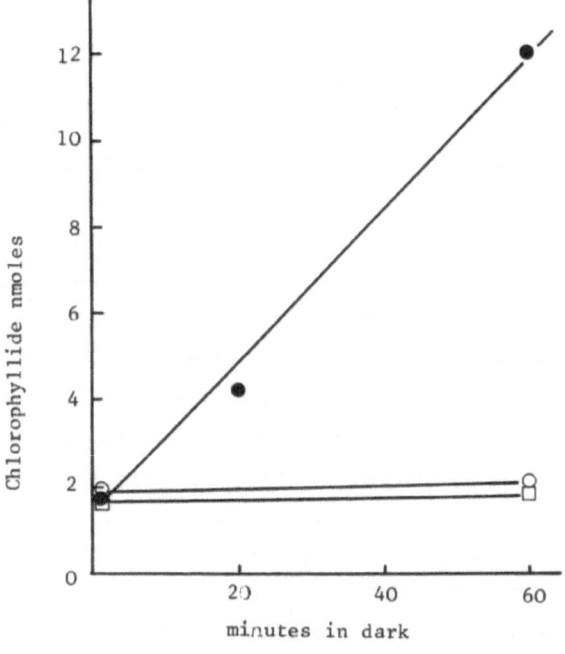

Fig.4. Chlorophyllide (nmoles) levels in reaction mixture as a function of time in dark

● + Pchlide + NADPH

○ - Pchlide + NADPH

□ + Pchlide - NADPH

Assuming a 7:1 protein : chlorophyll ratio for chloroplast membranes and a 10:1 ratio for pea tips it is possible to make a rough comparison between these rates for the light independent protochlorophyllide reductase enzyme in light grown peas and the light dependent enzyme in etiolated barley. This suggests (Table 3) that the light independent enzyme in developing pea chloroplasts has about the same activity as the light dependent enzyme in etioplasts obtained from dark grown barley which had been briefly exposed to light. Both rates are 4 to 5 times lower than the maximum rate reported for the light dependent enzyme.

TABLE 3.   Some estimated rates of chlorophyllide or chlorophyll formation in light and dark.

PEAS
chloroplast membranes in dark	0.29 nmols chlide/mg.Pro/min
shoot tips in dark	0.13 nmols chl/mg.Pro/min

BARLEY*
etioplast membranes in dark	1.0 nmols chlide/mg.Pro/min
etioplast membranes from plants exposed to light	0.2 nmols chlide/mg.Pro/min
etiolated leaves after 12 hours exposure to light	0.023 nmols chl/mg.Pro/min

* Mapleston & Griffiths (1980)

DISCUSSION

Our findings, that (1) chlorophyll synthesis continues when peas are transferred from light to darkness, (2) ^{3}H-ALA is incorporated in chlorophyll a and b in the dark, and (3) protochlorophyllide is converted to chlorophyllide in the dark, in vitro, in a reaction mixture containing chloroplast membranes and NADPH, indicate that pea chloroplasts contain an enzyme which does not require light to catalyse the reduction of protochlorophyllide. This enzyme is not present in pea etioplasts. As Popov & Dilova (1969) have shown and we have confirmed (Adamson, 1982a) many hours of light are required before the ability to synthesise chlorophyll in the dark is acquired by greening etiolated plants.

Although etiolated plants have provided a wealth of information about the different forms of protochlorophyllide which accumulate when angiosperm seeds are germinated in the dark, the manner in which these forms are interconnected and the characteristics of the enzyme which catalyse the photoreduction of protochlorophyllide (Griffiths, 1978; Griffiths & Mapleston, 1978; Mapleston & Griffiths, 1980; Oliver & Griffiths, 1980, 1981) they have also led to generalisations which have obscured the fact that many flowering plants do not require light for the reduction of protochlorophyllide under normal circumstances. These plants either have two enzymes catalysing the reduction of protochlorophyllide or a single enzyme which requires light to reduce protochlorophyllide in etioplasts but which can chemically reduce it under the conditions prevailing in normal chloroplasts. Until this is sorted out it is hard to see how the controversy concerning the relative contribution of the light dependent enzyme to chlorophyll synthesis in green plants can be resolved. Mapleston & Griffiths (1980) maintain that although light greatly reduces the activity and amount of the light dependent enzyme in dark grown plants, the enzyme remaining can still account for the rates of chlorophyll synthesis observed. Santel & Apel (1981) disagree. Detailed studies of the contribution of the light dependent and light independent enzymes to the reduction of protochlorophyllide in normally developing chloroplasts are urgently required.

Despite the difficulties of working with green plants we believe they are likely to be more rewarding than further work on dark grown plants greening under artificial (intermittent and long term continuous) light. These can produce puzzling results. For example, Bennett (1981)

observed that when etiolated pea seedlings were exposed to continuous
light for 24 hours and then returned to darkness, 38% of chlorophyll <u>a</u>,
74% of chlorophyll <u>b</u> and 84% of the light harvesting chlorophyll a/b
protein which accumulated in the light period were lost. He suggested
that breakdown of the apoprotein might be a mechanism whereby the plant
coordinates the pigment and protein components of the light harvesting
apparatus during normal growth. Believing that chlorophyll synthesis
only occurred in pea in the light and knowing that apoprotein synthesis
occurred in both light and dark he postulated that excess apoprotein,
which was not being incorporated into functional aggregates could be
degraded in darkness. He noted the wasteful nature of such a process and
commented on the need for further study.

We believe our findings suggest a different explanation. Four facts
need to be taken into account. (1) Under normal conditions peas contain
a light-independent protochlorophyllide reductase (evidence in this paper).
(2) Many hours exposure to light is required to induce or activate the
light independent enzyme in etiolated plants (Popov & Dilova, 1969;
Adamson, 1982a). (3) Newly formed thylakoids in partially greened
etiolated plants are unstable when returned to darkness. Increasing the
period of illumination increases their stability (Bennett, 1981). (4)
Chlorophyll is needed to stabilise the light harvesting apoprotein (Apel
& Kloppstech, 1980). We suggest that chlorophyll synthesis continues at
night in peas growing in a diurnal environment. This stabilises the
light harvesting apoprotein which is being synthesised at the same time.
There is no evidence that thylakoids are dismantled in the dark in young,
normally developing chloroplasts and some evidence to the contrary
(Adamson, 1982b). We attribute the instability of newly formed thylakoids
in partially etiolated plants returned to darkness, in part, to the
inactivity or low activity of the light independent protochlorophyllide
reductase. We interpret the direct relationship between membrane stability
in the dark and length of prior illumination to the effect of light in
inducing or activating the light independent protochlorophyllide
reductase.

This paper provides the first demonstration of a chloroplast enzyme
catalysing the <u>in vitro</u> formation of chlorophyllide from protochlorophyllide
in the absence of light. We believe this enzyme is widespread because it
is well known that certain green algae, mosses, ferns, gymnosperms and

angiosperms can make chlorophyll in the dark. It is likely that it is
a primitive enzyme because many prokaryotes (blue-green algae and certain
photosynthetic bacteria) do not require light for chlorophyll synthesis.

Since the light independent protochlorophyllide reductase is either
not present or not active in angiosperm seedlings which have not been
exposed to light and the precise conditions required to induce or
activate it in these plants are not known, it is essential that more work
be done on plants growing in diurnal environments. As this study has
shown, the greening etiolated plant is not always an appropriate analogy.

REFERENCES

Adamson, H. (1982a) Evidence for a light independent protochlorophyllide
    reductase in green barley leaves. In: Akoyunoglou, G. et al (ed.) :
    Cell Function and Differentiation, Part B. Progress in Clinical and
    Biological Research, 102B:33

Adamson, H. (1982b) Chloroplast development in green barley leaves
    transferred to darkness. In: Akoyunoglou, G. et al (ed.) : Cell
    Function and Differentiation, Part B. Progress in Clinical and
    Biological Research, 102B:189

Adamson, H., Hiller, R.G. & Vesk, M. (1980). Chloroplast development
    and the synthesis of chlorophyll a and b and chlorophyll protein
    complexes I and II in the dark in Tradescantia albiflora (Kunth).
    Planta 150:269

Adamson, H. & Hiller, R.G. (1981) Chlorophyll synthesis in the dark in
    angiosperms. In Akoyunoglou, G. (ed.) : Proceedings of the Fifth
    International Congress on Photosynthesis. V. Chloroplast Development.
    Philadelphia: p 213

Adamson, H., Packer, N. & Sanders, (198) Chlorophyll synthesis in the
    dark in peas. In Proceedings of the Sixth International Congress on
    Photosynthesis.

Anderson, J.M., Boardman, N.K. & Spencer, D. (1971) Phosphorylation of
    intact bundle sheath chloroplasts from maize. Biochim. Biophys. Acta 245:
    253

Apel, K. & Kloppstech, K. (1980) The effect of light on the biosynthesis
    of the light harvesting chlorophyll a/b protein. Evidence of the
    requirement of chlorophyll a for the stabilisation of the apoprotein.
    Planta 150: 426

Arnon, D.I. (1949) Copper enzymes in isolated chloroplasts. Polyphenol-
    oxidase in Beta vulgaris. Plant Physiol. 24:1

Baker, N.R. & Leech, R. (1977) Development of photosystem I and photo-
    system II activities in leaves of light grown maize (Zea mays). Plant
    Physiol. 60:640

Bennett, J. (1981) Biosynthesis of the light harvesting chlorophyll a/b
    protein. Polypeptide turnover in darkness. Eur. J. Biochem. 118:61

Boffey, S.A., Sellden, G. & Leech, R.M. (1980) Influence of cell age on
    chlorophyll formation in light grown and etiolated wheat seedlings.
    Plant Physiol. 65:680

Bradbeer, J.W. (1981) Development of photosynthetic function during
    chloroplast biogenesis. In: The Biochemistry of Plants. 8: 423. Eds.
    M.D. Hatch, N.K. Boardman, P.K. Stumpf & E.E. Conn, Academic Press

Castelfranco, P.A. & Beale, S.I. (1981) Chlorophyll Biosynthesis. In:
  The Biochemistry of Plants 8:375. Eds. M.D. Hatch, N.K. Boardman, P.K.
  Stumpf & E.E. Conn. Academic Press

Godnev, T.N., Shlyk, A.A. & Rotfarb, R.M. (1959) The synthesis of chlorophyll
  in the dark in angiosperms. Fiziol Rast 6:36. Plant Physiol. USSR Eng.
  Transl. 6:33

Griffiths, W.T. (1978) Reconstitution of chlorophyllide formation by
  isolated etioplast membranes. Biochem. J. 174: 681

Griffiths, W.T. (1980) Substrate specificity studies on PCR in barley
  (Hordeum vulgare) etioplast membranes. Biochem.J. 186:267

Griffiths, W.T. & Mapleston, R.E. (1978) NADPH-Protochlorophyllide Oxidoreductase. In. Akoyunoglou, G. et al Eds. "Chloroplast Development"
  North Holland: Elsevier Biomedical Press

Leech, R.M., Rumsby, M.G. & Thomson, W.W. (1973) Plastid differentiation,
  acyl lipid and fatty acid changes in developing green maize leaves.
  Plant Physiol. 52:240

Mapleston, R.E. & Griffiths, W.T. (1980) Light modulation of the activity
  of protochlorophyllide reductase. Biochem.J. 189:125

Oliver, R.P. & Griffiths, W.T. (1980) Identification of the polypeptides
  of NADPH-protochlorophyllide oxidoreductase. Biochem. J. 191:277

Oliver, R.P. & Griffiths, W.T. (1981) Covalent labelling of the NADPH-
  protochlorophyllide oxidoreductase from etioplast membranes with (^{3}H)-
  N-phenylmaleimide. Biochem J. 195-193

Popov, K & Dilova, S. (1969) On the dark synthesis and stabilisation of
  chlorophyll. Progress in Photosynthesis Research II:606

Robbelen, G. (1956) Uber die Protochlorophyll reduction in einer mutante
  von Arabidopsis thaliana (L) heyn. Planta 47:532

Santel, H.J. & Apel, K. (1981) The protochlorophyllide holochrome of
  barley (Hordeum vulgare L). The effect of light on NADPH-protochlorophyllide
  oxidoreductase. Eur. J. Biochem. 120:95

Treffry, T. (1970) Phytylation of chlorophyllide and prolamellar-body
  transformation in etiolated peas. Planta 91:279

Whatley, J.M. (1974) Chloroplast development in primary leaves of Phaseolus
  vulgaris. New Phyt. 73:1097.

EFFECTS OF LEAF AGE ON PROTOCHLOROPHYLLIDE AND CHLOROPHYLLIDE
FORMATION (A REVIEW)

Z. ŠESTÁK
Institute of Experimental Botany. Czechoslovak Academy of Sciences. Flemingo nam.2. Praha 6-16000 Czechoslovakia.

1. INTRODUCTION

The last steps of chlorophyll (Chl) biosynthesis have been
studied in detail since the end of the Second World War, thanks
to the development of exact spectrophotometric methods. The
spectral similarity of protochlorophyllide (PChlide) and chlorophyllide (Chlide) with protochlorophyll and chlorophyll, respectively, caused an ambiguity in the explanation of a probable
biosynthetic pathway. This has been solved by a combination of
spectroscopy with pigment separation between polar and non-polar
solvents, chromatographic techniques and radioactive isotopes
methods. Many studies in vivo have confirmed that the prevailing
sequence of the final steps is PChlide $a \xrightarrow{\text{reduction}}$ Chlide a
$\xrightarrow{\text{phytylation}}$ Chl a. The alternative pathway in which PChlide is
first esterified and the product is reduced, is certainly of
minor significance.

PChlide is probably esterified not by phytol, but by its precursor geranylgeraniol (26, 37), three stepwise reductions of
which lead then to phytol. To distinguish this different type of
esterification, the product will be called here PChlide ester
(PChlideE), in agreement with the recommendations of Kirk and
Tilney-Bassett (22) and Rebeiz and Lascelles (30). The term
protochlorophyll (PChl) is thus used as a term comprising PChlide
+ PChlideE, or where the nature of the pigment has not been
identified. As concerns the divinyl- and monovinyl-forms of
PChlide and PChlideE according to Rebeiz and Lascelles (30),
there is no need to distinguish them here as they have not yet
been discussed in the literature connected with leaf development.

Chl biosynthesis is mostly studied in etiolated plants, as
during normal development in light the major absorption of Chl

C. Sironval and M. Brouers (eds.); Protochlorophyllide Reduction and Greening. ISBN 90 247 2954 8

molecules overshadows the thousand times lesser absorption of
PChlides. The reduction of PChlide to Chlide and the final phytylation to Chl are accomplished in natural forms of these pigments
in vivo, marked by different pigment aggregation and formation
of various pigment-protein complexes, which are reflected in
spectral properties. Thus the above scheme may be complicated
by introducing PChlide 650, PChlide 635, Chlide 668, Chlide 678,
Chl 678, etc. (the figures give the wavelength of maximum absorption in the red part of the spectrum; sometimes the wavelength of
fluorescence maximum is also added).

In darkness, the formation of leaf tissue and its constituents,
including PChlide and the respective holochrome (probably with
a reductase function) is dependent only on metabolic transformations of reserve substances present in the seed. These formative
processes controlled by growth regulators result in some optimum
tissue structure containing PChl, the irradiation of which induces Chlide and Chl formation concurrently with an increase in
the activities of RNA polymerase, ribulose-1,5-bisphosphate
carboxylase and other enzymes; this enables the building of
chloroplast and thylakoid substructures necessary for accomplishing
optimum photosynthetic processes. Thus, 2-3 d old etiolated
French bean leaves contain 2-3 μm proplastids almost without
prolamellar bodies, while 7-9 d leaves contain 4 μm etioplasts
with large crystalline prolamellar bodies (25). The variance in
tissue age is observed also within the blade of one leaf, especially of monocotyledonous plants: leaf tip tissues are old, while
those at the leaf base are usually young, containing the meristems.
For example, PChlide content in darkness declines from leaf tip
to base in young maize (13) or barley (20) leaves. The lack of
irradiation at a proper time leads to sequential metabolic
losses in the etiolated tissues and, finally, to its deterioration.

Even if some papers on Chl biosynthesis still neglect the age
of leaves used for experiments, already the classical authors in
this field recognized that the presence of pigment-protein
complexes, kinetics and rates of reactions of Chl biosynthesis
depend on the development stage of the etiolated leaves (e.g.
36). The literature on the dependence of final steps and products

of Chl biosynthesis on leaf development is fairly voluminous
(for a general information on Chl biosynthesis see, e.g.,10,
11, 22, 30, 45).

2. PROTOCHLOROPHYLLIDE AND PROTOCHLOROPHYLLIDE ESTER IN THE DARK

PChlide is detectable already 24 h after the beginning of germination (2, 31), nevertheless, its later appearance was also
described (after 48 h - 27). In cucumber cotyledons its amount
rapidly increases namely during the 3rd d of dark growth and
reaches the maximum value after 5 d (31). In etiolated leaves
a rise in PChlide amount per leaf was observed to the age of 5 d
(barley - 38), 7 d (flax - 5) or 9-10 d (barley - 5; primary
leaves of *Phaseolus vulgaris* - 2, 15, 23), with a subsequent
decline. Calculation per fresh or dry mass may show a shorter
phase of rapid PChlide accumulation (3 to 4 instead of 6 d in
Phaseolus - 2, 29, 37; 5 d in wheat - 24; 2 d in cucumber cotyledons - 27) or only small differences with leaf age (pea, maize
- 29). The stimulation of PChlide formation by kinetin application
was most expressed in 5 to 6 d barley seedlings (38), but it was
not age dependent in radish cotyledons (12). In etiolated radish
cotyledons the amount of PChlide per fresh matter unit increased
between days 3 and 5, similarly as the amount of δ-aminolevulinic
acid dehydratase; both were lowered to less than half when cultivation proceeded in the presence of 4-thiouridine (35). The incorporation of δ-aminolevulinate in PChlide increased in barley
mutants ($tig-b^{23}$, $tig-d^{12}$) from the 4th to the 7th d very rapidly,
while in wild plants it was rather slow (17).

PChlide exists in vivo in two or three forms, absorbing   in
the red part of spectrum at 628, 632-637, and 650 nm. Their pho-
toconvertibility increases with increasing wavelength of maximum
absorption. According to Shibata (36), older etiolated leaves of
French bean contain a higher amount of the PChlide 636 form, non-
transformable during a brief irradiation, while in young leaves
the active 650 nm form prevails. Also Virgin (48) found in dark
grown barley leaves at the age of 3 d prevalence of PChlide 650,
while with increasing age (5 and 11 d) the 636 nm peak became
increasingly dominant. In dark grown wheat leaves the highest
relative amount of PChlide 636 was observed at the age of 8 d (24).

The results of Akulovich et al. (3, 4, 5) and Akulovich and
Raskin (6) are just the opposite: they found in etiolated 2 d
old flax cotyledons or barley leaves mainly PChlide 635-636,
while between the 3rd and 7th (12th in barley) d of etiolation
the formation of directly phototransformable PChlide 650-654
prevailed, and at the 12th d (between the 13th and 17th d in
barley) their ratio equilibrated to 1 : 1. The ratio PChlide
652/PChlide 635 increasing and declining during leaf development
(2 to 18 d) was observed also in *Acer* and cucumber (4).The ratio
of PChlide absorbances at 650 and 627 nm declined in radish
cotyledons from the 4th to the 6th d of growing in darkness and
then it continuously increased (12). Klein and Schiff (23) found
in 3-4 d primary leaves of *Phaseolus vulgaris* 50 % of PChl as
PChlide 628 and among the 50 % of transformable PChlides the
635 form prevailed, while between the 3rd and 7th d the preferential accumulation of PChlide 650 was accompanied by the enlargement
of prolamellar bodies. Hendry and Stobart (18) found in primary
barley leaves a ten times increment of PChlide 650 between days
6 and 8 [from 0.35 to 3.55 μmol kg^{-1} (fr.m.)], followed by a
decline to less than half at 10 d age. The maximum content of
both PChlide forms was found in 11 to 12 d old cucumber leaves and
8 to 9 d old barley leaves (5).

Cultivation under $\lambda > 700$ nm stimulated the formation of the
635 form (maximum in 5 d wheat leaves - 24). The ratio of individual forms may be detected on fluorescence emission spectra,
where the 655 nm peak reflects PChlide 650 (eventually PChlide
650 + PChlide 635) and the 630 nm peak reflects PChlide 628. The
ratio F_{655}/F_{630} in dark grown *Phaseolus vulgaris* leaves increased
from 0.9 (4 d) to 2.7 (15 d) and then declined (9), while in far-
red grown leaves of *Quercus robur* it increased from 0.22 (youngest
leaf) to 0.28 (3rd leaf) and then declined (8).

The addition of δ-aminolevulinic acid promoted formation of
PChlide 650 more in 8 d primary leaves of barley (by 87.5 %)
than in 7 or 10 d (ca. 22 % promotion) leaves; the increase was
explained by a suppression of a PChlide 650 breakdown mechanism
(in 8 d leaves with a rate constant 139 pmol nmol^{-1} h^{-1} and a
half life of 5 h) (19).

According to summarizing schemes (11, 22), PChlide 635 prevailing in 3 d seedlings of French bean is directly phototransformed after less than 2 s into Chlide 673, while in 7-14 d seedlings the transformation of PChlide 650 and PChlide 637 is started
even by a 1 ms flash, but it goes through the Chlide 678, Chlide
683, and Chlide 673 forms. Among the reasons for this ambiguity
of results is probably, in addition to variance in plant species
and methods of growing, the criterion of phototransformability.

Also the formation of PChlideE starts soon after the beginning
of germination but later than the formation of PChlide (after
24 h in cucumber cotyledons - 31). Its amount is much lower than
that of PChlide (ca. 1/10 in cucumber cotyledons 2-5 d old - 31;
ca. 1/6 in *Phaseolus* leaves 7-17 d old - 25), nevertheless the
PChlide/PChlideE ratio may decrease (31) or increase (25) with
tissue age. Isolation by high pressure liquid chromatography
shows that etiolated leaves contain always four types of PChlideE,
i.e. PChlide esterified either by geranylgeraniol, or dihydroge-
ranylgeraniol, or tetrahydrogeranylgeraniol, or phytol, but their
relative contents change with leaf age; thus in *Phaseolus vulgaris*
the relative content of phytol ester increases from 5 to 9 d
and then declines (probably due to a reversal of the reaction
sequence to dehydrogenation), while the relative contents of
other esters increase (37).

3. PROTOCHLOROPHYLL PIGMENTS AFTER IRRADIATION AND DURING DARK REGENERATION

The decrease in PChlide amount during ageing of an etiolated
leaf may be accompanied by a decrease (barley - 18) or increase
(wheat -24) in the ability to produce Chl upon irradiation. The
regeneration of PChlide and PChlideE per unit leaf biomass following a brief leaf irradiation is slower and has a longer lag
period in older leaves than in the younger ones (wheat - 43;
maize - 29, 42; *Phaseolus vulgaris* - 1, 2, 25, 42; cotyledons of
Pharbitis nil - 28; radish cotyledons - 12). The final amount of
PChlide resynthesized per area unit increases to a maximum and then
declines (3-6/8-13 d in *Phaseolus* - 2; 0-2-4 d in *Pharbitis* -28);
similarly per one leaf (wheat - 47). The duration of lag period of
PChlide regeneration is linearly related to the duration of the

Shibata shift, both increasing with age of *Pharbitis* cotyledons
(28).

During greening in the presence of δ-aminolevulinic acid more
PChlide is formed in the upper 20 mm than lower 20 mm of the
first barley leaf (39). The dark regeneration of PChlide is most
rapid in the base of primary wheat leaves (40), nevertheless,
the amount accumulated may depend also on the quality of acting
radiation; thus after irradiation by λ= 645 nm more PChlide was
formed in the second quarter of a young barley leaf blade than
in the top quarter, while after λ= 737 nm irradiation the top
quarter synthesized more PChlide (41). Spruit and Raven (42)
found differences in PChlide accumulation after 650 or 650+735 nm
irradiation in maize but not in French bean or pea leaves. The
amount of PChlide accumulated in the presence of δ-aminolevulinic
acid depended on the length of previous irradiation (10 min to
48 h): the largest age-dependent differences were found after
brief (10 min) irradiation (12 d wheat leaves formed less than
10 % of PChlide accumulated by 6 d leaves), while after 48 h
irradiation the respective difference was 16.7 vs. 30.4 μg PChlide
g^{-1} (fr.m.) (43).

4. PROTOCHLOROPHYLL PIGMENTS FORMED IN LEAVES GROWN IN LIGHT

Due to methodical complications, the amount of data on this
topic is rather limited. In primary leaves of field grown wheat
the amount of PChlide per leaf increases up to 11 d and then
declines, i.e. the phase of rise is more prolonged than in compar-
able etiolated leaves (13, 14). Maximum PChlide content per unit
area of primary leaves is in the second fifth of the leaf blade
from the base (19 d wheat - 40) or tip (maize - 44), the maximum
PChlide/Chl ratio being near to the leaf base. During darkening
of field-grown barley plants PChlide regeneration was more rapid
and lead to its higher amounts in upper than lower part of young
leaves (32).

5. CHLOROPHYLLIDE FORMATION

The conversion of PChlide to Chlide may proceed more rapidly
and be more complete after irradiation of young or mature than
old tissues. Thus in *Pharbitis nil* cotyledons the amount of PChlide

transformed to Chlide on irradiation increased to 2 d age and then declined (28). Nevertheless, in 2-3 d old primary leaves of French bean only 40 % of PChlide was reduced to Chlide, while in 7-9 d leaves the conversion was 80-90 % (25). Irradiation of young leaves of *Phaseolus vulgaris* containing a mixture of PChlide and PChlideE brought a mixture of Chlide and Chl, while in 7-9 d leaves, containing predominantly PChlide, Chlide 685 was mainly formed (25). In young primary leaves of this species, without prolamellar bodies, Chlide 672 originating from PChlide 635 was mostly formed on irradiation, while in older leaves the major pigment PChlide 650 was transformed to Chlide 683, with the following Shibata shift (23). In other experiments with French bean, after irradiation of 4 d etiolated leaves the Chlide absorption maximum slowly shifted from 677 to 680 nm, in 8 d leaves from 680 to 678 (during 5 s) and then to 682 nm (45 s), while in 21 d leaves from 690 to 692 (during 5 s) and then to 680 nm (45 s) (46). At temperatures between -45 and -90 $^{\circ}$C only the PChlide 650 (fluorescing at 655 nm) can be transformed into Chlide fluorescing at 688 nm, independently of leaf age (16). Half-time of the Shibata shift increased with leaf age (wheat - 24). Similarly in *Robinia pseudoaccacia* in 5 d leaves Chlide 673 was formed on irradiation, while in 20 d leaves the Chlide 680 prevailed (3). The photostability of both Chlide 684 and Chlide 673 as well as the rates of the Shibata shift and the late red-shift decreased with the age (5 to 11 d) of etiolated wheat leaves (7). The kinetic of 695 nm fluorescence emitted by Chlide at room temperature after first irradiation of etiolated primary leaves of *Phaseolus vulgaris* by a He-Ne laser differed with leaf age: a 2 s phase of rapid increase was followed by a smooth decline to a minimum after ca. 60 s in 6 d leaves, while with increasing age (10 and 14 d leaves) a second maximum (called "fluctuation A") was more and more expressed: it was caused by a Chlide fluorescence yield variation (21). The kinetics of fluorescence lifetime and intensity during greening of differently old etiolated leaves were similar in leaves of French bean and barley, but 3 d barley leaves behaved analogously as 10 d bean leaves, etc., probably due to differences in PChlide content per unit of leaf area (16, 46).

6. SUMMARY

The literature dealing with differences in formation of immediate precursors of chlorophyll, i.e. protochlorophyllide, protochlorophyllide ester and chlorophyllide, which depend on leaf age and various ages of tissues within one leaf blade, is reviewed. Synthesis of these pigments in etiolated tissues, in light grown plants, and their regeneration after tissue darkening are distinguished. Generally, rates of all respective synthetic processes and amounts of these precursors increase with tissue growth in darkness up to some leaf or cotyledon age and then decline. The eventual lag phase in their synthesis increases with tissue age.

REFERENCES

1. Akoyunoglou G. 1970. The effect of age on the phytochrome-mediated chlorophyll formation in dark-grown bean leaves. Physiol.Plant. 23, 29-37.
2. Akoyunoglou GA, Siegelman HW. 1968. Protochlorophyllide resynthesis in dark-grown bean leaves. Plant Physiol. 43, 66-68.
3. Akulovich NK, Godnev TN, Orlovskaya KI. 1970. [Features of spectral transformation of protochlorophyll(ide) holochrome of etiolated leaves during its formation.] Dokl. Akad. Nauk SSSR 191, 1406-1409.
4. Akulovich NK, Orlovskaya KI, Parshikova TA. 1973. [Interrelation of shape and function of forms of the protochlorophyll pigment in etiolated plants.] In Formirovanie pigmentnogo apparata fotosinteza, pp. 3-29. Minsk, Nauka i Tekhnika.
5. Akulovich NK, Orlovskaya KI, Parshikova TA. 1974. [Characteristic of processes of protochlorophyllide accumulation and formation of its spectral forms in developing etiolated plants.] In Shlyk AA, ed. Khlorofill, pp. 168-179. Minsk, Nauka i Tekhnika.
6. Akulovich NK, Raskin VI. 1971. [Formation of protochlorophyll-holochrome in etiolated leaves and its phototransformation in chlorophyll-holochrome.] In Problemy biosinteza khlorofillov, pp. 5-52. Minsk, Nauka i Tekhnika.
7. Axelsson L. 1977. The photostability of different chlorophyll forms in dark grown leaves of wheat III. Dependence on age of the plants. Physiol. Plant. 41, 217-222.
8. Axelsson L, Klockare B, Sundqvist C. 1981. Oak seedlings grown in different light qualities II. Photostability of early forms of chlorophyll(ide). Physiol. Plant. 51, 314-320.
9. Boardman NK, Anderson JM, Kahn A, Thorne SW, Treffry TE. 1971. Formation of photosynthetic membranes during chloroplast development. In Boardman NK, Linnane AW, Smillie RM, eds. Autonomy and biogenesis of mitochondria and chloroplasts, pp. 70-84. Amsterdam-London, North-Holland.

10. Bogorad L. 1976. Chlorophyll biosynthesis. In Goodwin TW, ed. Chemistry and biochemistry of plant pigments, 2nd ed, vol. 1, pp. 64-148. London-New York-San Francisco, Academic Press.
11. Bradbeer JW. 1981. Development of photosynthetic function during chloroplast biogenesis. In Hatch MD, Boardman NK, ed. The biochemistry of plants, vol. 8, pp. 423-472. New York-London-Toronto-Sydney-San Francisco, Academic Press.
12. Buschmann C. 1979. The influence of kinetin on the biosynthesis of chlorophyll. In Marcelle R, Clijsters H, Van Poucke M, eds. Photosynthesis and plant development, pp. 193-203. The Hague-Boston-London, Junk.
13. Chaïka MT, Savchenko GE. 1973. [Pigment metabolism in the course of development of a green leaf.] In Formirovanie pigmentnogo apparata fotosinteza, pp. 105-129. Minsk, Nauka i Tekhnika.
14. Chaïka MT, Savchenko GE. 1975. [Photoregulation of chlorophyll biosynthesis during chloroplast development.] In Fotoregulyatsiya metabolizma i morfogeneza rastenii, pp. 120-134. Moskva, Nauka.
15. De Greef JA, Caubergs R. 1972. Studies on greening of etiolated seedlings I. Elimination of the lag phase of chlorophyll biosynthesis by a pre-illumination of the embryonic axis in intact plants. Physiol. Plant. 26, 157-165.
16. Goedheer JC, van der Cammen JCJM. 1981. Protochlorophyll(ide) and chlorophyll(ide) fluorescence lifetime and other properties in etiolated and greening leaves. In Akoyunoglou G, ed. Photosynthesis, vol. V, pp. 39-44. Philadelphia, Balaban Int. Sci. Serv.
17. Gough SP, Kannangara CG. 1979. Biosynthesis of δ-aminolevulinate in greening barley leaves III: The formation of δ-aminolevulinate in tigrina mutants of barley. Carlsberg Res. Commun. 44, 403-416.
18. Hendry GAF, Stobart AK. 1977. Haem and chlorophyll formation in etiolated and greening leaves of barley. Phytochemistry 16, 1545-1548.
19. Hendry GAF, Stobart AK. 1977. Protochlorophyllide (P650) turnover in dark-grown barley leaves. Phytochemistry 16, 1663-1664.
20. Henningsen KW, Boynton JE. 1974. Macromolecular physiology of plastids IX. Development of plastid membranes during greening of dark-grown barley seedlings. J. Cell Sci. 15, 31-55.
21. Jouy M. 1982. Effect of age of etiolated leaves of *Phaseolus vulgaris* on the 695 nm fluorescence kinetics during first irradiation. Photosynthetica 16, 234-238.
22. Kirk JTO, Tilney-Bassett RAE. 1978. The plastids. Their chemistry, structure, growth and inheritance. 2nd ed. Amsterdam-New York-Oxford, Elsevier/North Holland Biomedical Press.
23. Klein S, Schiff JA. 1972. The correlated appearance of prolamellar bodies, protochlorophyll(ide) species, and the Shibata shift during development of bean etioplasts in the dark. Plant Physiol. 49, 619-626.
24. Klockare B. 1980. Far-red induced changes of the protochlo-

rophyllide components in wheat leaves. Physiol. Plant.
48, 104-110.
25. Lancer HA, Cohen CE, Schiff JA. 1976. Changing ratios of
phototransformable protochlorophyll and protochlorophyllide
of bean seedlings developing in the dark. Plant Physiol.
57, 369-374.
26. Liljenberg C. 1974. Characterization and properties of a
protochlorophyllide ester in leaves of dark grown barley
with geranylgeraniol as esterifying alcohol. Physiol. Plant.
32, 208-213.
27. Moran R, Arzee T, Porath D. 1980. Pigment accumulation and
plastid ultrastructural changes in dark grown cucumber
seedlings. In De Greef J, ed. Photoreceptors and plant growth,
pp. 237-240. Antwerpen, Antwerpen University Press.
28. Ogawa M, Tsutsui Y, Konishi M. 1978. Effects of illumination on absorption peak shifts in spectra of intact etiolated
cotyledons of *Pharbitis nil* II. Effects of leaf age on protochlorophyllide regeneration and the Shibata shift. Plant
Cell Physiol. 19, 127-132.
29. Raven CW. 1973. Chlorophyll formation and phytochrome.
Meded. Landbouwhogesch. Wageningen 73-9, 1-100.
30. Rebeiz CA, Lascelles J. 1982. Biosynthesis of pigments in
plants and bacteria. In Govindjee, ed. Photosynthesis, vol.
1, pp. 699-780. New York-London-Paris-San Diego-San Fran-
cisco-São Paulo-Sydney-Tokyo-Toronto, Academic Press.
31. Rebeiz CA, Yaghi M, Abou-Haidar M, Castelfranco PA. 1970.
Protochlorophyll biosynthesis in cucumber *(Cucumis sativus,*
L.) cotyledons. Plant Physiol. 46, 57-63.
32. Savchenko GE, Chaĭka MT. 1975. [Kinetics of dark protochlorophyllide accumulation in barley leaves at various developmental stages.] In Shlyk AA, ed. Biosintez i sostoyanie
khlorofillov v rastenii, pp. 83-103. Minsk, Nauka i Tekhnika.
33. Šesták Z. 1977. Photosynthetic characteristics during ontogenesis of leaves 1. Chlorophylls. Photosynthetica 11, 367-448.
34. Šesták Z. 1984. Chlorophylls and carotenoids during leaf onto-
geny. In Šesták Z, ed. Photosynthesis during leaf development,
in press. Praha-The Hague, Academia-Junk.
35. Shibata H, Ochiai H. 1975. [Studies on δ-amino levulinic
acid dehydratase during chloroplast development in radish
cotyledons.] Amino Acid Nucl. Acid 32, 16-24.
36. Shibata K. 1957. Spectroscopic studies on chlorophyll formation in intact leaves. J. Biochem. (Tokyo) 44, 147-173.
37. Shioi Y, Sasa T. 1983. Compositional heterogeneity of protochlorophyllide ester in etiolated leaves of higher plants.
Arch. Biochem. Biophys. 220, 286-292.
38. Shlyk AA, Averina NG. 1973. [Character of kinetin effect on
the process of protochlorophyllide accumulation in etiolated
and green leaves of barley.] Dokl. Akad. Nauk SSSR 213,
235-238.
39. Shlyk AA, Kostyuk NN. 1972. [Effect of δ-aminolevulinic
acid on chloramphenicol induced inhibition of dark accumulation of protochlorophyllide in green barley leaves.] Dokl.
Akad. Nauk SSSR 202, 707-710.
40. Shlyk AA, Savchenko GE. 1970. [Protochlorophyllide metabolism
in green leaves.] In Andreenko SS, ed. Fiziologiya i |biokhimiya

zdorovogo i bol′nogo rasteniya, pp. 185–197. Moskva, Izdatel′stvo Moskovskogo Universiteta.

41. Shlyk AA, Savchenko GE, Stanishevskaya EM, Shevchuk SN, Gaponenko VI, Gatikh OA. 1966. [Role of phytochrome in chlorophyll metabolism in a green plant.] Dokl. Akad. Nauk SSSR 171, 1443–1446.
42. Spruit CJP, Raven CW. 1970. Regeneration of protochlorophyll in dark grown seedlings following illumination with red and far red light. Acta Bot. Neerl. 19, 165–174.
43. Sundqvist C, Odengård B, Persson G. 1975. Light-stimulated accumulation of protochlorophyllide in leaves of different ages treated with δ-aminolevulinic acid. Plant Sci. Lett. 4, 89–96.
44. Tageeva SV, Savchenko GE, Semenova GA, Shlyk AA. 1969. [Dynamics of chloroplast organization and metabolism of protochlorophyllide in various parts of a maize leaf in darkness and light.] Fiziol. Rast. 16, 581–593.
45. Treffry T. 1978. Biogenesis of the photochemical apparatus. Int. Rev. Cytol. 52, 159–196.
46. van der Cammen JCJM, Goedheer JC. 1980. Kinetics of fluorescence lifetime during greening of etiolated bean and barley leaves of different age. Photobiochem. Photobiophys. 1, 329–337.
47. Virgin HI. 1961. On the formation of protochlorophyll in normal green wheat leaves of varying age. Physiol. Plant. 14, 384–392.
48. Virgin HI. 1975. In vivo absorption spectra of protochlorophyll$_{650}$ and protochlorophyll$_{636}$ within the region 530–700 nm. Photosynthetica 9, 84–92.

AUTHOR INDEX

Phycocyanin

- as internal standard for Chl fluorescence 255.

Phytochrome 243.

Picosecond spectroscopy

- apparatus for 99.
- of PChlide photoreduction 99.

Pigment protein complexes (see also PChlide and Chlide
protein complexes)

- effect of detergent on - 175
- in grana and stroma thylakoids 265.
- in isolated etioplast membranes 275.
- monomeric forms 255.
- oligomeric forms 255.
- organization 87-255.
- pigment composition 265.
- pigment-protein ratio 265.
- polypeptide composition 19-43-129-255-265.
- in PT and PLB fractions 53-69.
- with Zn-PChlide 175.

Pisum 353.

PLB (see Prolamellar bodies)

Polyacrylamide gel electrophoresis 19-43-53-69-129-265.
- of etiolated membranes 19-43-53-69-129.
- of greening membranes 19-43-129.
- of LHPC 265.

Polypeptide composition

- of chloroplast membranes 255-265.
- of digested membranes 43.
- of etiochloroplast membranes 19-43-129.
- of etioplast membranes 19-43.
- of grana and stroma thylakoids 265.
- of PChlide oxidoreductase 19-31-43-53-69-129.